A Guide to Library Service in Mathematics: The Non-Trivial Mathematics Librarian

FOUNDATIONS IN LIBRARY AND INFORMATION SCIENCE, Volume 30

FOUNDATIONS IN LIBRARY AND INFORMATION SCIENCE

A Series of Monographs, Texts and Treatises

Edited by
Thomas W. Leonhardt
Holt Library, University of the Pacific
and
Murray S. Martin
University Librarian and Professor of Library Science Emeritus, Tufts University

Volume 1.
The Mexican American: A Critical Guide to Research Aids
Barbara J. Robinson, *University of California Library , Riverside*
and J. Cordell Robinson, *California State College, San Bernardino*

Volume 2.
A Practical Approach to Serials Cataloging
Lynn S. Smith, *University of California Library, Riverside*

Volume 3.
The Microform Revolution in Libraries
Michael R. Gabriel, *Mankato State University* and Dorothy P. Ladd, *Boston University Library*

Volume 4.
China in Books: A Basic Bibliography in Western Language
Norman E. Tanis, David L. Perkins and Justine Pinto, *California State University Library, Northridge*

Volume 5.
Budgetary Control in Academic Libraries
Murray S. Martin, *Pennsylvania State University Libraries*

Volume 6.
Cost Analysis of Library Functions: A Total Systems Approach
Betty Jo Mitchell, Norman E. Tanis and Jack Jaffe, *California State University Library, Northridge*

Volume 7. *(Omitted From Numbering)*

Volume 8.
Management of a Public Library
Harold R. Jenkins, *Kansas City Public Library*

Volume 9. *(Omitted From Numbering)*

Volume 10.

Collection Development of Libraries: A Treatise
Edited by Robert D. Stueart, *Simmons College*
and George B. Miller, Jr., *University of New Mexico, Libraries*

Volume 11.

Introduction to Serials Management
Marcia Tuttle, *University of North Carolina Library, Chapel Hill*

Volume 12.

Developing Collections of U.S. Government Publications
Peter Hernon, *Simmons College* and Gary R. Purcell, *University of Tennessee*

Volume 13.

Library Management Without Bias
Ching-Chih Chen, *Simmons College*

Volume 14.

Issues in Personnel Management in Academic Libraries
Murray S. Martin, *Pennsylvania State University Libraries*

Volume 15.

Information Needs of the 80s:
Library and Information Services Role in "Bringing Information to People"
Based on the Deliberations of the White House Conference on
Library and Information Services
Edited by Robert D. Stueart, *Simmons College*

Volume 16.

ALMS: A Budget Based Library Management System
Betty Jo Mitchell, *California State University Library, Northridge*

Volume 17.

Options for the 80s: Proceedings of the Second National Conference of the Association of College and Research Libraries
Edited by Michael D. Kathman , *College of St. Benedict and St. John's University* and Virgil F. Massman, *James J. Hill Reference Library*

Volume 18.

The Library Services Construction Act:
An Historical Overview from the Viewpoint of Major Participants
Edward G. Holley and Robert F. Schremser, *University of North Carolina, Chapel Hill*

Volume 19.

Fundraising for Nonprofit Institutions
Edited by Sandy Dolnick, *Friends of Libraries, U.S.A.*

Volume 20.

Changing Technology and Education for Librarianship and Information Science
Edited by Basil Stuart-Stubbs, *University of British Columbia*

Volume 21.

Videotex and Teletext: New Online Resources for Libraries
Michael B. Binder, *Fairleigh Dickinson University Library*

Volume 22.

Research Libraries and Their Implementation of AACR2
Edited by Judith Hopkins and John A. Edens, *State University of New York at Buffalo, Library*

Volume 23.

An Encyclopedia of Film Festivals
Robert A. Nowlan and Gwendolyn Wright Nowlan, *Southern Connecticut State University*

Volume 24.

A Guide to the Literature of Electronic Publishing: CD-ROM, Desktop Publishing, and Electronic Mail, Books, and Journals
Michael Gabriel, *Northern Illinois University Libraries*

Volume 25.

Technical Services in Libraries: Systems and Applications
Edited by Thomas W. Leonhardt, *Holt Library, University of the Pacific*

Volume 26.

Collection Management: A New Treatise
Edited by Charles B. Osburn, *University of Alabama* and Ross W. Atkinson, *Cornell University Libraries*

Volume 27.

The Role of the American Academic Library in International Programs
Edited by Bruce D. Bonta, *The University Libraries, The Pennsylvania State University* and James G. Neal, *University Libraries, Indiana University*

Volume 28.

Academic Library Budgets
Murray S. Martin, *University Librarian and Professor of Library Science Emeritus, Tufts University*

Volume 29.

Academic Libraries and Training
Edited by Maryruth Phelps Glogowski, *E.H. Butler Library, State University College at Buffalo*

Volume 30.

A Guide to Library Service in Mathematics: The Non-Trivial Mathematics Librarian
Edited by Nancy D. Anderson, *Mathematics Librarian, University of Illinois at Urbana-Champaign* and Lois M. Pausch, *Geology Librarian, University of Illinois at Urbana-Champaign*

A Guide to Library Service in Mathematics: The Non-Trivial Mathematics Librarian

Edited by: NANCY D. ANDERSON
Mathematics Librarian
University of Illinois at Urbana-Champaign

LOIS M. PAUSCH
Geology Librarian
University of Illinois at Urbana-Champaign

JAI PRESS INC.

Greenwich, Connecticut *London, England*

Library of Congress Cataloging-in-Publication Data

A Guide to library service in mathematics : the non-trivial mathematics librarian / edited by Nancy D. Anderson, Lois M. Pausch.
p. cm. — (Foundations in library and information science ; v. 30)
Includes bibliographical references and index.
ISBN 1-55938-745-9 : $73.25
1. Mathematics libraries—United States. 2. Mathematics—Bibliography. I. Anderson, Nancy D., 1940- . II. Pausch, Lois M. III. Series.
Z675.M37G85 1993
026'.51—dc20 93-43482
CIP

55 Old Post Road, No. 2
Greenwich, Connecticut 06836

JAI PRESS LTD.
The Courtyard
28 High Street
Hampton Hill, Middlesex TW12 1PD
England

ISBN NUMBER 1-55938-745-9

Library of Congress Catalog Card Number 93-43482

Manufactured in the United States of America

To Felix Albrecht
Chairman of the Departmental Library Committee
Friend and Mentor

CONTENTS

ACKNOWLEDGMENTS

We wish to thank the following student assistants for their help in the preparation of the various bibliographies: Kara L. Baloun, Lou J. D'Andria, Desiree Dixon, Kevin Foster, Mike Napier, and Anne Ruttencutter of the Mathematics Library and K. L. Henebry, Derrick King, Lexis Robinson, and Scott Stortzum of the Geology Library.

Many thanks are extended to the mathematics librarians throughout North America who have, either in person, by writing, or via electronic mail, shared their thoughts over the years on improving service, working with mathematics faculty, answering reference questions, or solving problems: Sallie Barringer (Trinity University), Richard Funkhouser (Purdue), Joanne Goode (University of Kentucky), Janice Griggs (University of Minnesota), Kimiyo Hom (formerly at Berkeley), Carol Hutchins (Courant Institute), Dorothy McGarry (UCLA), Ella Ponomarenko (formerly at *Mathematical Reviews*), Berle Reiter (formerly at Michigan State University), Mary Ann Southern (Duke University), Martha Tucker (University of Washington), and Jack Weigel (University of Michigan). Their stories of early days on the job were, in part, the inspiration for this book.

Thanks go also to Hilda Britt of the Mathematics Department at the University of Illinois at Urbana-Champaign for word processing the many revisions to the list of journals.

We appreciate the financial assistance of the Research and Publication Committee, University of Illinois at Urbana-Champaign

Nancy D. Anderson
Lois M. Pausch
Urbana, July 1993

Chapter 1

INTRODUCTION

First of all, the editors would like to explain why we chose the title, "The Non-Trivial Mathematics Librarian." It is a play on words often used by the mathematical community. One of the first things a librarian is apt to hear as s/he works with mathematicians is the disdain expressed by them with regard to someone who does **trivial mathematics**: cranking out another theorem or supplying some new detail. The mathematics may be new, but so what? Thus, a **non-trivial mathematics librarian** would be one who would be innovative, creative, and provide essential services to those who come to the library.

This guide is designed for the librarian or information specialist who provides access to library resources for mathematicians. It will be especially useful for the librarian, new to the world of mathematics libraries, who does not have a subject background in mathematics (or in any other science), and, we hope, will help to overcome any "math anxiety" so that the new librarian can respond creatively and effectively to user needs. It will also be useful for the specialist librarian who is seeking to update his/her skills. As the authors see it, this book fills a niche between general works intended for librarians and specialist works intended for mathematicians.

There are chapters written by practitioners in a variety of library settings. Chapter 2, written by Anderson and Pausch, covers most aspects of mathematics librarianship in a departmental library in a research university. In Chapter 3, Thornhill writes on mathematics librarianship in a general science department of a large public library. Davidoff, in Chapter 4, provides guidance and direction for a government, technical, or special librarian. In chapter 5, the authors provide citations to references used in the preceding chapters as well as citations to works which will expand and extend discussion of the topics.

Since clientele and type of library impact strongly on services offered and emphasized, and on collection development and management, this approach has great value for the librarian seeking ways of dealing with the realities of his/her particular situation. However, even though the authors write from a particular viewpoint, all chapters can be scanned for valuable advice applicable to any setting. Earlier works by Pemberton, Dorling, and Schaefer are still useful but are outdated by at least ten years. Therefore, they do not cover such topics as online searching and CD-ROMs. These works also are meant as guides to reference sources while the present book includes many different aspects of librarianship, such as collection development, cataloging, and so forth. In addition, the earlier works were designed for the librarian who already understood the discipline of mathematics and had some knowledge of its literature. The authors hope that this book will serve as a stand-alone handbook that can be used by professional librarians to enhance their understanding of mathematics librarianship.

Later chapters in this volume consist of lists of books, journals, and reference works of particular interest to mathematicians and, therefore, to mathematics librarians. These lists can be used to check library collections for coverage. These listings are marked **(a)** for importance in an academic library, **(p)** for a public library, and/or **(s)** for a special library. At the beginning of each chapter, there is a note indicating the scope of each list.

Chapter 2

MATHEMATICS LIBRARY SERVICE IN ACADEMIC LIBRARIES

Nancy D. Anderson and Lois M. Pausch

INTRODUCTION

Mathematics librarians in academic libraries come from a variety of backgrounds. Some actually were math majors or did graduate work in mathematics at one time, some majored in Russian, some majored in some area of science, but many have no recent subject background in science or mathematics. Having gone through the interview process, the new librarian has grasped that the mathematics department is often passionately concerned about the library. The librarian will generally find a mathematics library with a good collection, but with procedures or practices that may need streamlining or modernization. Since this position was perhaps chosen because it offered independence and the opportunity to work with users who both need and want the librarian, the purpose of this chapter is to provide some insights into how to become a more effective academic math librarian.

CLIENTELE

Who Are They?

In order to better understand and appreciate the clientele, it helps to know something about their lives and work. Few mathematicians have written *about* mathematics, preferring to write mathematics. The authors of this chapter, however, recommend the following books to help broaden an understanding of mathematics and mathematicians. As mathematician Jenny Harrison writes: "It is only natural for people enamored of the beauty and perfection of mathematics to expect the mathematical community [read also mathematics library] to reflect, in its behavior and ideals, the perfection of mathematical thought." Jerry King attempts in his delightful book to illuminate the aesthetic delights of mathematics. Both Atle Selberg and Freeman Dyson describe vividly how they became mathematicians. Four additional excellent works are *The Mathematical Experience, The Mathematical Sciences, Mathematics as a Creative Art*, and *Mathematics: People, Problems, Results.*

Every mathematics department is different, although, at this time in the United States, most somewhat resemble that at the University of Illinois at Urbana-Champaign (UIUC). In 1989, the authors surveyed the faculty and found that male professors greatly outnumber the female (26/7), most are over 30 years of age (33/4), and most are pure mathematicians (45/17, the discrepancy in number of replies can be accounted for by noting that respondents were free to choose more than one area of study). Surveys have also shown that most research is being done in algebra (12) and analysis (12), next in combinatorics (5), number theory (5), geometry (8), and differential equations (4), then in logic (3), applied mathematics (3), topology (3), and finally in probability (2), control theory (1), numerical analysis (1), operations research (1), and programming (1). Faculty members can be described as pure mathematicians (with only 16 of 37 mentioning any interest in applied mathematics, as can the graduate students where only 8 of 29 described themselves as applied mathematicians.

Mathematicians who use the library can be characterized as very critical yet appreciative of library services, generally dismayed by the inadequacies of online catalogs (due in part to what they perceive as not very friendly interfaces) and very proud of the collection. They enjoy browsing through the collection and want all of it to remain in the Mathematics Library, regardless of age or use. Some faculty members make frequent use of the CD-ROM workstation, while others confuse it with the online catalog. The younger the academician, the more at ease s/he seems to be about using the CD-ROM workstation and the online catalog. Faculty members feel free to voice complaints about classification of library materials and expect the librarian to reclassify according to their judgment. This can be done more often now since changing call numbers in an online catalog is much easier than when catalog cards were printed and filed.

Mathematics faculty want materials received and processed as quickly as possible. A good working relationship with the person(s) in charge of acquisitions will often result in extra efforts being made to acquire materials as soon as possible. Also, colleagues at other mathematics libraries and the MathDoc Service at *Mathematical Reviews* may be e-mailed or phoned to supply an urgently needed article (when interlibrary loan cannot obtain it swiftly enough).

How Do They Use the Library?

In order for the librarian to understand how mathematicians use the library, it is important for the him/her to understand how mathematicians do research.

Frame writes, "The seed of an idea may be planted either in a mathematical talk, in a conversation with colleagues, or in reading a mathematical treatise or journal. Sometimes this seed takes root right away and grows into a mathematical idea worth publishing in its own right. Sometimes it lies dormant for a while and is only reactivated by new stimuli resulting from further reading or talking with kindred spirits. But the elaboration of the idea usually involves two essential ingredients: a mathematician must have periods of uninterrupted time to try out and develop his own ideas, and he must supplement this independent study by consulting the works of others. Seldom does a mathematician go to a library to read a book cover to cover. Usually it is to check one or two references, or to browse in the literature of a certain field to find what has been written on a specialized topic."

In the authors' study of use of the Mathematics Library at UIUC, it was learned that the faculty use the library more than the graduate students. Sixty-five percent of the faculty use the library 3 times a week or more. Most come to find a certain book or to scan the literature (100%), as do the graduate students (72%). Graduate students find their first library tour, held during their teaching assistant orientation, and a later individualized session at the beginning of their Ph.D. research, to be particularly valuable (31%). Both groups consider the help they receive at the reference/circulation desk to be most beneficial, although the graduate students are less likely to think they either need or can get assistance. The reasons given most often for requesting assistance were help in deciphering the online catalog or in locating a book/journal in the bookstacks. Neither faculty nor students consider it a high priority to consult the library staff on their research or to study in the library. The faculty consider the collection as the most important source of information.

The Kinds of Information They Seek

A number of librarians have written about the materials that should be in a mathematics library. Dorling, Pemberton, and Schaefer have written books, which though now dated, cover the entire spectrum of the literature, while Pausch and Stankus have written about particular aspects of the literature such as conference proceedings or scientific journals. Again, in the authors' survey of the Mathematics Department faculty, it was found that the faculty considered books, journals, and conference proceedings to be the most important information sources for their research. Graduate students, on the other hand, responded that dissertations and working papers were most important. (Several years before this survey the staff of the library began putting copies of dissertations on reserve shortly after their submission rather than waiting a year for the fully cataloged "official" copy to arrive. It was nice to have the survey corroborate experience.) In Steen's recent *Library Recommendations for Undergraduate Mathematics*, a committee selected works which should be available in any college or university library and which should reflect the interests of research mathematicians.

Importance of a Departmental Library

In 1963, Frame described the importance of an accessible mathematics library: "A mathematics library, adequate in coverage of books and research journals, and conveniently located near the offices of the senior mathematics staff, is a primary requisite of any mathematics department. It is an absolute necessity, if a department is to compete successfully in recruiting research mathematicians. It is as essential to the research worker in mathematics as the... laboratory is to the experimental physicist or chemist." Most first-rate mathematics departments insist on a nearby mathematics branch library. Thus, it comes as no surprise that nearly 30 years after Frame's work, Anderson and Rovnyak found in the American Mathematical Society's library survey that subject branch libraries comprise 83% of the total in Group I and 50% in Group II (the leading departments of mathematics in the United States). Of the 12 "best" mathematics libraries in the United States (Princeton, Harvard, Berkeley, Illinois, Michigan, Brown, Courant, Stanford, Chicago, MIT, Institute for Advanced Study, Yale, Cornell, Wisconsin, and Library of Congress—there are some ties), 11 are located in the same building as the mathematics faculty and 11 have a professional librarian at least part time.

In the authors' survey, 84% of the faculty and 72% of the graduate students responded that it was most important to have the Mathematics Library in the same building as the departmental offices even if not all materials related to their research were in that library. Further, the faculty members (87%) and graduate students agreed (the students were less vehement—72%) that the site of the library was more important than having all mathematics and related literature in one convenient location. The need for this kind of departmental library is borne out by reports from other universities. In the authors' experience, the most often asked questions are the ones from mathematicians at other universities asking how to establish a mathematics departmental library or from a science and technology librarian asking why mathematics faculty cannot be persuaded of the desirability of moving into "the" science library.

Mathematicians consider the physical environment of the mathematics library a very important quality, after the depth and breadth of the collection. Respondents to the AMS library survey noted that ease of use, access for browsing, current journal display, organization and arrangement of books, and hours open were very important for research. Faculty members often are quite vocal in demanding that the library provide areas in which to peruse newly received books and journal issues, away from the undergraduate students who use the library for a study hall or for access to reserves. It is sometimes difficult to provide this balance: on the one hand, the department encourages students to come into the library by placing items on reserve in the departmental library rather than in a central reserve unit, and, on the other hand, the faculty want quiet to review materials. These conflicting demands may be partially met by strictly enforcing quiet and by asking undergraduate students not to study in the areas set aside for New Journals and New Books. Undergraduates should be directed into those parts of the library which serve as a study hall. Faculty should be encouraged to use carrels in quiet areas where papers can be spread out.

RELATIONSHIP OF LIBRARY AND LIBRARIAN TO THE ADMINISTRATION AND THE INSTITUTION

How to Cope with Differing and Conflicting Aims of Your Library, the Library Administration, and Your Clientele

Rarely is an academic mathematics library in a stand-alone situation. It is usually part of a library system and thus must follow the same policies and procedures established for the library as a whole. The degree to which deviation from these policies and procedures can be tolerated depends on the enlightenment of the library administration, how persuasive a case is made to an immediate department head, and, in some cases, the old adage of what's not known won't hurt. It is certainly easier to manage a branch library far from the central administration of the library than one in the same building.

The departmental librarian must be a juggler: as a reference librarian, s/he must know the literature while as a manager, s/he must learn to balance budgets, hire, and train staff. These capabilities, however, are secondary to making the departmental library reflect and enhance the work of the mathematicians. One of the better ways of finding out what is needed is to schedule interviews with faculty members soon after being hired and then at regular, but not too frequent, intervals thereafter.

Staff should be encouraged to comment on and suggest improvements in service. They are a very important conduit to the people who use the library: they are the ones most likely involved with ready reference questions or circulation of books and journals. Chances are, if one mathematician complains or compliments, others will have remarked on the same thing over coffee or at a seminar. Listen attentively to staff, to faculty, to students, and to the library committee. Often a good response is to offer to help draft a letter to the library administration, requesting more funds or more service.

Role of Librarian as Advocate

Mathematicians seem to have a nationwide reputation with library administrators as "squeaky wheels." Because of this, many library directors will tend to dismiss comments on the need for greater funding of library collections and services arising from the Mathematics Department as just "more of the same." In many institutions, the library administration is at odds with the department over the need for a mathematics library. The mathematics librarian can do much to deflect this negative impression by championing the needs of the mathematicians with the library administration. The librarian can articulate, in ways understandable to another librarian, means by which policies can be altered or funds augmented or new services provided. When asked, the librarian should be able to justify requests for increases in materials budgets by reporting new course offerings, honors to individuals, or excessive price increases.

Staffing

Many departmental libraries may be characterized as being staffed by overachieving, recently graduated college students who want to take a few years off to earn some money before continuing on to graduate school or to another profession. As a consequence, frequent turnover may result in time-consuming and unending training, while procedures may be lost or forgotten from one individual to another. Thus, procedure manuals for each position are essential to the orderly operation of the library. There should be a manual for graduate assistants, library paraprofessionals, library clerks, and student assistants. Manuals may be kept in paper and/or electronic form with staff encouraged to update them on the computer as procedures change. Staff should also be requested to review the manuals before terminating employment so the library benefits from their experience.

Confidentiality of Library Records

Mathematicians and librarians are often at odds about the necessity of disclosing the name of an individual who has charged out a journal or book when another individual needs to take a quick look at it. Frame writes, "Seldom does a mathematician go to the library to read a book from cover to cover. Usually it is to check one or two references... [and] often it is vital to be able to check an idea at once." Yet librarians are constrained from releasing circulation information both by the section on Exceptions from Inspection and Copying of the federal "Freedom of Information Act" and by state law. In Illinois, the "Library Records Confidentiality Act," PA83-179, has, since 1984, provided a legal basis for privacy of patron registration. Many other states have similar laws.

However, since many faculty members are likely to want some way to circumvent these restrictions, one possible procedure is to set up a system for "waiving the right to privacy." Faculty members and graduate students can be given the opportunity, on an annual basis, to choose whether or not they wish to sign a waiver card allowing the library staff to release their names when requested by another faculty member or graduate student. A record, arranged by some identification number, is then kept of those who have agreed. For those who do not sign, library staff must call the borrower to return the book for a short period to allow the requester to get the information needed.

Hours of Opening and Student Assistants

Decisions on hours of service need to balance the differing needs of students and faculty. Generally, faculty like to have access for at least an hour after any afternoon colloquium. Graduate students (especially if they are teaching assistants) like the library to be open during the evening and weekends. Undergraduate students like the library to be open until midnight for studying. It may be necessary to make trade-offs, for example, to be open longer hours during the academic year by cutting back during holiday periods, interims and summer sessions.

The realization that graduate students will use the library in the evening when professional and clerical staff are not on duty means that student assistants must be able to supply much of the ready reference during those hours and must be skilled in using and interpreting the online catalog. Student assistants must be given extensive training on the contents of the desk reference collection, journal finding files, and the online and/or card catalog. Since these students are more likely to be trained by either clerical staff or some of the more experienced students, instruction may be uneven. In order to determine areas of training that have not been covered or need to be emphasized, an "exam" covering most of the situations s/he will encounter should be administered once a semester. Supplemental training is provided then in areas of notable weakness.

Keys to the Library

Keys to the library for afterhours access may be demanded when the library does not maintain long hours of opening. In the AMS library survey, 8 of the 12 "best" libraries reported that faculty do not have keys; these libraries are open an average of 90 hours per week when classes are in session. In 4 of the 12, faculty do have keys; these libraries are open an average of 55 hours per week. A recent survey of the "key problem" on PAMnet yielded responses from 21 mathematics libraries. More than half do not issue keys to the library. Of those libraries that do issue keys, two-thirds issue keys to faculty but not to graduate students. However, keys may be issued to the chair of the department and to the chair of the departmental library committee as a gesture of good will. Generally, issuance of keys means more problems with security and considerable reshelving of journals from after hours use and should be avoided, if possible.

Security

More than one-half the libraries in the PAMnet survey reported that they have security systems. Even in those instances where researchers had keys to the library, security systems cut losses considerably. Yet security systems are costly to install and, unfortunately, the funds for purchasing equipment are often more difficult to obtain than those for purchasing replacements. However, the mathematics department can influence the library administration's decision to install the equipment (through a letter-writing campaign). In at least one instance, faculty of a mathematics department raised the funds to purchase the security gates, provided that the library agree to pay for stripping the books.

Services Needed vs. Services Offered

Loan periods for materials may differ considerably for each kind of borrower and each type of material. For example, here at UIUC, the loan periods are as follows:

	Faculty	*Graduate Student*	*Undergraduate*
Books	16 weeks	4 weeks	4 weeks
Bound journals	2 weeks	2 weeks	1 day
Unbound journals	Overnight	Room use only	Room use only
New journals	Room use only	Room use only	Room use only

Materials may be renewed in a number of ways such as by telephone, by having borrowers note and return an overdue notice, or by calling a circulation telephone center, if one exists. Limits are not placed on the number of books an individual may charge out. In order to keep some control over excess circulation and to help keep down the number of books faculty and students may misplace in their offices, patrons should be asked to respond to overdue notices. It is not good practice to automatically renew books for any group of patrons, including faculty members. In many cases, fines cannot be assessed on faculty, so it is sometimes difficult to get a response on overdue notices.

Surveys to Determine Requests for Change

Mathematicians visit libraries at many different institutions over the course of their careers and frequently find procedures at other mathematics libraries more to their liking. This may result in requests for change in such things as the loan period (e.g., to loan books for 4 weeks instead of 16 or to not circulate journals at all) or shelving bound journals (some math libraries intershelve books and bound journals while others shelve journals separately by title). These requests should be taken seriously, with special attention given to the possible ramifications of such change. As an example, shortening the loan period increases the total amount of circulation and thus impacts on the workload of the staff. Reshelving to separate the journals and books would also cause a great deal of inconvenience and commotion if it were done while the library was open and with no additional staff. Any change should be made only after consultation with the library committee and perhaps after a survey has been done. If the results of the survey show a preponderance in favor of the status quo, those who requested the change can be expected to bow to the democratic process while the librarian will find out how pressing the demand for change really is. On the other hand, the authors learned through another survey that faculty wanted books/journals shelved higher off the floor. Since a shifting project was already in process, it was quite easy to respond to this request.

Communication with the Department

Effective and frequent communication with the faculty and graduate students is essential in maintaining rapport with the primary clientele. One suggestion for ways to bring this about is a regularly issued acquisitions list displayed prominently or distributed to those who indicate an interest. This list can be generated from records already in the online catalog. Additionally, new services, changes in hours or staff, and significant purchases can be announced in a departmental or library newsletter. Since newsletters often are issued at irregular intervals, memos can be distributed if more timely announcements are needed.

An effective departmental library committee is a real necessity. The UIUC Mathematics Department Library Committee consists of 12-14 faculty members, selected by the Department's Executive Committee, who advise the librarian on library matters. New members of the faculty are frequently appointed to serve on this committee as it is a good way to become acquainted with the library. Occasionally the Executive Committee has inexplicably appointed a faculty member who might be called a "problem patron." This has resulted in the "problem" becoming a better model for library behavior.

Duty on the UIUC Library Committee provides the faculty with the opportunity to note which books should be purchased by the library. Catalogs are circulated to the Committee with instructions to select items of interest. Committee members are urged to indicate "high" or "low" priority in case funds are inadequate to acquire all materials selected. Experience has shown that members of the Library Committee do not select as widely in their subject areas as might be expected so the librarian's judgement must often come into play. Remember, they are advisory to the librarian. The Library Committee is, of course, invaluable in offering guidance in journal cancellation.

The chair is the most important member of the library committee as s/he is the primary contact with the department. S/he can be an invaluable ally for additional funding requests, promotion of new services, and a focus for others' complaints and suggestions. Regular meetings with the chair to discuss budgeting, services, selection of foreign-language materials, and so forth, can be very rewarding. The chair of the committee is instrumental in bringing the concerns of the library to the department and of the department to the librarian and the library administration. Especially when one is a new mathematics librarian, the chairperson of the library committee will probably serve as the best source of information on how mathematicians work, how they regard the library and the literature, and how they can best be assisted with their teaching and research.

REFERENCE SERVICES

Paper Reference Sources

When a math librarian has little or no math background, any feelings of inadequacy for not knowing the language of mathematics will evaporate after the first week when the realization strikes that mathematicians may know their mathematics, but have real difficulty in using the library. And that's exactly what a librarian is good at! Most librarians have devised ways to help mathematicians find their way to the correct journal without having to ask can the library staff. One way is to have a public access Kardex file with a listing of *Mathematical Reviews* abbreviations of titles with local call number and location (if not in the mathematics library).

Another way is to develop a "quick ref" file at the reference/circulation desk, composed of 3 x 5 cards of catch words, mathematical terms, and acronyms usually in the words of the requester. This file will continue to grow and become very useful, especially for the new staff member or student assistant the first few times s/he is totally on his/her own in the library. Every staff member should be encouraged to contribute entries for topics on which 5-10 minutes were spent locating information. This file could also be easily be kept on a disc. When regularly scheduled hours of reference are offered, staff members should be encouraged to write down difficult-to-answer questions which can later be discussed in a staff meeting so all can learn.

It is fun and challenging to verify the citations mathematicians cannot find even after they have consulted the journal title abbreviation locators. *Mathematical Reviews* (MR) (see p. 122) or *Current Mathematical Publications* (CMP) (see p.p 120-121) are still the first places to look (depending on how recent the work is), keeping in mind that it is safest to start looking in *MR* about 12-18 months after date of publication. One caution: note that as *Mathematical Reviews* tries to keep production costs down, the percentage of titles in *CMP* that do not appear in *MR* has risen from 12% to nearly 20%. Occasionally, it will be necessary to search many years after the date of publication. Searching the cumulative indexes of *MR* speeds up the process, but, unfortunately, there are absolutely no plans for a new cumulation (covering 1986 to present) as everything is fully searchable through MathSci. Thus, having all the back files of *MR* on CD-ROM is essential for the verification process. If *MR* does not provide the answer, the *Zentralblatt für Mathematik und ihrer Grenzgebiete* (Zbl) (see p. 126) or *Referativnyi Zhurnal. Matematika* (RfZhMat) (see p. 123) should be consulted. *Zbl* is especially strong in European titles and is also available online while *RfZhMat* has excellent coverage of all titles. It is worth the effort to learn how to transliterate the cyrillic alphabet in order to access this reference work.

Mathematics librarians have their favorite bibliographies for accessing the older literature. For example, for the period 1868-1934, the *Jahrbuch über die Fortschritte der Mathematik* (see p. 122) is the best source to use. Several librarians have remarked on the usefulness of Tompkin's *MAST: Minimum Abbreviations of Serial Titles, Mathematics* (see p. 122). It helps with abbreviated titles, similar titles, acronym nicknames, and exotic foreign language titles.

Feelings of euphoria about providing effective reference service usually disappear with questions from math laypersons. However, all mathematics librarians have developed techniques for coping with such questions. Many report the utility of using the the *VNR Concise Encyclopedia of Mathematics* (see p. 102) or the MathSci keyword index to the AMS classification system (see *MathSci: UserGuide,* p. 117), or *Notes from MathSci* (Providence, RI: American Mathematical Society). It also helps to have the individual rephrase the question in order to give more clues as to what information s/he is seeking. If books in the library do not seem to contain the answer, do not hesitate to phone the reference department or anyone else in the university for help. An obliging mathematician may be asked for assistance (this is where knowledge of his/her research specialty really helps). One unforgettable question from a patron in a public library was what the geometric shape of an eggbeater was called. A geometer, when asked for help, worked several hours to find the answer to that question. Remember that others, besides librarians, know the answers. Mathematics librarians agree that providing reference assistance to mathematician and layperson alike has become a lot easier with electronic sources of information.

Online Electronic Reference Sources

Among the electronic databases available to the librarian are OCLC, RLIN, MathSci Online, MathSci Disc, SciSearch, CompactMATH, MATH database and other libraries' online catalogs (on Internet). It is very important for the librarian to learn to search effectively on these so that citations and other bibliographical data can be found in response to patrons' requests.

OCLC and RLIN are large-scale bibliographic utilities that are available for searching. They can be used for cataloging and for reference purposes. MathSci Online is a database maintained by the American Mathematical Society which can be accessed through BRS and DIALOG. More importantly, the AMS has come up with a new pricing structure for purchase of the tapes which will allow more libraries to mount MathSci as either part of the campus network or the Library's online catalog. These tapes will be available for a $2,000 yearly fee with backfiles priced at $10,000. MathSci Online includes data from *Mathematical Reviews* since 1940, *Current Mathematical Publications, Current Index to Statistics* (see p. 121), the Tukey *Index to Statistics and Probability* (see p. 121), *Computing Reviews* from the *Association for Computing Machinery* (ACM) plus the Stanford technical reports. For information prior to 1940, printed sources must be used.

Another aspect of online reference service is selective dissemination of information or SDI. For this, a patron-requested online search is placed with the vendor, who then runs it at regular, set intervals, and sends the results to either the searcher or patron depending on which vendor is used. Online SDIs are an efficient way to assist the user. However, each time the search is performed, a charge is made as in other online searches. This may prove to be too expensive for most mathematicians, causing them to bypass this type of service.

MathSci on CD-ROM is a very useful tool, but it does not include all of the information in the MathSci Online database. Only the *Mathematical Reviews* and *Current Mathematical Publications* data are available here. However, MathSci Disc does include all CMP items from 1985 on that do not appear in MR. This CD-ROM is currently produced by Silver Platter; it is updated semi-annually so that searching for current postings must be done in the print publications. Three discs are now available: 1940-1979, 1980-1987, and 1988-present. The librarian should compile a guide to searching this tool and make it available to users at the CD-ROM workstation. MATH is the online version of Zbl which cites worldwide publications in pure and applied mathematics.

Faculty and graduate students should be alerted to these services and programs should be devised to teach them to be effective searchers on their own. Both

Sichel and Murdock have described ways by which patron training on MathSci can be accomplished. This tool is licensed under agreement with the vendor either for single use or multiple use. Multiple use comes into play when there is a LAN on which the CD-ROM is networked. The licensing fee for networking is much higher and, of course, access will depend on being a node in the network.

MathSci Disc and MathSci Online were designed so that mathematicians could use them without the librarian as intermediary. It is entirely possible for them to search the online version from their offices, via DIALOG or BRS or their after hours services, Knowledge Index or BRS After Dark. However, many math librarians have difficulty in "selling" online searching and SDI because of cost recovery. It is not unusual for new math librarians to be really enthusiastic in trying to make this service attractive to their patrons and then find there is little response from the mathematicians due to cost recovery.

BIBLIOGRAPHIC INSTRUCTION

Faculty

While faculty will frequently listen in on the sessions provided for students, it is difficult to convince them of the need for individualized sessions on learning how to use the library more effectively. Newer, younger faculty members are more likely to be reached, often through announcements in the departmental newsletter or by handouts issued and distributed through the departmental mail system. Some will come in for sessions on using the MathSci Disc, on dial-up access to the online catalog from home or office, etc.

Library instruction should also be offered to new faculty at the beginning of the academic year and to visiting faculty. These individuals can be invited to any library-wide reception set up to welcome them to the university library and to demonstrate some services such as the online catalog. A follow-up session can then be held emphasizing the specific services of the mathematics library, and, again, a copy of the library guide should be handed to them.

Graduate Students

If the department has a graduate student orientation program, it makes good sense for the library to be a part of that program. A tour of the library, introducing them to the staff, the online catalog, online databases and CD-ROM searching, and to the way reserves are handled, is usually sufficient for this early session. One major emphasis at this time should be on the willingness of the library staff to help in answering questions. A guide to the library, if such is available, and other handouts relating to the use of the library or the catalog should be handed out.

When graduate students have completed sufficient study and have passed comprehensive examinations (if such are required before proceeding with Ph.D. dissertation research), the librarian should offer each individual an in-depth bibliographic instruction session to help her/him with dissertation-related research. At this time, the finer points of using the various automated systems such as the online catalog and MathSci Disc should be covered as well as use of subject headings and classification.

Undergraduate Students

Programs for undergraduate mathematics majors have been developed primarily for upper-level students. Sallie Barringer has described a successful program at Trinity University in San Antonio, TX, that includes a handout on information sources in mathematics, a presentation by a librarian which constitutes an assignment for the class of which it is a part, and several worksheets aimed at making sure the students use the instruction received.

PROVISION OF SPECIAL SERVICES

Display of New Materials

Journals

Mathematicians rely on librarians to give them some sense of order by which new journals are displayed. To accomplish this, library staff should try to check journals in daily within a reasonable time (in the case of the UIUC Mathematics Library, within two hours of receipt). A list of the current day's receipts may be typed and posted where the faculty is most likely to see it. The issues themselves can then be kept on a new journal shelf for a specified period of time (e.g., 1-2 weeks) for easy perusal. The day of the week the issue is received should be displayed prominently on each title so that those faculty members who come in regularly can quickly check to see what has arrived since the last time they were in the library. During this display period, the issues should not circulate except perhaps for photocopying (with the understanding that they will be returned immediately). At the end of the display, the still unbound journals may be shelved in a special area, arranged by journal title, where they will remain until there is a bindable unit plus one (in order to keep the most current issue in the library at all times).

Books

Completely processed new books can also be displayed in a new book shelf area, preferably for the same time period as journals. A sign-up slip should be inserted in each book. Before placing the new book in the display area, write in at the top of it the name of the person recommending purchase (if known). Books can be circulated in order of sign up, the loan period ranging from 1-4 weeks, depending on the number signing up. Those monographic materials awaiting original cataloging may also be displayed here as this accomplishes two things: (1) the book can be circulated when interest may be higher than it might be when it is finally cataloged; and (2) the number of names on the slip will give a sense of what priority the book has for cataloging.

Circulation Privileges for Short-term Visitors

One of the facts of life in most mathematics departments is the steady stream of visitors who come to give talks or otherwise interact with the faculty and students. These visitors come for varying lengths of time and should be allowed full use of the library. For the long-term visitor or for those designated as "visiting professors," there are usually established institutional procedures that will provide him/her with proper identification for use of the library. However, a visitor who intends to stay less than two weeks and is only interested in using the math library should be treated differently, in a less involved manner. In order to accommodate these short-term visitors, a set of special identification numbers can be set aside for assignment to them. Records should be kept noting departure dates and names of the "sponsoring" professors. Small cards with the special number can be given to the visitor together with copies of any guides to the library.

Other

Other special services can be offered, provided there is sufficient staff to perform them. Photocopies of tables of contents from journals held in other branch libraries can be made and posted near the New Journal Shelf. Faculty members can be encouraged to photocopy articles at little cost to them, using the library's photocopier, by having the mathematics department purchase a low cost copy card that is kept in the library for their use. A record is kept so that the department can charge back the cost of copies to the user.

If the library has mounted a current contents or other journal awareness service in its online public access catalog (OPAC), users should be encouraged to learn how to search, print out and/or download tables of contents or titles of interest. Those who are part of a campus network can send results directly to their own workstations.

COLLECTION DEVELOPMENT AND MANAGEMENT

Characteristics of the Mathematics Collection

What makes a good collection? One answer would be complete runs of mathematics journals, extending back into the nineteenth century. Since mathematicians often browse this older literature, every attempt should be made not to split serials. The importance of keeping the entire collection together cannot be overemphasized. In the AMS Library Survey, numerous respondents noted that nothing should be in remote storage.

Another response would be extensive holdings of foreign language materials, especially those in Russian. A mathematics library may rely on ordering some copies of Russian journals from MathDoc in Ann Arbor, but mathematicians will likely insist on having the important titles at hand.

Thirdly would be a good collection development policy which states the reasons for inclusion of some types of materials and not others. For example, the UIUC Mathematics Library does not acquire and retain preprints or technical reports. With the rise of networking of math preprints (e.g., Yale University's Instant Math Preprints), there is much less need to keep files of these materials. Usually, it is possible to satisfy a user request for a technical report with a published journal article.

The Mathematics Library at the University of Illinois is probably unique in having so many non-English language books; this reflects the attempt to acquire all the monographs reviewed in *Mathematical Reviews* (classed 00-65). In the early 1980s, funds from a Title II-C grant allowed this Library to begin a wide-ranging acquisitions program which has continued to the present time. Many mathematics libraries in the United States rely on interlibrary loan from the UIUC Mathematics Library for these materials.

In summary, books are important to mathematicians, more so than in the sciences. Good mathematics is written worldwide so the collection in an institution where English is the native language, while retaining an emphasis on English-language materials, should mirror the international character of mathematics research. Not all research will appear in "Eurojournals" nor will all be translated. In particular, Russian material will still be useful when patrons make the effort to either try to transliterate the Cyrillic themselves or to have someone who is more fluent translate it for them.

Who Plans and Controls the Budget?

It is extremely important that the mathematics librarian control the materials budget. Experience has shown that the library committee will often be so discriminating in its selection practices that funds will be underspent. Several years of this can lead to a permanent reduction in the budget. Also, the librarian is more likely to make sure the research interests of everyone in the department (or university) are met. Since the librarian cannot anticipate the research interests of those faculty the department might hire in the future, s/he has to acquire, at least minimally, in all areas of mathematics.

In the beginning, the new mathematics librarian will find that the library committee chairperson will be very demanding or solicitious about how and what is selected. By taking time to learn about the existing collection and faculty research interests and by working closely with the chairperson, the librarian can soon assuage his/her anxiety about collection building. Understanding the universe of mathematical publishing can be improved by scanning the books and videocassettes section of *Current Mathematical Publications.*

At many universities, the library administration will want the departmental librarian to plan and control the budget and the collection development officer will ask that the librarian project what needs will be. A good working relationship with departmental secretarial staff will help in gathering much of the information needed to fulfill these requests or will provide tips on who in the department can provide it. The more persuasive and cogent this statement of needs is, the better the opportunity that funding the mathematics collection will receive a high priority from the library administration.

If the library's bookkeeping system will allow it, the librarian should request a monthly budget statement, showing the rate of spending. It is good public relations to share this information with the library committee chairperson and department head. The more they know about funding shortfalls or difficulties, the more they will try to help. If at all possible, funds should be reported separately by type of material, that is, monographs, monographic series, serial continuations (e.g., annuals), and periodicals (i.e., journals) as price increases may vary considerably by type of material. The mathematics librarian should ask the library budget officer to explain the budgeting process and should bring duplicate payments to the attention of the person responsible.

How Materials Are Acquired

Since publishers often do not provide complete order information, staff members should be instructed on how to pre-order search under a variety of access points. Often this work can be done by student assistants. The results can later be checked by a library paraprofessional and/or the librarian before the orders are typed or input. The outcome will be fewer duplicate orders. An order file should be maintained, arranged by title (the title is less likely to change than the "main entry"—e.g., it is sometimes difficult to ascertain how conference proceedings will be cataloged). It is a good idea to place monographic series on standing orders because of their importance to mathematicians, and because claiming is usually easier and more efficient than ordering.

Another effective way to acquire books is to send a gift request letter to authors, editors, reviewers, institutions or other not-for-profit organizations. This can be remarkably successful in acquiring noncommercial publications. In some cases, the author will write back, advising how and where the publication is for sale. At other times, if the author resides in the former Soviet republics or the developing world, he may request that an American or western European work be sent in exchange. If the book being requested is not too expensive, and if the library administration will allow such purchases, complying with the request will result in getting a title wanted as well as aiding mathematicians where hard currency is scarce.

Sources of Information

The best sources of information on newly published materials are the "Books and Videocassettes Listed in This Issue" and "Serial Additions and Changes" sections of the biweekly *Current Mathematical Publications*, (see p. 121), the "Telegraphic Reviews" section of *The American Mathematical Monthly* (see p. 325), the "Buchbesprechungen—Book Reviews" section of the *Monatshefte für Matematike* (see p. 350), and the "Kritika i Bibliografii͡a: Novye Knigi po Matematik" section of *Uspekhi Matematicheskikh Nauk* (see p. 363). Steven Roman has come out with a new service, which seems promising as a real time-saver: no longer will each mathematics librarian have to scan numerous reviews, compiled from many different sources. *The Bulletin of Mathematics Books and Computer Software*, appears four times a year and contains reviews of math books and software, announcements, lists of "what's available" in various fields, a special column on advanced books written by Gian-Carlo Rota, and other information. Subscriptions to this newsletter are available at $19.50 for individuals and $49.50 for libraries from The Roman Press, 8 Night Star, Irvine, CA 92715.

On the other hand, if academic librarians do not wish to pay for selection information, there are other means of learning about new publications. A good source of information are the catalogs of such major publishers of mathematics/statistics materials as Academic Press, Birkhäuser, Dekker, Wiley, World Scientific, and so forth, which may be obtained by writing the publisher and asking to be put on the mailing list. Several international book vendors (Nijhoff, Harrassowitz, Casalini) supply forms for books which can be used for selection purposes (this is an excellent way to learn about many smaller European publishers). Harrassowitz, in particular, is very helpful in the identification of publications from eastern Europe and Germany. Finally, scanning the acquisition lists from other good mathematics collections (e.g., those from Courant Institute, University of Wisconsin, University of Michigan) can be helpful in the selection process. Usually, these may be obtained by writing the librarians at these institutions and arranging for an exchange of lists.

Russian Monographs

Until recently, the best way to acquire Russian research-level books was to establish a blanket order, especially for those published in Moscow, St. Petersburg, and Kiev. These plans worked well in library systems with a Slavic Library or Slavic bibliographer to provide help in identifying dealers (e.g., Kamkin in Rockville, MD). However, because of the changes in the Soviet Union, the effectiveness of such blanket orders needs to be evaluated. In the meantime, the library at *Mathematical Reviews* can probably provide the most reliable information on these publications.

Another source for these materials is a well-organized exchange program (see below for a fuller discussion of exchange programs). If the mathematics department has some materials to offer, many monographs and serial publications may be obtained in this manner. If the department does not publish a journal, then persuasion may be needed to convince the university library to purchase any mathematics journal to be used for exchange purposes.

Approval Plans

Participation in a library-wide approval plan can lower costs through higher discounts or speed up the acquisition of current American and British trade and university press books. These plans are set up through a vendor (e.g., Academic Book Center, Midwest Library Service, Blackwell North America, Scholarly Book Service, Coutts, etc.) using profiles specifying the needs of a particular institution and may be subject- or publisher-based. Research-level titles published in specific geographic areas such as the United States, Canada, and the United Kingdom including European publishers (e.g., Springer, Elsevier, etc.) whose titles are either published or distributed in the United States are most often those supplied. The plan can be set up to exclude certain types of materials (e.g., reprints, lower-level undergraduate texts, user manuals, books over a pre-set price limit, books published as part of a monographic series already on a standing order, etc.).

In the case of the UIUC Library, money for a publisher-based approval plan does not come from subject funds so departmental libraries benefit considerably. For example, in 1989-90, the Mathematics Library received an additional $17,000 worth of monographs. Having a central fund for this plan permits many librarians to protect a greater number of their serials by transferring monographic money to serials. So many trade books come in that a slackening in monographic acquisitions has gone unnoticed. Under this approval plan, the vendor also supplies announcements for publishers not currently on the approved list. This allows each departmental librarian to make the decision whether or not to purchase the title using their own subject funds.

Institutional Memberships

Institutional memberships are usually held by the mathematics department, but they can be extremely important to the mathematics library as another way to stretch the materials budget. The American Mathematical Society institutional membership offers discounts on a number of journals and monographs. However, there can be a down side to such a membersip. The AMS does not extend discounts to vendors, as of this writing, so that the library's acquisition department may be unhappy that they must order and prepay everything directly with the AMS. However, the prepayment policy will be reviewed in 1993. The AMS has improved its service to libraries, though, and now sends its renewal notices separately from journal issues. The Society for Industrial and Applied Mathematics (SIAM) also offers several options for discounts on publications. The Association for Symbolic Logic, the Institute of Mathematical Statistics, and the American Statistical Association are some other societies which offer institutional memberships that benefit libraries.

Firm Orders

All of the above are means to acquire material while expending as little of the acquisition budget as possible. When it finally comes to books and journals that must be purchased, the librarian must utilize whatever ordering system is in effect. Quite often, a large academic library will receive a good discount from vendors because of the total dollar volume and number of titles acquired in any one year. A good working relationship with the person(s) responsible for acquisitions is essential as mathematicians often want books "rush" ordered.

Once firm orders are placed and copies filed in a departmental library order drawer (or records are input into a database), they form the basis for keeping track of those titles that have been received and processed and those that are still outstanding. If, after a reasonable amount of time from ordering to hoped-for date of receipt has passed (e.g., 3 months for domestic and 6 months for foreign publishers) and the material has still not arrived, order copies can be used to claim or to cancel.

Out-of-Print Sources

Many mathematics collections already include a number of the important early mathematical works. These often are housed in rare book collections away from the mathematics library. To strengthen these collections, the identification of dealers who are helpful in the acquisition of these older materials is essential. **Bernard Quaritch, Ltd.** (5-8 Lower John Street, Golden Square, London W1R 4AU) is often the best place to start. Successful acquisition requires prompt perusal of their catalogs as well as a telephone confirmation of the order. Other dealers who occasionally offer mathematics books are **Gerhard Renner Antiquariat**, Auf dem Unteren Berg (Fuchsfarm), Postfach 16 48, 7470 Albstadt 2 Tailfingen, Germany; **William Patrick Watson** (97 Bishop's Mansions, Bishop's Park Road, London SW6 6DY); and **F.E. Whitehart** (40 Priestfield Road, Forest Hill, London SE23 2RS).

These out-of-print catalogs can also be used to identify materials shelved in the mathematics library that truly belong in the rare book collection. After checking the catalogs for items to purchase, other already-owned mathematics titles, especially those listed for sale at more than $500, may be checked for condition. If in good shape, the volumes can then be sent to the rare book collection for security and preservation.

It is more difficult to locate a source for the not-so-rare out-of-print book. These are often volumes which have been stolen or reported lost, are not "classics," and are more than 10 years old. In most cases, the acquisitions department staff should be asked to search through various out-of-print sources. Mathematicians have reported good success in obtaining this type of book from **Lange and Springer Scientific Booksellers, Inc.** (New York City, (212) 995-5250) and **Lubrecht & Cramer, Ltd.** (Forestburgh, NY, (914) 794-8539, Fax: (914) 791-7575). Firms that will search for out-of-print mathematics books are **Book Search International** ($3.00 for each title) (P.O. Box 120862, Nashville, TN 37212), **Elliott Books** (North Branford, CT, (203) 484-2184), **Lincoln Out of Print Book Search** (Scituate, RI (401) 647-2825), and **The Science Bookshelf** (525 Fourth St., Ann Arbor, MI 48103). Finally, it may be possible to replace the volume either from **Books on Demand** (UMI, Ann Arbor) or by photocopying the volume borrowed on interlibrary loan from some other library (after securing permission to photocopy from the publisher).

Gifts

At some point, every mathematician cleans out his/her office, retires, or dies. When any of these events happen, the professor (or his/her family) immediately thinks of the Mathematics Library!! If the mathematics library is new, gifts will generally be welcome. If it is an old, established library with a large and well-maintained collection, handling these gifts will require tact and diplomacy. It may never occur to the professor that the library may have already been offered, at least 10 times, a nearly complete run of the *Bulletin of the AMS*. However, it is not a good idea to alienate any donor because what is offered the next time might be something needed or wanted. Duplicates are difficult to give away without absorbing shipping fees. The many duplicates exchange services (e.g., there is one run by the Sci-Tech Division of SLA) take a lot of time to deal with, probably more than they are worth. There may be foreign graduate students in the department who would love to send these collections back to their home universities, but the professor giving these books and journals often wants to take an income tax deduction and that may not work if the materials go overseas. In addition, the student usually does not have the funds to ship the volumes.

So, what to do? First of all, ask the professor at the time the gift is first offered if there are any strings attached. Does s/he want any of it back? Can it be disposed of if it is not needed? Does s/he want an itemized list of what is kept? If the prospective donor indicates that s/he wants to take an income tax deduction or have the collection appraised, s/he should be given two articles to read: Dyal's "What old books are worth" and Payne's "A closer eye on appraisals" as well as a copy of "Instructions for [Internal Revenue Service] Form 8283." It will seem hardhearted to the giver, but the librarian cannot agree to appraise collections (it would be a conflict-of-interest). There is usually a bookdealer in the local community or in a nearby large city who will appraise the book/journal collection for a fee.

After agreement is reached, the donated titles should be searched in the catalog. If there is already a copy in the collection, if it is not currently being used, or if it is not a "classic," the gift should be rejected unless it is in excellent condition when it can then be used to swap with the library copy. These rejected gifts can be set out in a giveaway box and are usually taken within hours by math graduate students.

Gifts of journals require slightly different procedures. Often when the professor finds that the volumes are not needed, s/he will want to know the names of journal dealers who will buy them. A list of these dealers should be given to the professor upon request. Some dealers to be included on this list are **Abrahams Magazine Service**, 56 East 13th St., New York, NY 10003 ((212)777-4700); **Jerry Alper Inc.**, 271 Main St., P.O. 218, Eastchester, NY 10707 ((914) 793-2100, Fax: (914) 793-7811); **J.S. Canner & Co., Inc.**, 10 Charles St., Needham Heights, MA 02194 ((617) 449-9103, Fax: (617) 449-1767); **EDUCO Services International, Ltd.**, P.O. Box 226, 75 N. Kensico Ave., Valhalla, NY 10595 ((914) 997-7044, Fax: (914) 997-7995); **Alfred Jaeger, Inc.**, 66 Austin Blvd., P.O. Box 9009, Commack, NY 11725-9009 ((516) 543-1500, Fax: (516) 543-1537; **Kraus Periodicals**, Route 100, Millwood, NY 10546 ((914) 762-2200, Fax: (914) 762-1195. The donor should be encouraged to try a variety of dealers as the prices s/he will be offered by each dealer will vary considerably.

It is preferable to ask the professor to consider donating his/her journals to a library overseas as mathematics libraries in the developing world or in the former Soviet Union are in desperate need of good books and journals. The Smithsonian Institution Libraries (SIL) have a program whereby they send materials overseas if the donor will wrap and forward them to the Smithsonian. This is a forwarding service for educational institutions so the department or the library must agree to forward the materials for the individual professor. More information about this program is available by writing to **Smithsonian**

International Exchange Service, SIL, 1111 North Capitol Street N.E., Washington, DC 20560 ((202) 357-1964). The American Mathematical Society's Committee on Service to Mathematicians in Developing Countries is trying to get a more robust program going to distribute materials to universities in need.

Friends of the Library and Fund Raising

As university libraries have received less public funding, librarians have had to look elsewhere for support. A library development officer can be a valuable asset for fund raising and in the planning and development of a Friends group. This group may publish a newsletter in which librarians ask for help in purchasing reference materials or equipment that the budget will not cover. Remarkable success can be achieved; the main problem is that departmental libraries must often take turns advertising their needs.

The advantages of fund raising for mathematics libraries have been known for many years. For example, here at UIUC, a professor from another institution made a large bequest to the Mathematics Library in memory of his long-dead advisor in the Mathematics Department. The income from this bequest helps not only to extend the meager monographs budget, but is also used to attract other donors to the Library. The widow of a former head of the Mathematics Department was solicited to give the initial subscription to MathSci Disc; she has maintained her interest in this venture and recently contributed some of the funding for the acquisitions of back files on MathSci Disc. This relationship demonstrates two rules of successful fund raising: the first gift is difficult to obtain and subsequent gifts depend on the donor being thanked and informed of the value of the gift.

Cornell recently received widespread publicity for its fund raising efforts in which mathematics faculty members are persuaded to teach a course "for free," the savings in salaries going toward an endowment for its mathematics library. At the University of Illinois, the Mathematics Department often designates its library as beneficiary when a donor is solicited. Also, for the past two years, faculty have proven to be valuable allies in mounting a fund raising campaign to collect $10,000 annually so journals would not have to be cancelled. Many of the contributions are in the form of monthly payroll deductions. Anyone giving over $350 is given the privilege of selecting a journal which will have a bookplate with his/her name on it when the journal is bound.

Exchanges

An effective exchange program is the best way to make a good mathematics library better. First, exchanges bring in serials which would otherwise be unobtainable. At the University of Illinois, the *Illinois Journal of Mathematics* has been exchanged for years with institutions in the former Soviet Union, Eastern Europe, China, and Latin America. At present, around 200 serials are received on this exchange program. It follows, then, that any mathematics library with a good collection is almost always one which maintains a good exchange program. These publications are generally those that are not available through subscription agents so the only way to get them is on exchange. If a department does not publish a mathematics journal, it may be worthwhile to purchase a mathematics journal to use for exchange purposes. With the breakup of the Soviet Union, exchanges still work but it is anticipated that more Russian mathematics journals will be published commercially in the future.

The second reason to maintain exchange agreements is the savings in the cost of journal subscriptions. The authors estimate that over $15,000 worth of journals are received on exchange, comprising approximately one-third of the journals in the collection. In many cases, the exchange is not equal in value as the *Illinois Journal of Mathematics* is worth more than some of these irregularly published journals. The Department, which subsidizes the publication of the *IJM*, supports these unequal exchanges as it believes that American mathematics departments should support scholarship in developing countries.

Journal Cancellation

Cancellation of journals does not make anyone happy. Humanities and social science librarians see their monograph funds reduced in order to partially protect science serials. Science librarians do not think this is enough and have effectively argued that the ratio of expenditures for serials and monographs in science should range between 70/30 and 90/10. Devin and Kellogg have published citation data which validates the impression widely held by librarians of what the ratio should be for their collections. For mathematics, the ratio is calculated at 77/23, serials to monographs.

Some librarians report that mathematics faculty members are reasonable and understand the reasons behind almost yearly journal cancellation efforts. This is not always the case: it may be necessary to survey and amply document the rationale behind each prospective cancellation. When duplicate titles are cancelled, there is usually little outcry from the faculty. However, when unique titles begin to be cancelled, faculty members need to be involved and informed as to the whys and wherefores of the choices made.

Librarians should work closely with the faculty on determining which titles should be cancelled (lists can be proposed and distributed by the librarian to which faculty members respond or detailed use studies can be undertaken). Be advised that the reaction of the mathematicians is often to refuse to answer any survey and to complain to the campus administration, the library director, and the librarian. Agreement on which titles to cancel can be difficult to achieve.

Some titles that can be considered for cancellation are the very expensive non-AMS translation journals, the eastern European titles, or the ones which have increased their prices well above the average for the year. Mathematicians are beginning to respond to announcements from librarians about outrageous price increases and are reconsidering to which journals they will submit articles. In the short run, this concern may translate into a greater willingness on their part to cancel a particular title. The tendency of many faculty members, on the other hand, is to protect titles of immediate interest in their particular fields. Thus, in order to make informed decisions on journal cancellations, the librarian should rely to a certain extent on departmental input from at least the library committee. Various types of use surveys may need to be undertaken such as counting volumes left after photocopying or before reshelving, or counting the number of users from the sign-up slips in recently received issues. Faculty members can be asked to list their "top 20" journals or to rank, from not important to very important, all journals received on subscription. Searches on MathSci Online or Disc can be used to determine in which journals faculty members are publishing and the frequency with which each person publishes. A search on the CD-ROM version of *Science Citation Index* (Philadelphia: Institute for Scientific Information) can be done to determine in which journals faculty papers were cited. All this data can then be used to rank each journal in priority order. For example, a title may receive a ranking of 1 and thus will be the first to go in any future cancellation project. After all ranking is completed, the library committee can be asked to comment on the rankings and may choose to modify the list. They may decide to retain all newly published titles in order to have time to judge the value of these new journals. All members of the faculty should be shown the final list of titles to be cancelled before it is forwarded to the library administration. This lengthy procedure may help to explain the whys and hows of cancellation in a way that will be acceptable to mathematicians.

Electronic Journals

Mathematics is at the forefront of developing electronic research journals. There is both a department and standing committee of the American Mathematical Society concerned with the development and future of electronic journals. John Franks spoke of the role of libraries in this rapidly evolving field. Firstly, libraries will be subscribers as these journals will be cheaper to publish than the paper versions many of us are now being forced to cancel. Secondly, mathematicians hope that libraries will be the ones to archive these electronic journals. There is certainly historic precedent for this; libraries are viewed as more dependable than publishers who may go out of business.

Librarians will need to respond to these new ventures: they will need to figure out how to deliver this information to the faculty offices, necessitating new modes of interaction. They will also need to interact with the computer experts at their institutions in order to provide additional services (e.g., linkages to other journals).

Maintenance of the Collection

Preservation

Deterioration of the collection is cumulative and, in old, established collections may be quite noticeable. If the collection is housed where there are insufficient environmental controls, the process is hastened. As an example in one such library, a study by Chrzastowski and a group of UIUC librarians shows that 37% of the volumes in the Central Bookstacks are seriously deteriorated (embrittled paper). This finding is also reflected in the Mathematics Library collection. Without adequate financial support, little can be done other than to replace volumes by in-house photocopying and microfilming; at least in this way, the intellectual content is retained. Less damaged volumes can be boxed or paper laminated by in-house conservation staff and rarer volumes transferred to the rare book collection which may have better temperature and humidity controls.

Kenney and Personius wrote about a promising development (now an actuality), at least for mathematics, which was tested at Cornell University where the department and the library collaborated with Xerox Corporation to record deteriorating books as digital images and to reproduce, on demand, high quality paper copies. Titles were selected by a faculty advisory committee and represent those printed before the 1920s. Information on the 500 reproduced books or the project is available from the mathematics librarian, Steve Rockey, at the following e-mail address: DIGITBKS@MATH.CORNELL.EDU. The reprints produced by Cornell are both very high quality and low priced (less than 10 cents/page, plus $7.00 for binding and $3.00 for postage and handling) so they are "selling like hotcakes" to individual mathematicians and libraries. Finally, the Cornell list can be used to determine the percentage of the books held by a library that will need to be replaced.

Shelfreading/Inventory

Shelfreading and inventory are related and necessary to the efficient operation of the library. The difference is mainly in the number of workers involved and the frequency of the operation. Because users often seem to have a different notion of how a book numbering system works, they should be asked NOT to reshelve volumes browsed. A designated place can be set up where the volumes can be deposited until reshelving is done by a library staff member. Even if this is done, misshelving will remain a problem. During the academic year when the problem is likely to be the worse, a student assistant should be assigned on a daily basis to read the shelves. A complete "reading" (checking for books shelved in correct call number order) should be completed at least once a year. In some areas, such as analysis and applied mathematics, should be read each semester.

Inventory is still a useful tool to determine the extent of loss in the collection. Ideally, it should be done every 5-10 years. Two staff members take the shelflist card files into the stacks and match the books with the cards. Subsequent checking will determine that some are in circulation, and that many are just missing. Where a card shelflist no longer exists, parts of the collection known to be in high demand can be selectively inventoried by asking for a printout of particular call number ranges from the office responsible for the online catalog and then following the steps noted above. Either way is very labor- and time-intensive. Consider doing it only when a particular area of the collection gets very messy, faculty complaints become overwhelming, or additional student assistants can be hired.

Weeding and Transferring to Remote Storage

The problem with weeding the collection is that decisions based on circulation will not be accurate in that browsing use by patrons will not be counted. About the only items that can be withdrawn from the collection without consultation are multiple copies of old texts. Objections will probably not be raised if earlier editions, teaching of mathematics materials, and older titles in applied mathematics (as these do become dated) are transferred to another site. If space limitations are particularly severe, some serials may be selected for remote storage, for example, those for which there is no current subscription. Other than these few categories, broad guidelines should be solicited from the library committee on materials which could be shelved elsewhere. For example, volumes that have not circulated in ten years, duplicate copies, or volumes that are falling apart are often good candidates for remote storage. Student assistants can be assigned to retrieve the designated volumes. When this step is completed, the librarian should look at what has been found and further winnow the selection, based on his/her knowledge of the politics of the department or of unrecorded anecdotal use. A list can then be made of what is left and distributed to faculty members and graduate students, asking for comment. Finally, those titles on which there is agreement can be transferred to whatever other site is available.

Cooperative Arrangements and Resource Sharing

Since the 1980s, when funding for materials budgets started to decline at the same time as science journal prices began to escalate, collection development officers at many research university libraries have been encouraging greater cooperation among subject specialists in the development of collections. Despite special meetings that have focused on cooperation among research libraries (one such was sponsored in 1985 by the Council on Institutional Cooperation (CIC, made up of the Big 10 universities plus the University of Chicago)), little enthusiasm has been shown by mathematics librarians for cooperative collection management. One reason for this lack of enthusiasm is that document delivery services need to be vastly improved. Realizing that regional institutional initiatives are greatly needed to make any cooperative collection development program work, the CIC is looking into using its communication linkages to enhance telefacsimile transmission of journal articles.

Another reason for lack of cooperation is that mathematicians maintain that browsing access to at least the last 20-30 years worth of literature is needed in their local library. Mathematics librarians report that whatever funds they receive from their own institutions must go to supporting the research interests of their clientele even if this means that all the collections come to more closely resemble each other. On the other hand, several mathematics librarians admit that a hidden cooperative arrangement exists because they feel free to buy fewer foreign-language materials since they know that the University of Illinois would be acquiring them (if those titles were reviewed in *Mathematical Reviews*).

BIBLIOGRAPHIC CONTROL

In any discussion on the bibliographic control of mathematics materials in a major academic library, it is appropriate to note that, to most mathematicians, the library classifications used (either the Library of Congress or the Dewey Decimal Classification) are not very useful. Because Dewey must fit all of mathematics, whether theoretical or applied, in the 10 digits of 510 and then does not make use of two of those (517 and 518), many of the classes are seen by library patrons not to fit the literature. Bartle suggested almost 20 years ago that the level of specificity of Dewey numbers be greatly simplified so that the average cataloger or librarian without a mathematics background could understand and use the schedules with more precision. Donovan, Hunt, and Mack recommended studying the possibility of some cross classification using the Dewey base numbers for the large areas of study (i.e., algebra, geometry, etc.) and then adding on numbers from the *Mathematical Reviews* classification (MRC). In his article, Bartle lists several ways in which library classification could be made to work better both for librarians and for patrons. Although the papers cited referred to earlier editions of Dewey, many librarians and mathematicians agree that these remarks can also be applied to the current edition.

No attempt has been made to follow most of these suggestions although to one who has ever had responsibility for cataloging and classifying mathematics materials, they appear to be eminently reasonable. However, more authors and/or publishers are listing the major and minor *MRC* classification numbers somewhere in the early pages of books and other types of publications to assist librarians in classifying them.

The last major revision of the QA schedule in the LC classification occurred in 1973. At that time, major portions of the schedule were assigned new numbers and some portions were reassigned numbers that had been in use for other topics in mathematics. The Library of Congress Office for Subject Cataloging Policy is responsible for changes in the LC classification and has stated that any changes or revisions follow as much as possible the MRC classification and other reference works when interpolating new class numbers into the existing schedule. However, these changes and revisions are made only when identified by a cataloger who may or may not have a working knowledge of mathematics. Again, decisions are made without direct contact with experts in the field.

Donovan, Hunt and Mack published a concordance between the 18th edition of Dewey and the *MRC* classification in 1975. In each edition of the *MathSci Users Guide* (see p. 117), a section of concordances between the LC classification and *MRC* and vice versa is included.

Keeping in mind that the two classification systems are meant to provide major subject coverage at the whole book level and that *MRC* was devised to provide access at article level (and, therefore, is not suitable for use under current cataloging standards), the use of multiple subject headings becomes the only way to provide access to the other topics a book might cover. For most large academic libraries, the Library of Congress Subject Headings (LCSH) serve as the major source for headings. It is widely noted in the literature, and in discussion, that LC subject headings as a whole, but most especially in the sciences and mathematics, are often archaic and inadequate. Some improvement has been seen in recent years with the LC decision to change many inverted headings so that, for example, instead of Numbers, Theory of, the heading now reads Number Theory. LCSH still reflects the general nature needed to cover headings for all types of libraries and, therefore, are not changed quickly enough to keep up with the literature in the sciences.

The result is that bibliographic control of materials in mathematics and statistics continues to be less than satisfactory. Until such time as the major producers of classifications and subject headings develop good working relationships with experts in the many fields of mathematics for the development of more suitable classifications and subject headings, the situation will remain a problem for mathematicians and a difficulty for librarians.

PERSONAL DEVELOPMENT

Continuing Education

Updating skills or learning about other aspects of librarianship can occur in a variety of settings. Among these are participation in one- or two-day short courses, offered for credit, which precede annual library association conferences. Time management and expert systems are some of the topics that are offered and taught by nationally recognized professionals. There are also workshops, sometimes one day or longer, that are presented at the local, state, and regional levels by library and other education-related organizations.

Communication with Library Colleagues

The academic mathematics librarian should be a member of both the Special Libraries Association (SLA) and the Association of College and Research Libraries (ACRL). Membership in each association costs around $100 a year. If only one can be afforded, SLA probably offers more opportunity for interaction with peers. Then it will be necessary to rely on the library administration or colleagues to provide access to the publications of ACRL.

ACRL is an association within the American Library Association (ALA) and places the academic library setting in perspective, while SLA brings together many of the most active mathematics librarians in the country. The Science and Technology Section of ACRL focuses on issues of interest to academic science librarians while the Science & Technology Division of SLA focuses on issues of interest to science librarians in any special library (academic, corporate, or governmental). Funding is usually not available to attend both ALA (and thus, ACRL) and SLA annual conferences so the best way to learn of developments in both associations is to form mutual discussion groups with colleagues who are members and who can report on each association.

The Special Libraries Association has several subdivisions of interest to the mathematics librarian. A personal membership includes one subject division and one geographic chapter. Additional chapters and divisions may be joined for a small fee. Thus, it is possible to belong to a state chapter, the Physics-Astronomy-Mathematics Division and the Science & Technology Divisions. Chapters offer monthly or quarterly meetings, usually after work, with a speaker on some topic of management or technology. If situated close to several state chapters, it is possible to shop around to see which one to join since one state chapter may have more interesting programs or more academic library members than the other.

The Physics-Astronomy-Mathematics Division of SLA (P-A-M) was formed nearly twenty years ago. Although the Division is not very big, with perhaps around 150 members, it has become a very important vehicle for communication between academic librarians around the country. It is quite easy to get to know and communicate with the 20 or so librarians who can be counted upon for assistance with obscure reference questions, informal interlibrary loan, information on "customer" behavior, and other aspects of working life. The *P-A-M Directory* lists these librarians.

From the beginning, the P-A-M Division has emphasized subject-oriented workshops on one of the three specialties, that is, mathematics or physics or astronomy. These workshops usually last for 1 1/2 hours with around 30 librarians in attendance, some of whom are math librarians, some math-physics librarians, some science librarians, and all variations in between. Some participants are new to the profession, some have been coming to these meetings for two decades. All seem to have some area of expertise and all are extremely generous in sharing their knowledge. As a result, each workshop participant learns something that enables him/her to give better service upon return to the library. Some topics are discussed year after year: bibliographic instruction, serials cancellation, or what the American Mathematical Society or Mathematical Reviews is doing that affects librarians. In the past few years, topics have included the AMS Library Survey, acquisition of Russian materials, e-Math and how it could be beneficial to libraries, remote storage, and mathematical statistics core journals. The discussion leader solicits topics first through an announcement in PAMnet and then announces those to be discussed in the May issue of the *P-A-M Bulletin*. It is often very difficult to complete the discussion within the time allotted so informal followups often occur nightly at the division suites during the conference.

Recently the P-A-M Division began sponsoring a vendor update so that online/database vendors or publishers could inform librarians about new products, journals, and books coming out in the next year. While many in the audience will have already visited their booths in the exhibition area, this gives all attendees a chance to hear more balanced, in-depth presentations. While announcements of new journals are always greeted with little enthusiasm, the interaction between librarians, vendors and publishers allows librarians to learn much about the publishing world and its particular challenges. The vendor update also gives the AMS a forum to explain its products.

The Science & Technology Division also offers programs of interest to the mathematics librarian. One of the largest divisions in SLA, this division sometimes co-sponsors programs with the P-A-M Division. The mathematics subject specialist in a large research university library reference department will certainly want to participate in this division. Mathematics librarians may wish to join this division as well as P-A-M as it is a good way to become more involved in projects of wider appeal.

PAMnet

PAMnet is an electronic discussion forum on Bitnet for astronomy, mathematics, and physics librarians. Although it was started by the P-A-M Division, one does not need to be a member of the Division to join. In fact, less than half of the nearly 100 subscribers are P-A-M members. Devoted users of e-mail will find this listserver an effective way to communicate informally and frequently with other librarians, publishers, researchers, etc. Not only is PAMnet more effective than the *P-A-M Bulletin* in making timely announcements, it is extremely useful in securing advice from colleagues. Recently, a mathematics librarian wanted to know what policies there were on issuing library keys to faculty and graduate students. Another time, PAMnet featured a discussion of the relative merits of the two competing mathematics CD-ROM products. The person asking for information summarizes and shares the responses with the others. It is a great way to ask for help with a reference question, to locate a missing issue of a journal, to warn others of imperfect copies of publications or to note a lag in publishing issues of a journal. It has really opened the world to mathematics librarians: they are no longer all alone in a one-professional library. For more specific information on PAMnet, consult the November 1991 issue of the *P-A-M Bulletin* or read the overview provided by Stern in *Special Libraries.*

Other Electronic Mail Sources

There are several other electronic sources being utilized by mathematicians with which math librarians should be familiar. Included in these are publication services such as IMP (Instant Math Preprints) from Yale University and MP ARC which is an electronic archive for research papers in Mathematical Physics.

Another, e-MATH, an electronic initiative by AMS in electronic communication and publishing, is being developed for mathematicians and is a node on the Internet. It is available to anyone with Internet access at no cost; users do not have to be AMS members: telnet e-math.ams.org or telnet 130.44.1.100, login: e-math, password: e-math. Librarians will find using e-MATH to access the membership directory or the MRC classification scheme easier than searching for their paper copy. Another application currently provides access to the *MR* document delivery system on a fee for service basis. A bulletin board is being planned which may prove valuable in the exchange of information between librarians and mathematicians. For further information, or for assistance in accessing and using e-MATH services, send email to: SUPPORT@E-MATH.AMS.COM.

Professional Literature

A personal membership in the Special Libraries Association includes a subscription to a number of publications of interest: the monthly *Special*

Libraries with news of the association and articles which are of interest to the mathematics librarian, the monthly newsletter *Specialist* with more timely announcements, and the quarterly *P-A-M Bulletin* and *Sci-Tech News* plus newsletters from the chapter(s) joined. These should be scanned except for the *P-A-M Bulletin,* which should be read cover-to-cover as soon as received.

A personal membership in the American Library Association and thus the Association of College and Research Libraries includes subscriptions to the monthly *American Libraries* and *College & Research Libraries News,* both lively professional news journals, and to *College & Research Libraries* where scholarly articles appear. The Science and Technology Section of ACRL produces two publications: the semi-annual newsletter, the *STS Signal* and a new, irregularly published electronic newsletter, *Issues in Science and Technology Librarianship.* The premier issue (December 1991) of the electronic publication contained articles on science literacy and reports from the Coalition for Newtworked Information and from a 1991 preconference on serials cancellations. Call or write the ACRL Office for more information on this publication. All of these should be scanned for articles of interest.

A special interest electronic publication, *Newsletter on Serials Pricing Issues,* supplies extremely current information on a topic which occupies a major part of any academic librarian's time. This is issued at frequent intervals as news becomes available. Address subscription requests to TUTTLE@UNC.BITNET.

It may be that the library administration or another branch library will subscribe to other library-related journals which publish articles that may be of interest. If possible, the table of contents of such titles as the quarterly *Science & Technology Libraries* (Binghamton, NY: Haworth Press) and *Journal of Academic Librarianship* (Ann Arbor, MI: Mountainside Publishing) should be scanned. Other special interest journals, for example, those on collection management, serials, bibliographic instruction, are listed in the references and further readings chapter.

The monthly *Notices of the American Mathematical Society* (Providence, RI) is highly recommended. This is "the" news journal for mathematicians and provides valuable insights into the concerns and directions of their professional society. Recent articles on the workings of *Mathematical Reviews* were of particular interest to mathematics librarians. Issues of the *Notices* and *The Mathematical Intelligencer* (New York: Springer-Verlag) are readily available in the library; take the time to scan them when they arrive.

CONCLUSION

The academic mathematics librarian will enjoy the excitement of working in two worlds for some time to come. S/he will balance the benefits of working with advances in information technology, (e.g., electronic mail, TeX, online bibliographic databases), with the assurance of knowing that tradition (e.g., conventional forms of publication, personal skills in offering access) will still be highly valued by mathematicians. Confidence in his/her abilities to provide the very best service, meeting the expectations of a critical yet appreciative group of library users, means that the mathematics librarian and the departmental library will continue to occupy an important role in the intellectual life of the department.

Chapter 3

MATHEMATICS LIBRARY SERVICE IN PUBLIC LIBRARIES

Robert Thornhill

INTRODUCTION

The place of mathematics materials in public library collections has received very little attention, perhaps because many people, including librarians, think math is esoteric and difficult. Librarians may not particularly like or feel comfortable with mathematics; many patrons probably feel the same. This, coupled with little education or background in the subject, means that librarians pay too little attention to mathematics collections and service. They simply consider mathematics materials with the rest of the science collection and do not give them very high priority. Yet people do need and use math materials, and their importance is increasing. Considerable attention has been paid recently to the innumeracy of the American population. Public libraries could be a major factor in alleviating this problem. That will require developing and maintaining good math collections and making them readily accessible to patrons. The aim of this chapter is to assist librarians with developing and providing access to mathematics collections in the public library setting.

Public libraries are probably more diverse than any other type of library. Their sizes range from small to very large more so than nearly any other type of library. Their budgets show an equal range. One finds them in every setting, from the largest cities to the smallest hamlets. They serve a more varied group of patrons, of all ages, educational levels, cultural and ethnic backgrounds, and socioeconomic status than either academic or special libraries. The more general and varied needs of these populations must be reflected in more general and varied collections. Collections must be highly diverse in both mathematical topics and level of difficulty included. Discerning the needs of this variety of patrons and locating and acquiring appropriate materials presents different problems than those faced by librarians in academic, school, and special libraries. Public librarians must consult a wider range of selection tools and sources. Reference and other services to patrons of elementary school age through adults of any background and educational level also require special attention. Given these difficulties and that math is a difficult and perhaps unpopular subject for many people, it is not surprising that it is often neglected.

PATRONS

Public libraries, by definition, are established to serve the public, that is, everyone in their communities. Their patrons usually include children and young adults, college and university students, adults needing help with remedial education, professional people, job seekers, "street people," and retired people, to name a few. That results in an extremely diverse patron community, with equally diverse needs and expectations of the library. All these people need and use math to some degree, from professionals who use it at a sophisticated level to "math phobics" who try to avoid it altogether. Patrons using mathematics materials and services often fall into several broad categories.

A very large patron group in most public libraries are students in elementary and high school. Their mathematical knowledge varies directly with their age and grade level, but their interest, ability, and mathematical sophistication can be at almost any level. They have in common that they all are learning math. These students will be primarily interested in help with what they are learning in school, understanding concepts and solving problems. Some may be interested in looking further into the field, at biographies of notable mathematicians, the history of mathematics, or general works discussing the ideas of mathematics or "what math is all about." These may be highly motivated students who want to learn more than they are given in school. They come to the library freely, motivated by interest and a desire to learn. For these students, a good math collection is especially important. Students want material in all the areas of math likely to be covered in their classes: arithmetic, algebra, geometry, finite math, analytical geometry, trigonometry, and, perhaps, beginning calculus. Most students today use calculators and computers as a matter of course and will want material on their use. A great many will dislike math and be apprehensive of it. Some may be suffering from math anxiety and need material to help them overcome it. Others may enjoy mathematical games and puzzles as a recreation. Whatever their need, they will probably need assistance from the catalog or a librarian to find and perhaps make use of the material; probably few will attempt to find what they want by browsing. They may, in fact, expect more help than the librarians can give, hoping to have problems solved for them or lengthy explanations because they "didn't understand the teacher." Since few librarians are also mathematicians or teachers of mathematics, they may find it difficult to provide this kind of assistance. Students will refer to materials for a formula or method of calculation, a definition, or an explanation of some specific concept. Problem-solving books, handbooks, books of tables and formulas, mathematical dictionaries, and workbooks will answer most of their requests. Students will probably tend to use this reference material, especially expensive handbooks and books of tables, in the library more often than not and will probably need help from librarians to find and use it. However, there is no reason why students (and others) cannot check out and take home most problem-solving and workbook-type books as well as narrative works intended to be read. In general, students will probably be frequent, though perhaps not enthusiastic, users of the math collection.

College and university students have much the same needs as lower grade students and use the math collection in much the same way. They are generally more sophisticated library users, however, and may require less assistance from librarians. Since colleges and universities nearly always have good libraries, these students probably make less use of public libraries. They need material at a higher level of difficulty than elementary and high school students, and in areas such as the calculus, analysis, and topology not found in lower grade curricula. Students at this level may also be more interested in mathematical philosophy and logic, history, and general mathematical concepts. They may want works dealing with applied math in fields such as engineering, computer science, business, and other scientific and technical disciplines. Though they may request reference service less often, when they do, it will probably be at a more demanding level for the librarian. They will also be more likely to request other services such as computer searches and interlibrary loan of unavailable material.

Another group of students who often make use of public libraries are those in vocational and technical schools. Sometimes these schools do not have libraries or have only limited ones and their students must rely on public libraries. Vocational students may be more reluctant to request a librarian's help, perhaps owing to less familiarity with libraries generally. They may not ask for services like computer searches or ILL unless the librarian suggests it. These students may require special attention from librarians since they are apt to have less educational experience. They need material dealing with practical, applied math, for carpenters, electricians, mechanics, plumbers, and so forth. Workbooks and study guides are useful for them, as are basic handbooks and compendia of math tables. Works with clear explanations of calculus are valuable. Basic textbooks are also useful. These students will probably want books to check out more than those for use in the library, so a good circulating collection is important. These patrons may not be as numerous as other groups but public library materials and services are especially important to them.

Teachers as well as students make frequent use of public libraries, of course. Mathematics teachers in a library's service area may become quite familiar with the public library's math collection and services and direct many students to use them. As users, they need the same materials as their students as well as others concerned with math education. They may also have strong interests in the philosophy and history of mathematics and in biographies of eminent mathematicians. Popularizations of mathematical themes may also interest them for recreational reading. Most of them will be knowledgeable library users and may make regular use of ILL, online literature searches, and even selective dissemination of information (SDI). They are likely to request searches of the ERIC database, for example, and to be interested in the mathematics periodical literature. Teachers and librarians are natural allies in education, and a cooperative, mutually supportive relationship is good for both.

Needing most of the same materials as students and teachers are patrons pursuing a self-study program in mathematics. This group includes those studying for employment and promotional examinations, for the GED, and for civil service exams as well as those educating themselves for the pleasure of learning. It also includes those patrons, both youth and adult, engaged in what can be called remedial learning, making up deficiencies in their education. Often these patrons are strongly motivated and make heavy use of a library's collection and services. They may spend long hours in the library studying, especially if unemployed or retired. Often they are very specific in their needs and express them precisely. They may, for example, request books by title and author rather than by topic. Like students and teachers, they will want a good selection of handbooks and compendia of mathematical or statistical tables. Those studying for exams will want test study guides in their fields, since math is included in virtually every occupation. Remedial learners will want review books and materials designed especially for them as well as textbooks. Some of these independent study patrons will make extensive use of ILL and online searching while others will not, depending on the object of their study. They will call on librarians to some extent for help with specific questions, but more often for guidance to the kind of materials they need. For these patrons, libraries take the place of schools, and a good collection and service are exceedingly important.

Another important group of patrons seeking math information is comprised of professional and technical users of math such as engineers, architects, poll-takers, business men and women, and electrical, construction, medical, and chemical technicians and workers. Anyone who uses math in his or her work or trade fits into this category. These patrons need specific information dealing with their work, formulas and methods of calculations, for example. Handbooks and texts geared to specific fields serve these people well. Often these books are classed in the various subject areas, but math librarians should be familiar with them since they deal mainly with practical, applied mathematics. These patrons seldom browse the collection and are not typically interested in aspects of math outside their expertise. They rarely read general works of mathematics, including popularizations. They usually come directly to the reference service point and ask for the material they want. Sometimes they know the specific source, sometimes they request a book that will help them solve one specific problem. Most of the time, they do not need to check out the book but want to use it for brief reference in the library. A good collection of handbooks and texts in the various technical fields is necessary to serve these patrons.

Finally comes the largest and most varied group of math patrons, the "general readers." It includes all patrons who do not fit into one of the other, more clearly defined groups. Such patrons may be recreational readers who enjoy books with mathematical themes, or someone with an occasional need for a formula or calculation, or someone wanting to interpret tables of data or graphs from a newspaper or magazine article. At some time, nearly every library patron will need mathematical information, and it can be anything at all involving mathematics and numbers. It is difficult to generalize about such a group. They may ask for what they want or they may find it themselves by browsing or consulting the catalog. They may inquire for a specific formula or calculation by telephone. Librarians must be prepared to answer questions and assist patrons to find answers to the problems they present. Often, patrons themselves are confused or unclear about what they need or do not know how to express those needs clearly. This can be challenging for librarians and may make great demands on their knowledge and skill. Some experience and education in math can be valuable to librarians dealing with many math-related requests. The kinds of information needed are often found in introductory textbooks or in general works on broad mathematical topics, but general users will make use of the entire collection.

THE COLLECTION

In general terms, the traditional philosophy of public library collections is that they contain the right material for each and every patron. Public libraries are open to everyone, therefore the collection must contain material for everyone. This means that the mathematics collection must be of sufficient size and scope that everyone in the community can find the mathematical information needed, regardless of level of education or prior mathematical knowledge. This is a laudable goal, but one that is not always attainable in practice. The realities of budget and space limitations usually mean that libraries fall short of this ideal. Thus librarians must get to know their patrons well enough to develop collections that will satisfy their needs as far as possible and must utilize online searching, interlibrary loan, and alternative services when the collection is not adequate. They must define the purposes of the math collection within the framework of the library's stated mission and goals and establish standards and guidelines for collection development to satisfy those purposes.

The purposes of a mathematics collection in a public library are several. First, it is educational. In a broad sense, all mathematical books are educational in that one learns from them, even popularizations and puzzle books. More narrowly, textbooks, study guides, and problem-solving books are designed for education. Most communities today have school libraries so that curriculum support is not a function of the public library. Sometimes, however, school libraries do not exist or are inadequate and the public library fills this role. In this situation, supporting the math curriculum is an educational purpose of the library. Another is to provide materials that will stimulate the growth of mathematical knowledge in the community, enabling people to better deal with aspects of their lives involving mathematics and fostering an appreciation of the utility and beauty of math.

Another purpose is to provide factual technical information needed by people to accomplish specific tasks. Architects and engineers, for example, use mathematics in all aspects of their work and may need materials not available in their personal libraries. People doing surveys and polls may need statistical information to analyze their data. Homeowners planning an addition may need to calculate amounts and costs of materials required. Investors may want to calculate profits, losses and yields on their securities. Thus, the need for formulas and methods of calculation pervades everyone's daily life. For most individuals, if the specific mathematical information is not available to them in their home or office, the public library can be the most convenient place to obtain it.

Few people think of mathematics books as entertaining or recreational reading. Yet books about mathematics can be read, just as biography or history or science can be, for pleasure, for ideas, for entertainment, for an understanding of civilization and culture, for an appreciation of the scientific and humanist contribution to western culture of mathematical thought. Many people enjoy recreational mathematics, problem solving, puzzles, and mathematical games. Others derive pleasure from "figuring the odds" and gambling. Some people are attracted by the peculiarities and mystical association of numbers and their lore. Providing materials to satisfy these entertainment needs is another purpose of the public library math collection.

Mathematics books for children are especially important in public libraries. A good collection should contain books that are instructional, recreational, and entertaining. Math books are published for all ages of young people, beginning with simple counting books for very young children. Not all math books for children are textbooks and workbooks. Some introduce abstract mathematical concepts in nonmathematical ways (no signs and symbols) that are fun and entertaining, notably some of the books by Masaichiro and Mitsumasa Anno. Abstract mathematical concepts are often difficult to explain and illustrate at a child's level of understanding, and the Annos' books do this very well. The *Annos' Mysterious Multiplying Jar* (New York: Philomel Books,

1983), for example, introduces factorials. Presenting mathematics in such entertaining, enjoyable ways can foster a lifelong interest in and understanding of mathematics. Children's librarians need to pay special attention to finding and purchasing books with mathematical themes.

Collection Development

In order to satisfy these purposes, libraries must establish criteria and policies for developing the mathematics collection. These guidelines must address questions of the size and scope of the collection, the kind of materials to be included, the knowledge level of materials collected, the quality of the materials, and the level of collecting intensity and collection density. The ALA *Guidelines for Collection Development: Guidelines for the Formulation of Collection Development Policies* (Chicago: American Library Association, 1979) provide essential guidance in formulating and implementing collection development policies and procedures. Several public libraries, including the Enoch Pratt Free Library of Baltimore and the Fairfax County Library in Virginia, have published their collection development policies. Other public libraries may want to use these as models in formulating their own policies. *Policies and Procedures for Selection of Instructional Materials* by the American Association of School Librarians (Chicago: American Library Association, 1989) may also be of some use. The literature of librarianship contains myriad articles and books on collection development. A great many of these works deal with collection development and book selection in public libraries although not specifically with mathematics. Virtually nothing has been published on mathematics collections in public libraries specifically, but much of what has been published about school and academic libraries applies to public libraries as well. In formulating library policies and criteria, librarians will have to rely on their knowledge of their patron communities. Written statements of collection development policies should explicitly address the mathematics collection.

The ALA *Guidelines for Collection Development: Guidelines for the Formulation of Collection Development Policies* (Chicago: American Library Association, 1979) set forth a hierarchy of levels of collection density and collecting intensity. These levels "describe the range and diversity of titles and forms of materials." From the highest to lowest, the levels are comprehensive, research, study (further broken into advanced study and initial study levels), basic, and minimal. Even the smallest public libraries should have mathematics collections at the minimal level. Larger libraries should have collections at the basic and initial study levels, adequate to support independent and undergraduate college level study. Large public libraries need collections at the advanced study level which can support advanced undergraduate, master's level and sustained independent study. Only a few, the largest urban public libraries which function as research libraries, may want collections at the research level. Professional mathematicians and researchers at this level are nearly always associated with universities and research institutes and rarely need research collections in public libraries. It is a good idea to incorporate a statement of the collection level into the collection development policy.

In addition to the level of collecting intensity and collection density, the detailed policy statement for mathematics should indicate languages, chronological periods, geographic areas, forms of material, and "the library unit or selector with primary selection responsibility for the field" (see *Guidelines* above). In American public libraries, it is highly unlikely that mathematics books in any other language but English will be collected. Foreign language math books are difficult to obtain, expensive, and would receive very little use. Exceptions may occur in places with very large populations of recent immigrants, where there might be a demand for a small collection in specific languages. There are many works from France, the former Soviet Union nations, and Germany available. Larger libraries with higher level collections may find imported books translated into English worthwhile. All chronological periods are worth collecting in mathematics. As for form of material, almost all are printed books and periodicals, although some instructional material in other forms may be available. Policy statements sometimes contain references to the quality of material as well. Probably very few librarians have a mathematical background sufficient to evaluate most mathematics books as to content. Materials should, however, be of acceptable technical quality, readable, and typographically attractive. In looking at any title for selection or retention in the collection, librarians must decide for what users it is intended, apply an appropriate standard of usefulness, and make a decision as to the suitability of that work. No item will be useful for all patrons, but each title should be utilized by a significant group of them.

The question of whether or not to include textbooks in the collection must be answered in each library based on user needs and general policies. Most public libraries do not include them, since supplying school textbooks is not one of their purposes. A few widely accepted basic textbooks, however, are a worthwhile addition to a math collection. They provide much basic knowledge not easily found in other places. They are also good reference sources for the librarian answering "ready reference" questions. In addition, they are good supplementary texts for students using other tests for their courses or for self-study. Answer books, study guides, teachers' guides, and other supplementary material associated with particular textbooks are usually not appropriate for public library collections.

Collection development implies more than simply book selection. It involves collection evaluation and deselection (or "weeding") as well. Standard methods of evaluation apply. Comparing the collection to standard lists or to the catalogs of comparable libraries with "good" collections is a frequently used technique. Several lists are available, such as those published by the Mathematical Association of America and the *Public Library Catalog* (New York: H.W. Wilson Company, 1984). Counts of the numbers of titles in the subdivisions of the math classifications (510-519 for Dewey and QA for Library of Congress) may also be useful. Numbers should be reasonably evenly distributed among the various classifications, although those topics in demand should be emphasized. Librarians should also pay attention to patrons' requests. Unusually frequent requests for unavailable material on a particular topic may mean that the collection is inadequate in that area. Whatever the method of evaluation used, there is no substitute for librarians' knowledge of, or familiarity with, their collections and of their patrons in assessing the collection.

Unlike some fields, mathematics books generally do not quickly become obsolete. Older works are usually just as valuable as new ones. Deselection is usually done more on the basis of book condition or replacement by a new edition than on obsolescence of content. The collection development policy statement should contain criteria for deselection as well as for the selection of new material.

An established program of collection evaluation and management is important and should be described in the library's policy statement. If the library has an established evaluation program, it should be followed for the mathematics collection. If there is no such program for the library, the mathematics librarian should develop and implement a plan of regular and systematic inspection and review of the collection. For small collections, this will not be a difficult task, and librarians can accomplish it without a great expenditure of time and effort. For larger collections, it may be necessary to use a formal manual of procedures, standard forms, and keep perpetual written records. Whatever the method used, it should be a permanent, ongoing part of the mathematics librarian's task to insure the best and most useful collection possible.

Selection Tools

Finding materials in mathematics presents no greater difficulty than finding materials in any other scientific or technical subject. There are a number of selection tools available. Bibliographies are some of the most important (see pp. 80-86). For retrospective collection development, many of the standard lists are useful, including *Books for Public Libraries* (Chicago: American Library Association, 1982), *Public Library Catalog* (New York: H.W. Wilson Company, 1984), *The AAAS Science Book List: 1978-1986* (see p. 80), *The Best Science Books and A-V Materials for Children* (Washington, DC: American Association for the Advancement of Science, 1988) and others. There are some lists dealing exclusively with mathematics. In 1992, the Mathematical

Association of America revised its out-of-date basic library lists on undergraduate mathematics: *Library Recommendations for Undergraduate Mathematics* (see p. 85) and *Two-Year College Mathematics Library Recommendations* (see p. 86). An eighth edition of William L. Schaaf's *The High School Mathematics Library* (Reston, VA: National Council of Teachers of Mathematics, 1987) was recently released. Those lists oriented to school or academic libraries will be useful for public libraries. M.E.B. Martin has prepared *Mathematics for the Non-Mathematician: An Annotated Booklist Based on the Collections of the Halifax City Regional Library* (Halifax, NS: Halifax City Regional Library, 1985). The works it contains are an introduction to mathematics "prepared for the general reader who wishes to get a feeling for mathematics and an understanding of the fundamental ideas of modern mathematics." Although brief, it is excellent, and all public libraries probably should own all the books it includes. *Books in Print* (New York: R.R. Bowker) and *Children's Books in Print* (New York: R.R. Bowker) are, of course, indispensable.

For recently published materials, there are perhaps fewer resources for locating mathematics books than for books in other fields. Publishers' catalogs and vendors' lists are, of course, available. *Publishers Weekly* (New York: Cahner) lists new mathematics books as they are published. The American Mathematical Society distributes its lists to libraries.

Reviews of mathematics books are relatively scarce. *Science Books and Films: The Quarterly Review* (Washington, DC: American Association for the Advancement of Science) includes math books. *Science News* (Washington, DC: Science Service) has short reviews of mathematics and science books of interest to the general reader. Other sources of reviews are the mathematics periodicals, but most public libraries subscribe to few or none of these. *Mathematics Teacher* (see p. 348), for example, has reviews of new books of interest to students and teachers. The Telegraphic Reviews section of *The American Mathematical Monthly* (see p. 325) is also a source for good reviews. Periodicals in other scientific and technical disciplines often carry reviews of applied mathematics books, and some may be found in *New Technical Books* (New York: New York Public Library). *American Reference Books Annual* (Englewood, CO: Libraries Unlimited) includes new mathematics reference books. The more general review media include only a limited number of mathematics books. *Choice* (Chicago: American Library Association), *Library Journal* (New York: R.R. Bowker), *School Library Journal* (New York: R.R. Bowker), *American Libraries* (Chicago: American Library Association), and *Booklist* (Chicago: American Library Association) contain almost no reviews of mathematics books. Reviews of some general, popular works may appear in *The New York Times Book Review* (New York: *New York Times*), *Newsweek* (Los Angeles: Newsweek, Inc.), and other popular publications. Teachers, statisticians, and others who use mathematics professionally are also good sources of information about new books, as are patron requests for mathematics books, but these cannot be counted on for regular information. Librarians who select mathematics books must rely more on simple lists and catalogs than on reviews as a rule.

Acquisition

The budget for a mathematics collection depends entirely on the library's budget, which depends on the level of community support for the library. In larger libraries, there may be a separate line item for mathematics in the budget. Smaller libraries may include it in the total amounts for science or nonfiction. Every public library has its own method of allocating money among the different subject areas. The amount apportioned to mathematics depends on the allocation criteria, patron and collection needs, and on the person(s) doing the allocation. Where no one has a strong interest in mathematics and the mathematics collection, and allocation criteria are subjective, it is apt to be slighted. For this reason, criteria that are as objective as possible are preferable.

Libraries usually have no difficulty procuring math books through their regular vendors. If some titles cannot be obtained from a vendor who deals mainly with public libraries, they can usually be had from another who primarily serves academic and/or school libraries. For better discounts, some titles may have to be purchased directly from the publisher. This is true, for example, for those published by the Mathematical Association of America or the American Mathematical Society. Buying directly from the publisher does not present a problem in most libraries.

Some mathematics books may be acquired as gifts and donations. More often than not, these are old textbooks from college or high school and are not useful or suitable for the collection. Acceptance of donated books will be governed by the library's policies, and if the policy permits, old textbooks should be refused. If they must be accepted, they may be sold at used book sales or simply discarded. Occasionally, however, useful popular or technical math books may be received. Technical or business people may donate older editions of handbooks, books of tables, and so forth. Since these books are often very expensive and the information in older editions is still valid, these may be worthy additions to the collection, especially if the library does not own the newer editions. Librarians should be cautious in accepting most gifts and donations, however. If a particular work is not useful to the donor, it may not be useful to anyone else either.

Scope of the Collection

The scope of the collection will be mainly determined by the needs of the patrons, and should contain something from all the areas of mathematics:

general works, introductory works, arithmetic, algebra, geometry, and so forth. It should also contain material at all levels of difficulty likely to be used. Librarians will want to emphasize those areas and levels most needed by their patrons, but at least a few titles should be available in each area and at a level usable by the majority of patrons, in even the smallest public libraries. There should also be at least a few volumes of mathematical history and biography. There should be a minimum of one or two mathematical periodicals as well. Although perhaps not absolutely essential, a few of the great classics of mathematical literature are also important. A few of the works of Euclid, Newton, Leibnitz, Russell, Whitehead, and Ramanujan add greatly to both the breadth and depth of a mathematics collection. Research monographs, conference proceedings, and advanced professional-level treatises are of little use in public libraries and should be excluded from all but the very largest collections. The size of the collection will be determined by the interaction of demand for mathematics material and the library budget, but the numbers of books should be adequate to reasonably satisfy patron needs.

General and introductory works are especially important in public library collections. They are intended for a wide spectrum of readers and are readily understandable by most people regardless of mathematical background. The books listed in *Mathematics for the Non-Mathematician* (Halifax, NS: Halifax City Regional Library, 1985) are good examples of this kind of work. There are many others. *Mathematics: An Introduction to its Spirit and Use: Readings from Scientific American* (San Francisco: W.H. Freeman, 1979) gives an overview of the study of mathematics. *Mathematics in the Real World* (Boston: Birkhäuser, 1979) depicts the world of mathematics as it exists today, its usefulness and application in other fields, and implications for future development. *Mathematics Today! Twelve Informal Essays*, edited by Lynn A. Steen (New York: Springer-Verlag, 1978) is similar and stresses the relevance of mathematics for everyone. More recent titles of high quality are *The Mathematical Experience* by Philip Davis and Reuben Hersh (see p. 73), *Mathematics: The Loss of Certainty* by Morris Kline (see p. 152), *Mathematics and the Unexpected* by Ivar Ekeland (see p. 151), and *The Mathematical Tourist: Snapshots of Modern Mathematics* by Ivars Peterson (see p. 150). Examples of other books on more specialized topics but intended for the educated nonmathematician are Douglas Hofstadter's *Gödel, Escher, Bach* (see p. 147), and *Chaos* by James Gleick (New York: Viking, 1987). Books of this type may be the most used in the mathematics collection.

History and biography are also important in the collection. *An Introduction to the History of Mathematics. Sixth Edition* by Howard W. Eves (Philadelphia: Saunders, 1990) and *History of Mathematics* by Florian Cajori (see p. 232) are two fine general histories of the field. A shorter work suitable for younger readers is *A Concise History of Mathematics* by Dirk J. Struik (see p. 228). Another general history intended for general readers, students, and teachers is *Historical*

Roots of Elementary Mathematics by Lucas N. Bunt and others (Englewood Cliffs, NJ: Prentice-Hall, 1976). *Men and Institutions in American Mathematics* (Lubbock, TX: Texas Tech University Press, 1976) is a collection of conference papers on different historical aspects of mathematics in this country and emphasizes the twentieth century. Another history with a somewhat different approach is *Mathematicians and Their Times: History of Mathematics and Mathematics of History* by L.C. Young (New York: Elsevier North-Holland, 1981). In addition to biographies of noted individual mathematicians such as Bertrand Russell, Alfred North Whitehead, Ramanujan, and Ada Lovelace, there are a number of collections of short biographies. A classic one is Eric Temple Bell's *Men of Mathematics* (see p. 138). Another, aimed at people unfamiliar with mathematicians, is *Makers of Mathematics* by Alfred Hooper (New York: Random House, 1958). A recent work in this area is *Mathematical People: Profiles and Interviews* edited by Donald J. Albers and G.L. Alexanderson (Boston: Birkhäuser, 1985). Two works dealing specifically with women mathematicians are *Math Equals: Biographies of Women Mathematicians and Related Activities* by Teri Perl (see p. 138) and *Women in Mathematics* by Lynn M. Osen (see p. 240). A useful source for further biographical information about women in mathematics is *Women of Mathematics: A Bibliographic Sourcebook* edited by Louise S. Grinstein and Paul J. Campbell (see p. 139). For information on black mathematicians, see *Black Mathematicians and Their Works* edited by Virginia K. Newell and others (Ardmore, PA: Dorrance and Company, 1980). These are only a few of many works available. Librarians can find others by consulting one of the guides to mathematical literature, *Books in Print*, or other references sources to scientific biography.

Works covering individual aspects of mathematics are also necessary in the collection. There should be at least a few titles available in each of these areas: arithmetic, algebra, trigonometry, analytical geometry, finite math, calculus, set theory, number theory, analysis, logic, topology, and statistics and probability. A great many publishers put out books by many authors in each of these areas, and it is difficult to recommend any individual titles as many are of good quality, giving clear and concise explanations with attractive typography. Many, but not all, are textbooks. Some are intended for general readers or for those studying independently, especially those who are learning and developing mathematical skills at a basic level. The most useful textbooks are those at an introductory level which present a broad, general survey of one particular area of mathematics. If there is need and the budget permits, more advanced and specialized works can be added in any area. Ching-Chih Chen lists many worthwhile ones in the chapter on monographs and treatises in her *Scientific and Technical Information Sources* (Cambridge, MA: MIT Press, 1987).

All of the above works, except periodicals, should be available for the patron to check out. A minimum reference collection for use only in the library consists of a dictionary of mathematics, a concise encyclopedia of mathematics, the *CRC Handbook of Mathematical Sciences* (see p. 107), the *CRC Standard Mathematical Tables* (see p. 127), a compendium of functions such as the *Handbook of Mathematical Functions with Formulas, Graphs, and Tables* (see p. 112), and a compendium of statistical tables, of which many are available. A book of conversion tables is also a basic necessity. There are several good mathematical dictionaries available, including *Dictionary of Mathematics* by T. Alaric and William Millington (see p. 89), *Facts on File Dictionary of Mathematics* (see p. 91), and William Karush's *The Crescent Dictionary of Mathematics* (see p. 88). A more recent one is Douglas Downing's *Dictionary of Mathematics Terms* (see p. 89). The *VNR Concise Encyclopedia of Mathematics* (see p. 102) is a good, brief encyclopedia. Larger libraries will want to add more examples of these works as well as more specialized mathematical dictionaries, encyclopedias, and compendia of tables and functions. Many are available in the various areas of mathematics, especially in applied mathematics and statistics. A notable example is the multivolume *Encyclopedia of Statistical Sciences* (see p. 100).

Beyond these basic types of reference works, there are others needed in large mathematics collections. Guides to the literature are very useful. Two representative examples are *Use of Mathematical Literature* edited by Alison R. Dorling (see p. 105) and *Guide to the Literature of Mathematics and Physics Including Related Works on Engineering Science* by Nathan G. Parke (see p. 104). There are at least two guides to the various tables used in mathematics: *Guide to Tables in Mathematical Statistics* by J.A. Greenwood and H.O. Hartley (see p. 128) and *An Index of Mathematical Tables* by A. Fletcher and others (see p. 128). The basic abstracting and indexing service for mathematical literature is *Mathematical Reviews* (see p. 122). An electronic equivalent called "MathSci" is available online from the commonly used vendors of electronic information services. For statistics, there is *Statistical Theory and Method Abstracts* (see p. 126). For those needing directories, there is the *World Directory of Mathematicians* (see p. 142) published by the International Mathematical Union and distributed in the United States by the American Mathematical Society, which also publishes the *Mathematical Sciences Professional Directory* (see p. 141). Dated but still of some use is the American Mathematical Society's *Directory of Women in the Mathematical Sciences*, (see p. 140) as is the Mathematical Association of America's *Guidebook to Departments in the Mathematical Sciences in the United States and Canada* (see p. 141). A major treatise which should probably be assigned to the reference section of the library is *Fundamentals of Mathematics*, in three volumes, edited by H. Behnke and others (see p. 147). There are also reference works in statistics, logic, history, and other specialized areas of mathematics that are useful additions to extensive reference collections. Librarians can find these and many other reference sources in mathematics in Eugene P. Sheehy's *Guide to Reference Books* (Chicago: American Library Association, 1986), Ching-Chih Chen's *Scientific and Technical Information Sources* (Cambridge, MA: MIT Press, 1987), *Finding Answers in Science and Technology* by Alice Primack (New York: Van Nostrand Reinhold, 1983), and the other guides to science and technology reference sources that are available. All librarians concerned with science and technology should be familiar with these.

No collection is complete without periodical literature, and there are many mathematical periodicals published. Most of them are at the research level, intended for professional mathematicians, and not suitable for most public libraries. Those appropriate for a public library have been listed in Chapter 8 of this book. It is important to examine representative copies of journals and magazines before ordering subscriptions, as titles are not always descriptive. Those titles selected should be appropriate for the needs of the users and for integration with other materials in the collection both in subject matter and level of difficulty. For effective use, the collection must also include indexes for the titles chosen.

REFERENCE SERVICES

Effective reference service in mathematics requires that librarians have a thorough knowledge of the collection and the various reference materials giving access to it as well as other sources for finding relevant materials not in the mathematics collection. It also requires being able to communicate effectively with patrons at varying levels of mathematical knowledge in order to determine precisely what the patron needs. Librarians who provide mathematics service do not have to be mathematicians or mathematics teachers, but they should have some understanding of mathematical language: alphabetic, numeric, and symbolic. The job also requires patience and understanding of patrons who dislike or fear the subject. This is true of other subjects as well, of course, but anxious patrons seem to be encountered more often when dealing with mathematics. Once the information has been located, the librarian must often assist the patron in making use of it. Only when the patron has what he or she requires is the service completed.

Librarians giving mathematics reference service in public libraries should be prepared for an almost endless variety of questions. According to Katz, questions may be categorized into three types. Ready reference involves responding to factual, easily answered questions. Somewhat more difficult are specific search questions which require pulling together a number of sources on a given topic. The third type may be termed research questions where the inquiry is more scholarly and may involve locating resources outside the library. The level of difficulty is inversely related to the frequency of encounter. Research questions are much less common than relatively simple factual ones.

Examples of the first type of question are requests for the formula for the circumference of a circle if the radius is known or for the date of birth of a prominent mathematician. "What are fractals and why are they important?" is an example of the second type. Research into the applications of Markov models in various fields is a question of the third type. Most public library collections would contain the answers to the first two questions and the librarian could supply them without difficulty. The answer to the third might involve searching print and online indexes and requesting books and articles through interlibrary loan.

Patrons often formulate questions in terms of general topics and prescribed sources. Rather than accept these questions at face value, librarians will use reference interview techniques to determine the actual information the patrons need. It often helps to know the context of the patron's question, or why the information is needed. If a patron asks for information on fractals, for example, it helps to know whether s/he has any knowledge of what fractals are or has ever worked with them. The way librarians phrase their questions is also important. They should ask "open" questions to elicit the most useful response from the patron. Questions beginning with the words who, what, why, where, or when are apt to be open and get the best response from patrons. "Closed" questions, those beginning with is, was, do, or does, often lead to games of twenty questions that result in frustration for both the librarian and the patron. This applies to all reference questions, including mathematical ones. Questions such as "What are you trying to calculate with this formula?" and "Where did you read about this mathematician?" will probably elicit more information than those with simple yes or no answers.

Reference librarians assisting patrons with mathematical questions must be thoroughly familiar with the many guidebooks, compendia of tables, directories, guides, indexes, and other reference sources as well as with the books in the general math collection. Since patrons usually require very precise information and because of the complex nature of the field, math reference may require more time and painstaking effort than reference work in other fields. Giving a patron the wrong formula or method of calculation, or a formula with a single incorrect operational sign, may have serious negative consequences for that patron. The greater the librarian's knowledge of reference sources and the collection, the less likely this is to happen.

As a general rule of thumb, ready reference, factual answers can usually be found in handbooks, tables, and introductory textbooks. More extensive questions may involve consulting the catalog, various mathematical indexes, and other sources to find books and periodicals on the subject. Advanced research questions may require librarians to search not only any of the previously mentioned sources in greater depth but sources in other fields and one or more online databases as well, and to assist the patron in using the sources and perhaps guiding him/her through the process of bibliographic research in mathematics. Librarians must also be familiar with the different branches of mathematics and the terminology used in them. If a patron is interested in chi-square, for example, librarians should know that it is a term used in statistics rather than algebra or geometry. Looking up an unknown term in a mathematical dictionary usually gives some clue as to where it is used. It is not always necessary to find answers to mathematical questions in mathematical sources. For some questions where a lengthy answer is not required, librarians should not neglect general reference works such as encyclopedias. *Americana* (Danbury, CT: Grolier), *Britannica* (Chicago: Encyclopaedia Britannica), and *World Book* (Chicago: World Book) have good articles on many mathematical topics. Older editions of encyclopedias may still be useful. The old eleventh edition of *Britannica* has some excellent articles on mathematics.

Many public libraries provide telephone reference service, although generally only brief, factual answers can be given by telephone and many libraries restrict the time allowed for telephone patrons. Patrons at the reference desk do not like to wait while a librarian is answering a telephone question. Some mathematical questions are very difficult to answer by telephone, especially those involving formulas with symbols that neither librarian nor patron may know. In such cases, the librarian should advise the patron to come to the library where they can determine the formula exactly. Each library will have policies for telephone reference service that should be followed in answering mathematics questions as well as for answering questions in other fields.

SPECIAL SERVICES

Searching online databases is a service provided by most public libraries today. Mathematical information can be found in MathSci Online, a database of mathematical literature, as well as in numerous other scientific and technical databases. They are available through several database vendors used by public libraries. Searches are done using the commands and protocols of the vendors, who also supply information about searching individual databases. O.T. Hawkins reviewed online physical sciences and mathematical databases in three issues of *Database* (see p. 75) in 1985. Once citations are obtained from a database, the articles can be requested through interlibrary loan if the library does not have the periodicals needed.

Another service offered by many online services vendors is selective dissemination of information (SDI). Few public libraries provide this expensive service unless the patron agrees to pay the charges.

Some other services may be of interest and use to mathematics patrons. Public libraries commonly display new books in browsing sections apart from the rest of the collection. Displaying new mathematics books in this way can stimulate interest in the subject and perhaps increase use of mathematics materials. Acquisition lists are another way of letting people know about new books. They are perhaps more often seen in academic than in public libraries but are an effective way of calling attention to recent acquisitions. If the size of the library and number of acquisitions warrant, a separate list could be produced for mathematics books. Setting up separate, small collections of mathematics books for special purposes is another service some libraries may want to provide. Books useful for students doing math homework is one possible collection. Another is a special collection of material for students doing mathematical science fair projects. The value of setting up these collections would depend on the need for such services.

Most public libraries include video cassettes in their collections. Not a great many are available dealing with mathematics, but there are a few. The Public Broadcasting Service's PBS Video sells a series called *Futures* (Alexandria, VA: PBS Video) with teacher Jaime Escalante, known from the movie "Stand and Deliver." This is a series of motivational programs for students in grades 6 through 12 that shows the relevance and fun of mathematics. While instructional, they are lively and entertaining. PBS Video has recently released a second series called *Futures^2* that continues where the first series left off. These would be excellent additions to any public library collection. There are other instructional videos in mathematics available from educational producers, but these are usually too costly for most public libraries.

CONCLUSION

Mathematics collections and services in public libraries have received little attention from either professional librarians or library users. No studies have been done, but it seems probable that both collections and services in most public libraries are adequate at best, and are often only minimal. Yet all patrons need mathematical information at some time, and the public library is the most convenient source for most of them to use. This chapter has pointed out some aspects of working with mathematics collections and providing mathematical services in public libraries. Given the widespread problem of innumeracy in the United States, these collections and services are becoming increasingly more important.

Chapter 4

MATHEMATICS LIBRARY SERVICE IN SPECIAL LIBRARIES

Gary Davidoff

INTRODUCTION

Congratulations. You have convinced everyone, up through the Chief Executive Officer, that you are the perfect person to handle the job, even though you have not had a mathematics class since high school algebra. Yes, your history degree demonstrates your ability to act and think critically. And, yes, your MLS demonstrates your thorough grounding in the tools of your trade. But today you have to talk to mathematicians, who, second to philosophers, with whom they are kindred spirits, seem to interact very strangely with the real world. Well, take heart; this could be your first day in a chemistry library.

The universe that is covered by the name "government and special libraries" is especially broad when compared to other types of libraries. One travels from the solely-for-profit business to the professional not-for-profit society to the government research and development (R&D) organization, drowning in red tape (though, luckily, not here at Argonne).

There are, of course, no tried and true rules of how to succeed in your organization. What works in one library might well be destructive in another. In the discussions that follow, you must decide what is appropriate to your own situation. A summary of my experience might be helpful in this evaluation.

Argonne National Laboratory is an R&D facility owned by the United States Department of Energy and operated under contract by the University of Chicago. It is a multipurpose laboratory covering a broad range of energy research. The facility itself is spread out over a large area. As a result, the site is served by 10 libraries tied together by centralized processing (library support services). Each library is situated in the cluster of buildings in which related research is conducted, so that we have a Materials Science Library, a Chemical Engineering Library, and so forth. I am in charge of the Mathematics/

Computer Science/Nuclear and General Physics Library and have a half-time clerical assistant. I have been pushing the professionalism of the information specialist at Argonne for eight years. Prior to that, I was the Science Librarian at Juniata College, a small liberal arts college (we must all start somewhere).

A few thoughts before we begin to deal with your terror. As most psychiatrists would agree, very few problems or fears are unique. There are many people who have walked into a new situation, unsure of themselves or of their ability to survive. Most do survive. And just as you will be facing people who speak a strange language, it will amaze you to learn how poorly they speak your language. Since most scientists still assume all we (librarians*) do is shelve and file, early success will win most of them to your side. And what is most beautiful is that only a minority of your patrons will ever be aware that you are capable of making errors. As you speak to other information specialists, they will gladly regale you with stories of their mistakes. It comes with the job.

PRECAUTIONS

I know this seems like a strange place to start, but you can save yourself future hassles by proper preventive medicine. The list might be endless, but we will start with two that will demonstrate the idea.

One day, shortly after your have started, a user will walk in and tell you, "Do this; after all you are here to serve me." My response, when I was actually told this, was "No, I am here to work with the scientists, not to serve them. Now what is the problem?" (Well, actually, I was not quite that glib, but it was the gist of my answer.) From the very start, you must be a professional and a colleague. Those librarians I know who did not assert this at the beginning had to fight to achieve that status. It took time, and some humiliation. Of course, if it is your boss, bow from the waist, and jump. Use common sense.

If it is a new library you are walking into, you have to decide on a circulation procedure. At Argonne we have no due dates for books, with recall after two weeks if another scientist needs the book. However, there is no going back. Transition to a two-month loan period would create a ruckus I would not want to face. The point being, certain decisions are essentially one way. Protect yourself by starting with the conservative position. Move to the more liberal position when you have enough experience to understand its implications. You gain users' appreciation and avoid adverse publicity for the library.

I will touch on these ideas repeatedly. To a large extent, you control how you are perceived.

CLIENTELE

From the first day you start this job, you will be involved in an ongoing process that never ends. The day you have finally pinned down the interests and needs of every person who uses your library, a new project will be initiated by management, bringing in new people, changing the orientation of your regulars, and calling forth new materials and services from yourself. To participate fully in this process, you must know who your clients are. Are they applied or theoretical scientists? Are they involved in bringing products to market or in long-term research? Are they involved in their own project area or do they principally serve other projects (such as statisticians might)? Also, it is not enough to react solely to users' questions. You must be prepared to anticipate new directions. However, you do not need to know all these at once; but where to start?

Be assertive. As people become familiar, ask them what they do. Do not worry about understanding everything (anything). Find out if it is theoretical or applied. Are they involved in a particular project? Do they usually find the materials they need in the library? Are they interested in journals, books, reports? The majority of people who use the library regularly are glad to help you out. Over time you will be able to classify the users.

Do they get together for coffee at a regular time during the day? Drop in for your own cup. Often you can hear about new initiatives that may be coming up. You may also learn about the power structure of your organization. It is always important to avoid making the wrong enemies. Here you can ask people in small groups about the library. On an individual basis, they will often just tell you how wonderful this library is, but in a small group, they may end up being more critical (in the positive sense).

Look at the organization of your company. Are your users arranged in groups? If your organization is more of a business, it may be arranged by projects. If it is more R&D oriented, it may have a divisional arrangement (like that of universities). In either case, there will be units that hold meetings. As you gain people's confidence, ask to attend some of the general meetings. Caution: avoid having them perceive you as a spy. Be placed on the mailing list for divisional reports such as five-year plans.

Are there general interest seminars offered? Just appearing at these enhances your professionalism. You gain reality outside of the circulation desk. Attend some technical seminars, standing close to the door so you can leave early, but even five minutes of introduction to each seminar will add up to useful knowledge. Do any of the groups offer training? At Argonne, the Mathematics and Computing Sciences Division offers a training program in parallel computing. While I do not intend to become a programmer at this level, I have attended part of the three-day training session as an observer, just to gain a better understanding of what that group does.

Are you willing to do research? Use your library tools to track down books, articles, and reports by your users (or you can ask them directly—I have always been too embarrassed). Where do they publish? What kinds of references do they use? Are their references current? Do you need to push them to make better use of the library's resources? Do they seem oblivious to the uses of the patent literature? You will find out that most users wear blinders; they do not realize there is more to the library than they are already aware of.

RELATIONSHIP TO THE REST OF THE ORGANIZATION

Of necessity, I must paint with a fairly broad brush. In any real situation, there are people who must simply be obeyed: no questions, no arguments. You may know you are right but, by definition, they are right. These people, however, do not affect your general strategy, and I will ignore their influence.

At some point in time, it is necessary to make a philosophical decision as to why you are there. You might be there to ensure that books are shelved, cards filed, and the library always ready to be used. I operate under the assumption that I am there to get information to those who need it, even if they do not realize such information exists. This requires me to be proactive. I initiate interactions with almost eveyone who enters the library and especially if I know the person, even if they seem to know what they are doing—often they do not. So, the only reasons I do not give and do everything for everybody, are budget, staffing, and space. One is thrust into a decision-making process because one must continually balance everyone's needs.

One of the first formal acts I would recommend is the establishment of user committees, as advisory groups. It is important not to be perceived as being arbitrary, consistently making unilateral decisions. No matter how wise those decisions might be, there will always be those who are affronted by having no say in the process. I need to emphasize the advisory nature of these committees. The groups' tendency will be to tell you what to do. Always thank them for their advice, tell them you will take their advice into consideration, and so forth, but remember, you are the information professional. Now, with all these warnings under our belts, how do we form the committees, and what do they do?

As noted before, your organization will be divided up into smaller units in some fashion, whether project-oriented or divisional. Decide which groups appear to be fairly permanent and approach the leaders/chairpersons about having people serve on the committees. Make clear in these initial discussions (memos and follow-up memos are good) what these advisory committees are for.

And, what are these advisory committees for? Certainly not for the everyday running of the library. But there are many decisions you will have to make, especially at the beginning, where their support and advice will be helpful in such things as circulation practices, library hours, type of catalog, and so forth. You will also turn to them for help for ongoing practices: collection development, weeding, journal purchasing and cancelling, and so forth. They also serve as important conduits for requests such as new journals or new services. They are also informal conduits for requests in the other direction—to management. The committee members may become advocates for higher library budgets, for larger space allocations, or for equipment needed. These things cost money and it is important that management perceive that it is their scientists and profit-producers who need the extra services and not the librarian trying to build an empire. And you will find that you are often faced with contradictory demands. Your users often want immediate access to materials, which generally translates to more books and journals, with concomitant costs, or access to a wider range of materials than a research library can supply, which translates to greater staffing needs. Users of the report literature want more money and space allocated to reports, while journal users fight replacement of hard copies with microfiche and book users want everything, regardless of duplication of information. The committees offer you a place for educating users about new services and technologies that affect these decisions and to learn about the underlying needs and fears driving those demands.

Now, you may be walking into a preexisting library, with set procedures, or a new but, as yet, unused facility, or a library that exists only in your mind. The first case allows you to go along with the status quo at least until you are comfortable making decisions. The second and third require immediate decisions as to how you will interact with the rest of the organization. Remember, as I have said before, consider these decisions in terms of reversibility. You can always become more liberal without much complaint.

Access is an important problem in research libraries. Mathematicians often need 24-hour access to the facilities. They include statisticians working on special demographics projects for marketing to your night owl who just does not work during the day. The only argument I can make for limited access is security. Theoretically, your presence will inhibit theft. However, you will find that, between meetings, helping individuals back in the stacks, and other calls on your time that take you away from your desk, your presence is not much of a deterrent. Second, and I consider this more important, you have been hired as a professional information dispenser, not as a policeman. As long as access in the off hours can be limited to organization personnel, consider having 24-hour access. This will require circulation procedures that do not require your presence and a photocopier left on (which is a problem if you charge back for copying). There may be areas of your collection (limited access company reports, books that will be stolen if not kept under lock and key), and pieces of equipment (personal computer, and information thereon) that need to be made secure.

Personnel costs are high. As a result, you will be unlikely to be able to offer service outside of the working day. As long as this is clearly stated, there should rarely be problems. I firmly believe one should not try to make up for this lack oneself. Ultimately, that is the path to burnout. If your organization finds it needs service during off hours, then it should hire the personnel.

Organizations usually consider all records and files generated within the organization its property. It may include library records within its purview. The Special Libraries Association and the American Library Association have both come out in defense of the confidentiality of such records. However, in contrast to this, is the need to facilitate the availability of materials for the mathematicians. If one user can be sent directly to another to look at a book, less time is wasted. But do you have the right to share such information? It is very fine line to walk.

PHYSICAL LAYOUT

There are books dedicated to this topic that really do know what they are talking about. I have never had to construct a library from scratch and would not begin to advise how to do it yourself. What I will offer are some ideas to keep in mind as you make your own decisions.

Research libraries are usually small enough that the reference desk should be placed where everyone who enters the library must pass it. You will get to know the users much faster, and, most importantly, you have the chance to offer help to anyone who looks the slightest bit confused.

Confusion is also the reason that you want the reference books, abstracts, and indexes placed near the reference desk and the aisles placed so that you can see down them. The prior set-up in my library had the reference books near my desk, but the shelves blocked my view. The abstracts were placed at the back of the library. This was frustrating because I could not see the people who needed help. The mathematicians wasted time, often settled for less useful resources, or gave up. We are in the process of shifting books and moving desks and the situation has improved immensely.

If possible, have your photocopier near the bound and unbound journals. If these two areas are split in your library, you can only accomplish one. (I consider placement of the unbound, current journals more important.) This simplifies the problem of keeping the most current journals from wandering about the library so that they stay displayed in one area. As a majority of the users will consider scanning the latest journals one of the most important functions of the library, this last is important.

Plan for the new technologies. Online catalogs will require terminals and outlets. CD-ROM workstations require space and printers. Reader/Printers for microfilm require darker areas. All these should be fairly near your work area as they will require attention: users needing help, replacing paper, and so forth.

To reiterate, the theme of all these suggestions is that we do our job best when we can aggressively assist our users and, simple as it may sound, line of sight is the greatest aid you can have.

REFERENCE SERVICES

Why do I harp on aggressive service so much? When I showed up in my library, I had two barriers to overcome. I was new and the users were used to the librarian I had replaced. They did not believe I would be able to offer service as good as hers. Also, those who were not used to working with the librarian were afraid to interrupt my important duties: shelving and filing. As you overcome these initial barriers you will find they will turn to you more quickly. Do not worry about your lack of mathematical language. Most of the questions you will have to answer will be of the more generic library type: journal articles are found using the abstract journals; books are found using OCLC and *Books in Print* (New York: R.R. Bowker). The *Mathematical Reviews* classification scheme (MRC) (see *MathSci User Guide*, p. 117) can be used as an aid to understanding the questions.

Regardless of format, there are two major abstract/index sources that I use: *Mathematical Reviews* (*MR*) (see p. 122), and *Science Citation Index* (*SCI*) (Philadelphia: Institute for Scientific Information). In larger libraries, or those with larger budgets, *Physics Abstracts* (*PA*) (London: Institute of Electrical and Electronic Engineers), along with *Chemical Abstracts* (*CA*) (Columbus, OH: American Chemical Society), or *Biological Abstracts* (*BA*) (Philadelphia: Biosciences Information Services), would be useful for those mathematicians doing applied research. *Zentralblatt für Mathematik und ihrer Grenzgebiete* (see p. 126) is a mathematics abstract journal that has especially good coverage of European journals, especially those from Eastern Europe and the former Soviet Union. It is available, also, on STN as the MATH database. Ultimately, experience will teach you which is appropriate for the task at hand. The decision you must make is that of format. They are all available online (DIALOG, STN, etc.) and *Computer and Mathematics Search* is available on BRS. Several of them are available on CD-ROM.

The advantages of the printed format are: 24-hour access by multiple users, the ability to browse (*MR, PA, CA*, and *BA* are arranged by subject) and, for the mathematicians, there are no additional charges (the cost is paid up front by the library). Disadvantages to the user are poorer searches and the cost in time to the researcher (at the $100-$200 an hour they receive, including benefits, this is quite substantial). You have the problem of the cost of these indexes and the storage problem. You might find it almost impossible to justify the cost of the lesser used indexes (*PA, CA, BA*).

Online searching and CD-ROM are what you will find most pratical. I am certain you have seen their virtues extolled before. Also, as CD-ROM is still in its infancy, I will not compare its benefits to online searching. I believe the two are complementary and will be for some time. *MR* is an abstract that cries out for computer access. Very specific and broad information requests are both difficult to pull out of the printed format. There are also some databases that are not available online. The NTIS database is the best source for scientific report literature. One CD-ROM product I would like to mention is not available online. It is called *Computer Select*. It covers approximately 40 journals full text and indexes about 150, with an orientation toward software and hardware. The reviews are very good, the product is very easy to use, and there is really no equivalent print or online product.

At some early point, but before budget time, you will have to decide who pays for online searches. If you cover all, or just a set amount, of each search, this must appear in your budget. The problem is deciding how much to budget (it is painful to stop subsidies in the middle of the fiscal year because you have run out of money). An additional consideration is documentation of your support of research. By charging back, there is quantitative evidence of library use by different units of the organization. Charging back also prevents abuse of library resources by any one unit. The general search (everything on nonlinear partial differential equations) will help break the bank.

Somewhere along the line you will begin to consider bibliographic instruction in end-user searching for some subset of your users. Prior to CD-ROM, my suggestion would have been quite succinct: do not bother! Most users would not search often enough to maintain their skills and, of those who might, very few would take training to update their skills. Also, they would lose all the benefit of the skills you have acquired, which will be considerable. Finally, since a significant proportion of online searching is done for new lines of research, you may not become aware that your researchers are proceeding in this new direction. This will detract from your ability to support this research in a timely manner. Now that that is said and done, CD-ROM changes everything. Those users who are interested will have considerable chance to practice. Having had the chance to refine their search on the CD-ROM, they would be able to apply the same search online. Unfortunately, the user will not make use of your knowledge, not will you be aware of his new line of research. However, the user saves money and avoids having to interact with another human being. Essentially, the CD-ROM technology is too new to properly assess its impact on end-user online searching.

Now we know that your literature searching will be so successful that you will be driven to an extensive interlibrary loan program. In fact, this is a direct result of the success of online searching. I still have to explain to users that the databases covers far more material than I own in the library. This will require you to network. Consider local, state, regional, and personal networks.

Contact your local Special Libraries Association chapter, the public library system nearest you (they are increasingly becoming multitype library systems) and the state library, among many. Print sources you might consider are the *Encyclopedia of Information Systems and Services* (Detroit, MI: Gale Research, 1985) and the annual P-A-M (Physics-Astronomy-Mathematics division of SLA) *Membership Directory* (Washington, DC: Special Libraries Association, 1992). Do not be afraid to call up someone in P-A-M and just ask for help.

As part of your ILL program, you will be constantly be looking for means to speed up the process. While books may be an intractable problem, there are several means available to you for journal articles. First, consider purchasing a FAX machine. You will find these in an ever increasing number of libraries and it can bring the turn-around time down to under a day. The dark side of the FAX is the poor quality of the product. New FAX boards for PCs are a possible way to overcome this problem. Most of the articles you find online are also available for purchase online. The American Mathematical Society is an online document supplier. Turn-around time on these purchases can be very short; however, the sooner you want it, the more you pay.

It is important that you supply good ILL service from the start, and maintain it. The timeliness and quality of Argonne's ILL service has improved manyfold in just the last five years. However, I still deal with the old prejudices—"ILL takes so long; they can never find what I want," and so forth.

COLLECTION MANAGEMENT

It is important to develop a framework for your acquisition and weeding of materials. Both budget and space constraints require that your acquisitions be limited and your weeding be continuous. My philosophy at Argonne is that this library supports active research. This fact alone helps define what I purchase and what I weed. For example, I purchase very little that deals with the history of mathematics. While some scientists may develop an interest in this field, they will not have the same time-driven needs as others and can easily wait for ILL. I also believe that there is little chance that history of mathematics will ever become an initiative at Argonne, so these materials are prime candidates for weeding. I do maintain a small collection for those scientists with a personal interest, but its space requirements are minimal.

An interesting aspect of the mathematics literature is the way some of it becomes written in stone. I will try to explain this with an example. In the late 1950s and early 1960s there were many books published on linear algebra. Since then, and especially with the advent of computers, the field has moved on and changed. But the fact and techniques included in these books are still valid, and useful. As it was a hot topic at the time, many linear algebra books were purchased. As I reviewed these books from our current perspective, the authors tended to say the same thing in only a limited number of ways. I was able to weed most of them without affecting research. This also affects my acquisitions policy. Even when a topic is "hot," I try to limit my purchases so that each book contributes something new. There are some books that are so useful that many scientists will want to keep one at their desks. At this point, I buy one, put it on reference, and strongly recommend that the scientists buy their own copies. I buy very few books on programming languages or computer software. The number of books available is limitless and users need them at their desks.

Alternatively, there are books whose information is so out of date that it is dangerous to still own them. Statistical techniques have been replaced by newer, more accurate ones. The older books should be considered for weeding to prevent their inadvertent use as not all users bother to check for publication date.

As a librarian in a research library, you will have to consider the level at which the books are aimed. Do you purchase only those materials that are aimed at graduate students or researchers? This assumes use of the library only by mathematicians. If other scientists will be using the library, do you need more elementary materials so they can cross over to new fields?

Language of publication is another concern. There is more than enough published in English to absorb any budget. However, there is other work being done in the world. Are any of your users fluent in other languages? While being aware of research being done elsewhere is important, is the money wasted if no one will use the materials?

How does one find the materials to purchase? You will probably want to use several different methods, depending upon the items being purchased and your budget. If the budget is small, you would have to rely upon reviewing publisher's catalogs, reviews, and mathematicians' requests. *Current Mathematical Publications* (see p. 121) is a good source of titles. If your budget is large enough for substantial purchases, you should consider an approval plan through a vendor (e.g., Blackwell North America or Ballen) Approval plans will cover most of the technical publishers in which you are interested and will give you access to their books either through the book itself or through an announcement. Depending upon the company, you may supply them with a subject list based upon their own thesaurus or one based upon the Library of Congress classification. It simplifies purchasing (only one vendor to deal with) and enables you to scan a much wider range of materials. There is also a discount on book price when you deal through a vendor. There will be series where you know you will buy everything (e.g., *Lecture Notes in Mathematics* [New York: Springer-Verlag]) and it might be cheaper to buy these directly from the publisher.

There are publishers, such as the American Mathematical Society, who do not participate in the approval plan. You will have to order from these publishers directly. The simplest method is a blanket order: you promise to buy everything they publish. This can become expensive and run out of control. I would suggest that in a research library it is more effective to use standing orders and to peruse the catalogs. Be wary of the symmetry argument: because you own the first 75 numbers of a series does not mean you should continue to purchase every one. The prime directive is need, not symmetry.

Journals are a different problem. There are several unavoidable truths, or at least the scientists will insist they are truths. The price of current subscriptions will go up, and at a rate that exceeds inflation or, more importantly, your budget. Second, all the new journals are extremely important and their purchase cannot be put off, if even for a year. Next, we need to keep all our current subscriptions. One of the mathematicians might need to look at an article from one of the journals at some indefinite time in the future. Last, we cannot afford to get rid of any of our old runs. One of the mathematicians might need to look at an article from one of the journals at some indefinite time in the future (yes, it is meant to be repeated).

Actually, I hope you will find that I have overstated the case. If you have developed a good collegial relationship with your users, the problems of purchase, maintenance, and retention will be simpler, though never simple. It is very important to have your users intimately involved in these processes. This is one area your library user committee can be valuable.

Many, if not most, libraries nationwide are facing a severe budget crunch when it comes to new journals: more new journals mean fewer books, and sometimes a moratorium on book purchases. One way librarians are dealing with this is the one-for-one rule: if a group wants a new journal purchased, they must allow you to cancel a current subscription. Some libraries require cost equivalence of the two journals. At Argonne we require a written justification (the library supplies the form) for every new title: why is the journal important and who will use it.

We have just instituted a new procedure that we hope will be successful. As new major initiatives are submitted for funding, we work with the groups involved to include library support as part of their requests.

As mentioned above, your first actions set the stage for what can follow. Do not let the users cow you at first and agree to purchase everything they request. If you do play doormat, you will find it much more difficult to implement whatever plan you finally decide upon. And, if you stand up for yourself and for your library, you will find yourself being treated as more of a professional.

You still may find it necessary to cancel journal subscriptions because of rising costs. The choices of how many, and which titles to cancel must be done with the advice of your user groups. This will help minimize the effect the cancellations will have on the scientific work of your users. A useful byproduct is the support this gives you in facing future criticism because no matter how careful you are, at least one of the titles cut will turn out to be important to some group. You should attempt to gather data: circulation records, sign-up sheets at the photocopier, surveys, or marking of spines. Surveys may range from a questionnaire attached to each title over the course of several months to a complete list of subscriptions with questions about use and need. For my older collection, I put colored dots on the spines (of bound journals) as they are reshelved. Visually, it becomes easy to identify those journals that are reused frequently.

The outcome of these activities is usually an annotated list which is submitted to the user group. I then explain that it is their choice: they recommend the cuts made, or I will. Making "no cuts" is not an acceptable choice.

Keep in mind that success is often a matter of power. It is necessary to have management (at least your immediate supervisors) on your side because users may be willing to go up the chain of command to find someone who will support their position. You must keep your superiors informed about what you are doing, and the complaints they may expect to receive. Anticipation will make it easier for them to supoort you.

Formalized resource sharing with another library is another way to limit subscriptions. You may each agree to collect in certain areas or just divide up the titles on a weighted need and cost basis. If you can guarantee a 24-hour turnaround on articles it may be easier for users to accept cancellation. This should be formalized because, as personnel change, or just with the passage of time, verbal agreements are changed or forgotten.

BIBLIOGRAPHIC CONTROL

As in all good theatre, there is conflict. But I see it not between the major cataloging systems. The battle is cataloging scheme versus subject headings.

To resolve this conflict I offer my Theory of Cataloging: All good (useful) books are, by definition, checked out. **Lemma 1:** Browsing is bad because users will not find the good books (they are checked out). **Lemma 2:** The shelflist is an acceptable way to browse, so a classification scheme is needed. If you disagree with these as a given then, please, read a book on cataloging schemes; I cannot compete with the specialists. Otherwise, proceed.

If the library already uses a major classification scheme (e.g., Library of Congress [LC], Dewey Decimal Classification [DDC], or the Mathematical Reviews Classification [MRC]), there is no need to change. There is good documentation for future cataloging and the time involved in reclassifying and retraining the users is not productive. Before we automated, I found it almost impossible to get any of my users to use the shelf list (it was available alongside the author/title and subject indexes) and I discouraged browsing on the shelves, so the differences between the classification schemes was unimportant in my situation. What is important is the ability to use other people's intellectual efforts as you copy the cataloging from OCLC or RLIN or....

Possibly you have inherited a nonstandard classification scheme. Then, face the music, bite the bullet, and forget time and money constraints: you must recatalog. Even if the prior librarians have documented their system, the effort needed to maintain it, and to incorporate the changes required by new directions in mathematics, may approach the ridiculous. If you do not make the effort, personal idiosyncracies may lead to so many internal inconsistencies as to make the system useless for anything other than assigning a place locator on a shelf.

Finally, you may be entering a new library or one that has been run without a classification scheme (e.g., a reading room). In your studies, or discussions with other librarians, you may have in mind a system you believe is better than the others. But, take these factors into consideration. Most of your mathematicians will have strong academic backgrounds and most of academia uses the LC scheme. If users find the shelving more familiar, they will be less likely to need help. A second factor is that the Library of Congress should exist as a functional unit for longer than you need to be concerned with. Therefore, cataloging utilities will continue to offer LC cataloging and we need not worry about availability or drastic changes in the system.

If I seem to ignore the theoretical aspects of cataloging, that is because I do ignore them. The subject index should be the means by which the users find the books they need. It gives multiple access and, again, eliminates the problem of books that are checked out. And, as you move toward an online catalog, subject headings become easier to search (you no longer will need an exact match—plural vs. singular, regular vs. inverted headings).

From my last sentence you may detect a strong feeling toward the online catalog. Your first day on the job you may not feel secure enough to request an immediate transition to an online system. But, as you settle in, the advantages will become evident and the only question will become, "When can I convince management to fund the project."

There are two considerations: one I call the technical imperative, and then there is everything else. As information technology continues to change, the librarian has to struggle just to stay aware of what is becoming available. If you do not begin to use the new technologies, your ability to understand what is available diminishes. You owe it to yourself, as a professional, and to your organization to be able to employ this technology. This applies to online catalogs, and personal computers, and CD-ROMs and networking and.... You do not need to be on the cutting edge—very few people can be there, or it is no longer the cutting edge—but you do need to move forward with the profession.

And now for everything else. You will find an online catalog useful when a mathematician only remembers it is a new book with the word optimization in it, or it is a 1986 conference on numerical methods. Possibly you have a collection of titles you are thinking of ordering, and you want to see if you own them yet. An online catalog simplifies and improves efficiency. Users may access the system from their offices. It is simpler for users to produce printouts during subject searches. Records may more readily be downloaded from cataloging utilities. I almost forgot the most important point, at least as it relates to your sanity. You no longer have to file catalog cards.

What are the problems with an online catalog? There are hardware (microcomputer, minicomputer or mainframe) and software (turnkey systems only) costs. The hardware must be available when the library is open. There must be enough terminals available and there must be someplace to put them. Someone must be responsible for system maintenance. You will also find strong resistance to change—even from scientists. When we automated, there were many people wanting to revert to the old days.

Most important, if you cannot purchase an online system that will do the job properly and completely, then you should not go online. At Argonne, our records have been in digital form for a long time, since the 1960s. However, we only printed out a comfiche catalog and did not use the data to produce an online catalog. Although the records were in digital form, there was no simple and cheap way of incorporating them into a searchable system. The system we would have ended up with would have been so poor that too many users would have walked away not knowing that the material they wanted was available. We waited until we could purchase the proper system before we decided to go online.

SPECIAL SERVICES

The services you will be able to offer your users will greatly depend upon such factors as the size of your library and staff, the support services available (e.g., centralized technical processing in a multilibrary system), the number of patrons you support, and so on.

You will almost certainly want to display your new acquisitions (books, journals, etc.) The users who want to regularly peruse these materials have a legitimate need to do so. These journals should be displayed for a regular time interval so that users can be assured of not missing materials.

If possible, there should be some printed output of new books and reports. You can safely ignore journals, except that you might want to announce new journal subscriptions. If your organization has electronic mail, use it to distribute the list—it is efficient and very professional appearing. You might consider using FAX if your users are physically dispersed.

Where there is an interest, encourage use of selective dissemination of information (SDI) for online literature systems. If lack of interest shows lack of knowledge, rather than lack of need, spread the word. Online SDIs are an efficient and nontime intensive method of helping your users. New online services are making this process even easier. DIALOG has introduced direct mailing of searches. The results will be mailed directly to the user without any intervention on your part. Another service that has been introduced on DIALOG is subaccount billing. This considerably simplifies your billing process which is probably the most time intensive part of the process.

And then there are the services that sound wonderful but may not be feasible, given time and personnel constraints. As you get to know your users you might consider photocopying the Tables of Contents (TC) from journals for their convenience. Now, while this might be simple for a few titles for a few people, success could be dangerous. Consider the worst possible scenario: photocopying the TC of every journal for every user. Weigh the number of journals you get and the expected popularity of the program with your users and compare it to the amount of time you are willing to spend on the project before you begin. If you find it necessary to cancel a popular service because of time constraints you may create ill-will. On the other hand, it could always be used as a tool for getting more support.

You may consider automatic routing of journals to users, but try not to do so. Automatic routing has two major problems. When the journal is new, and the most valuable, it will be unavailable to most of your users. Journals tend to sit on users' desks and you cannot count on their timely return. Second, as the scientists' interests change, they rarely inform you they are no longer interested in receiving the journal, so lists only get longer, they rarely shrink. Worse, the scientists will not inform you when they leave the organization (the library just might be the last thing on their minds) and you will then be routing the journal to a ghost with the journal most likely fading from sight for quite a while. However, if the person requesting the routing is high enough in the hierarchy, show your flexibility and route the journal. Use this as an opprtunity to build goodwill upstairs and also as an opportunity to expand library services.

You will probably find that, when you first walk into your new library, your mind will be so very fertile. You will have ideas about this and about that, and you will want to implement everything immediately: pause, count to 10, and then take the time to reconsider. Any service you offer will take time and, as I have noted before, if it is popular, it may become written in stone. Even if you find the activity takes up an inordinate amount of your time, you may no longer have the choice of backing off.

Also, save some of your ideas for later. The psychological effect of a series of innovations spaced out over a considerable period of time is much greater than the effect of the same innovations all offered at one time. This spacing also gives you the chance to evaluate the cumulative effect these changes have on your available time. With infinite time, money, and support you could afford to implement all your ideas. Since these are finite, careful planning is needed to maximize your effectiveness in the time you are willing to give to your job.

PERSONAL DEVELOPMENT

Warning, this section is being written by a groupaphobe: a person who has always found it impossible to belong to or work with groups. Networking and active membership in an association have almost always been anathema to me. With this in mind, my advice is to network and participate as much as you can.

Networks and association participation are important in several ways. They are methods of forming very strong support groups which can be very helpful in times of adversity. They also provide ways of increasing your ability to do your job. The advice and information other librarians can supply can solve specific problems/questions as well as keep you informed of new areas in which to improve your knowledge. It is very hard to function as a professional soley on your own. The breadth of knowledge needed is too large to absorb all of it.

The Special Libraries Association is the major association for special librarians, whether academic or corporate. You might also consider joining the American Mathematical Society. It has a strong interest in mathematics librarianship and the mathematics literature.

I find attending classes and workshops almost irresistible. They help keep my brain active and my mind interested. And this, perhaps, is the most important reason for working on personal development. You many have hopes of working in this field for 30, 40, 50 years. Boredom and burnout are quickly attained if you limit yourself solely to what goes on in your little enclave. And how quickly you will find yourself a relic of the past.

CONCLUSION

Although, ostensibly, the purpose of this chapter was to cover all mathematics special libraries, you may have noticed that it did not. It was written from my viewpoint, filtered through the library system I work within. What I have attempted to do is to raise the questions you will have to ask. Your answers may, and will, be different, depending on your situation within your organization. But, I hope, the questions will be the same. Possibly, one should treat this chapter as you would your grandmother's recipe: a dash of automation, a pinch of professionalism, and 3 shakes of confidence. Just remember, if you look like a professional, feel like a professional, and think like a professional, then you must be a professional.*

And as Lewis Carroll, mathematician, writes in *Alice's Adventures in Wonderland* (New York: Holt, Rinehart and Winston, 1985, p. 115): "Begin at the beginning," the King said, gravely, "and go on till you come to the end; then stop."

**Note*: I use the terms librarian, information specialist, or any other permutation, interchangeably. What matters is that we are professionals.

Chapter 5

REFERENCES AND FURTHER READINGS

"Accessing PAMnet," *P-A-M Bulletin*, 20(1): 2 (August 1992).

Anderson, Nancy D., "Alternatives of public funding of university science and technology libraries," *IATUL Quarterly*, 4:215-220 (December 1990).

Anderson, Nancy D. and James L. Rovnyak, "Mathematics research libraries: A 1990 snapshot," *Notices of the American Mathematical Society*, 38(10): 1258-1262 (December 1991).

Ardis, Susan B. and Karen S. Croneis, "Document delivery, cost containment, and serial ownership," *C&RL News*, 48(10): 624-627 (November 1987).

Auld, Larry, *The University of Illinois Mathematics Library: A Preliminary Historical Survey, 1868-1970* (Urbana, IL: Graduate School of Library Science, University of Illinois, 1975), 22pp.

Barringer, Sallie H., "Introducing undergraduates to mathematics information resources," *Research Strategies*, 7(3): 129- 133 (Summer 1989).

Bartle, Robert G., "Comments on the Dewey Decimal Classification scheme for mathematics," *P-A-M Bulletin*, 7(3): 32 (February 1980).

Bartle, Robert G., "One mathematician looks at the classification of mathematics," in *The role of classification in the modern American library* (Urbana, IL: Graduate School of Library Science, 1960) pp. 96-97.

Battiste, Anita L., et al., "The University of Florida's moving experience," *C&RL News*, 50(6): 467-471 (June 1989).

Bostic, Mary J., "Gifts to libraries: Coping effectively," *Collection Management*, 14(3/4): 175-184 (1991).

Chrzastowski, Tina, et al., "Library collection deterioration: a study at the University of Illinois at Urbana-Champaign," *College & Research Libraries*, 50(5): 577-584 (September 1989).

Cziffra, Peter, "A computer-assisted stackshift in a university science library," *Sci-Tech News*, 46(1): 4-6 (January 1992).

Committee on Support of Research in the Mathematical Sciences, *The Mathematical Sciences: A Collection of Essays* (Cambridge, MA: MIT Press, 1969), 271pp.

Davis, Philip J. and Reuben Hersh, *The Mathematical Experience* (Boston: Birkhauser, 1981), 440pp.

Devin, Robin B. and Martha Kellogg, "The serial/monograph ratio in research libraries: Budgeting in light of citation studies," *College & Research Libraries*, 51(1): 46-54 (January 1990).

Diodato, Virgil, "The use of English language in non-U.S. science journals: A case study of mathematics publications, 1970-1985," *Library & Information Science Research*, 12(4): 355-371 (October-December 1990).

Dionne, Richard J., "Science libraries at a crossroads," *American Scientist*, 76(3): 268-272 (May/June 1988).

Donovan, Peter W., David C. Hunt, and John M. Mack, "The Dewey Decimal scheme and mathematics," *Wilson Library Bulletin*, 48(3): 220-222 (November 1973).

Donovan, Peter W., David C. Hunt, and John M. Mack, "Letter to the editor," *Notices of the American Mathematical Society*, 20(3): 139 (April 1973).

Donovan, Peter W., David C. Hunt, and John M. Mack, "Mathematics in a major library using the Dewey Decimal Classification," *Australian Academic and Research Libraries*, 6: 87-91 (June 1975).

Dyal, Donald H., "What old books are worth: When price, not preservation, is paramount, it's best to set 'em straight," *American Libraries*, 16(11): 774-776 (December 1985).

Dyson, Freeman J., "Willard Gibbs and the teaching of science," *Proceedings of the Gibbs Symposium* (Providence, RI: American Mathematical Society, 1990), pp. 269-276.

Edmonds, A. and C. Hutchins, "The IMA library survey," (unpublished manuscript, 1991).

"European Mathematical Council survey of European mathematical periodicals," *Notices of the American Mathematical Society*, 33(6): 905-909 (November 1986).

Fedunok, Suzanne, "Mathematics," in *Scientific and Technical Libraries. v.2 Special formats and subject areas* (Orlando: Academic Press, 1986), pp. 152-160.

Ferguson, Anthony W., Joan Grant, and Joel S. Rutstein, "The RLG Conspectus: Its uses and benefits," *College & Research Libraries*, 49(3): 197-206 (May 1988).

Frame, J. Sutherland, "Departmental libraries," in *Buildings and Facilities for the Mathematical Sciences* (Washington, DC: Conference Board of the Mathematical Sciences, 1963), pp. 79-90.

Franks, John, "Electronic research journals: What will they be like." Session on Emerging Electronic Publishing Medium: User Perspectives. Special Libraries Association Annual Conference. Cincinnati, June 9, 1993.

Garfield, Eugene, "Journal citation studies. 36. pure and applied mathematics journals: What they cite and vice versa," *Current Contents*, 5(15): 5-10 (March 8, 1982).

Gibbs, Beth Liebman, "Subject specialization in the scientific special library," *Special Libraries*, 84(1): 1-8 (Winter 1993).

Gödert, Winfried, "Subject headings for mathematical literature," *Journal of Documentation*, 36(1): 11-23 (March 1980).

Gould, Constance C. and Karla Pearce, "Mathematics," in *Information Needs in the Sciences: An Assessment*, (Mountain View, CA: Research Libraries Group, 1991), pp. 63-68.

Halmos, P. R., "Mathematics as a creative art," *American Scientist*, 56(4): 378-389 (1968).

Hardesty, Larry and Jamie Hastreiter, "Mathematics periodicals in selected liberal arts colleges," *Collection Building*, 9(1): 3-11 (1988).

Harrison, Jenny, "The Escher staircase," *Notices of the American Mathematical Society*, 38(7): 730-734 (September 1991).

Hawkins, O. T., "A review of online physical sciences and mathematics databases. Part 1: Physics, mathematics, and computer science," *Database* 8: 14-18 (February 1985).

Hurlbert, Janet McNeil and Richard A. Weida, "Developing a BI project for mathematics research in the small college library," *Research Strategies*, 9(4): 189-194 (Fall 1991).

Janusz, Gerald J., "Reviewing at *Mathematical Reviews*," *Notices of the American Mathematical Society*, 38(7): 789-790 (September 1991).

Katz, William A., *Introduction to Reference Work. Sixth Edition* (New York: McGraw-Hill, 1992), pp. 12-15.

Kenney, Anne R. and Lynne K. Personius, "Update on digital techniques," *Newsletter*, Commission on Preservation and Access, 40: 1-6 (November-December 1991).

King, Jerry P., *The Art of Mathematics* (New York: Plenum Press, 1992), 313pp.

Kister, Jane E., "From published paper to *MR* review: How does it happen?" *Notices of the American Mathematical Society*, 36(10): 1362-1365 (December 1989).

Kulldorf, Martin, "A survey of statistical journals," *The IMS Bulletin*, 21(4): 399-407 (July/August 1992).

Kuzma, Taissa T., "Computerization of Mathematical Reviews into Mathfile," *Science & Technology Libraries*, 3: 49-62 (Fall 1982).

Kuzma, Taissa T., "From Mathematical Reviews to MathSci," *Notices of the American Mathematical Society*, 37(9): 1260-1263 (November 1990).

Kuzma, Taissa and Elizabeth Wolf, "MathSci: From data to database," *Notices of the American Mathematical Society*, 38(4): 306-310 (April 1991).

Ladner, Sharyn J. and Hope N. Tillman, "Using the Internet for reference," *Online*, 17(1): 45-51 (January 1993).

Library of Congress. Subject Cataloging Division. *Library of Congress Subject Headings*, 13th ed. (Washington, DC: Cataloging Distribution Service, Library of Congress, 1990) 3v.

Line, Maurice B., *Academic Library Management* (London: The Library Association, 1990), 266 pp.

Marks, Kenneth B., et al., "Longitudinal study of scientific journal prices in a research library," *College & Research Libraries*, 52(2): 125-138 (March 1991).

"Mathematics classification and subject cataloging at the Library of Congress," *P-A-M Bulletin*, 18(4): 7-12 (May 1991).

Mathematics Subject Classification (Providence, RI: American Mathematical Society, 1991), 35pp.

MathSci User Guide. Second Edition. (Providence, RI: American Mathematical Society, 1990), various pagings.

McKinlay, John, "Dewey and mathematics," *Australian Academic and Research Libraries*, 4: 105-111 (September 1973).

Mischo, William H., et al., "University of Illinois at Urbana- Champaign," in *Campus Strategies for Libraries and Electronic Information* (Bedford, MA: Digital Press, 1990), pp. 117-141.

Mischo, William H., David Stern, and Timothy W. Cole, "Use of microcomputer workstations to enhance access to library collections," *Science & Technology Libraries*, 11(1): 31-45 (1990).

Moline, Sandra R., "Mathematics journals: Impact factors and cents per thousand characters," *The Serials Librarian*, 20(4): 65-70 (1991).

Murdock, Martha, "Mathematics made easier on disc," *CD-Rom EndUser*, 1(10): 54 (February 1990).

Pasterczyk, Catherine E., "Checklist for the new selector," *C&RL News*, 49(7): 434-435 (July/August 1988).

Pausch, Lois, et al., *Report on the Pilot Study on the Value of Conference Proceedings in the Scientific Literature* (Chicago: Association of College and Research Libraries, Science and Technology Section, 1986), pp. 74-75.

Payne, John R., "A closer eye on appraisals: The new tax law and its effect on gifts, donors, and appraisers," *C&RL News*, 46(2): 52-56 (February 1985).

Raeder, Aggie W. and Karen L. Andrews, "Searching library catalogs on the Internet: A survey," *Database Searcher*, 6: 16-31 (September 1990).

Roberts, Elizabeth P., "ILL/document delivery as an alternative to local ownership of seldom-used scientific journals," *The Journal of Academic Librarianship*, 18(1): 30-34 (March 1992).

Rockey, Steven, "Library "publishes" out-of-print mathematics monographs," *ARL: A Bimonthly Newsletter of Library Research Issues and Actions*, 161: 11 (March 2, 1992).

Sanders, Nancy P., Edward T. O'Neill, and Stuart L. Weibel, "Automated collection analysis using the OCLC and RLG bibliographic databases," *College and Research Libraries*, 49(4): 305-314 (July 1988).

Sapp, Gregg, "Subject coverage in general and specialized mathematics journals," *The Serials Librarian*, 16(3/4): 93-117 (1989).

Selberg, Atle, "Reflections around the Ramanujan Centenary," *Collected Papers. Volume I.* (Berlin: Springer-Verlag, 1989), pp. 695-701.

Shkolnik, Leon, "The continuing debate over academic branch libraries," *College & Research Libraries*, 52(4): 343-351 (July 1991).

Sichel, Beatrice, "Training endusers on MathSci disc," *Special Libraries*, 82(4): 282-287 (Fall 1991).

Slutsky, Bruce, "How to avoid science anxiety among science librarians," *Science & Technology Libraries*, 12(1): 11-18 (1991).

Smith, Warren D., "How to buy a math and computer science library," (Unpublished manuscript, 1990).

Stankus, Tony, *Scientific Journals: Improving Collections Through Analysis of Publishing Trends* (New York: Haworth Press, 1990), 205pp.

Stankus, Tony and Virgil Diodato, "Selecting multispecialty mathematics research journals via their underlying subject emphases," *Science & Technology Libraries*, 4(1): 61-78 (1983).

Stern, David, "An alternative national electronic mail network for libraries," *Special Libraries*, 79(2): 139-142 (Spring 1988).

"Survey of American research journals," *Notices of the American Mathematical Society*, 36(9): 1193-1198 (November 1989).

Weigel, Jack, "Mathematics," in *Selection of Library Materials in the Humanities, Social Sciences, and Sciences* (Chicago: Collection Management and Development Committee, Resources and Technical Services Division, American Library Association, 1985) pp. 289-299.

Williams, Lynn B., "Subject knowledge for subject specialists: What the novice bibliographer needs to know," *Collection Management*, 14(3/4): 31-47 (1991).

Woolf, William B., "The Mathematical Reviews database: The power of modern technology," *Notices of the American Mathematical Society*, 37(2): 132-134 (February 1990).

Note: The editors realize that not all readers may have access to these references at their institutions. Thus, readers may request photocopies from: Nancy D. Anderson, Mathematics Librarian, University of Illinois, 1409 West Green Street, Urbana, Illinois 61801.

Chapter 6

REFERENCE MATERIALS FOR THE MATHEMATICS LIBRARY

Willis F. Cunningham, Mary B. Harper, and Allen D. Merry*

CONTENTS

CRITERIA FOR SELECTION

The titles on this list were selected from the literature (reviews in *Mathematical Reviews, The American Mathematical Monthly, Choice, Booklist, ARBA*, Eugene P. Sheehy's *Guide to Reference Books*, C. D. Hurt's *Information Sources in Science and Technology*, and Ellis Mount's *Using Science and Technology Information Sources* and from the reference collection in the Mathematics Library, University of Illinois at Urbana-Champaign. The list below is exhaustive and includes many out-of-print titles, because they are worthwhile additions to any mathematics reference collection. Appropriate titles for academic libraries are coded (**a**), for public libraries (**p**), and for special libraries (**s**). Coding was done in collaboration with chapter authors.

**Note*: Prepared by the Graduate Assistants of the Mathematics Library, University of Illinois at Urbana-Champaign under the direction of the librarians.

BIBLIOGRAPHIES

AAAS Science Book List: 1978-1986. Ed: Kathyrn Wolff et al. Washington, DC: American Association for the Advancement of Science, 1986. xvi + 568pp. $25.00. ISBN 0-87168-315-6.

Complete entries including prices and LC numbers. Annotations are descriptive and often offer evaluative comments. Arrangement by broad DDC subject categories. Serves as a supplement to the AAAS science book list (1970) and to the first supplement (1978). **(p)**

Annotated Bibliography of Expository Writing in the Mathematical Sciences. Matthew P. Gaffney, Lynn Arthur Steen. Washington, DC: Mathematical Association of America, 1976. xi + 282pp. o.p. ISBN 0-88385-422-8.

Classified, occasionally annotated listing of articles (some books) useful for university level teaching. Seven major subject headings are divided by degree of difficulty plus related entries. Author index has complete citations. **(a)**

Annotated Bibliography of Films and Videotapes for College Mathematics. David I. Schneider. Washington, DC: Mathematical Association of America, 1980. vi + 107pp. $10.50. ISBN 0-88385-434-1.

Films are listed by distributor. Includes a catalog of mathematics TV courses, references sources, film rental libraries with their mathematics holdings, and subject and title indexes. **(p)**

An Annotated Bibliography of Works on Babylonian Mathematics. Joran Friberg. Göteburg, Sweden: Department of Mathematics, Chalmers University of Technology, 1980. 18pp. Price not available.

A comprehensive work on the subject. **(a)**

Bernoulli Numbers: Bibliography [1713-1990]. Ed: Karl Dilcher, Ladislav Skula, Ilja Sh. Slavutskii. (Papers in Pure & Applied Mathematics; 87). Kingston, ON: Queen's University, 1991. iv + 175pp. $14.25.

A new and enlarged edition of the original which was published in 1988. Includes articles known to the authors prior to June 30, 1991. Lists 1,956 publications by 839 authors and is extensively indexed. **(a)**

Bibliography and Research Manual of the History of Mathematics. Kenneth O. May. Buffalo, NY: University of Toronto Press, 1973. 818pp. o.p. ISBN 0-8020-1764-9.

Consists of two parts. First a manual on retrieval, storage, and analysis of information. Part 2 a bibliography on history of mathematics with more than 30,000 entries arranged under five broad headings. Standard work for academic and research collections. **(a)**

Bibliography for Dynamical Systems. Kenichi Shiraiwa. Nagoya, Japan: Dept. of Mathematics, College of General Education, Nagoya University, 1985. 390pp. Price not available.

This is revised from the 1981 edition and contains lists of writings about dynamical systems and related topics. **(a)**

Bibliography of Basic Texts and Monographs on Statistical Methods, 1945-1960. Second Edition. William R. Buckland, Ronald A. Fox. New York: Hafner, 1963. vii + 297pp. o.p.

Lists 200 books. English-language entries on statistical methods and applications arranged by broad subject categories. Excerpts of reviews included. Author index. **(a)**

Bibliography of Early Modern Algebra, 1500-1800. Robin E. Rider. Berkeley: University of California, 1982. 171pp. $8.00. ISBN 0-918102-08-1.

Sketches development of algebra and algebraic theory and influence of publication practices during the period. Arranged chronologically with detailed author index. **(a)**

Bibliography of Mathematical Works Printed in America through 1850. Reprint Edition. Louis Charles Karpinski. Ann Arbor, MI: University of Michigan Press, 1940. xxvi + 697pp. $60.00. ISBN 0-405-12553-4.

Chronologically arranged coverage of over 1,000 mathematical books printed in America. Indexed by author and anonymous titles, topical, and publisher/printer indexes. **(a)**

Bibliography of Mathematics Published in Communist China, 1949-1960. Comp: C.K. Tsao. Ann Arbor, MI: Books on Demand UMI, 1961. 83pp. $31.00 (hbk.); $25.95 (pbk.). Order #: 2012295.

As the title says, this is a bibliography of mathematics published in the Peoples' Republic of China, 1949-1960. There is a second part available from AMS that is a report on Chinese mathematical dictionaries by S.H. Gould (1969. ISBN 0-8218-0016-7. $9.00.). **(a)**

Bibliography of Multivariate Statistical Analysis. T. W. Anderson et al. Huntington, NY: Krieger, 1977. x + 642pp. $49.50. ISBN 0-88275-477-7. Reprint of 1972 edition.

Exhaustive compilation of literature on multivariate analysis to 1966 for articles, and to 1970 for monographs. Provides author and subject access to some 6,250 papers and books. Useful background information. **(a)**

Bibliography of Non-Euclidean Geometry. Second Edition. D. M. Y. Sommerville. New York: Chelsea, 1970. xii + 410pp. $27.50. ISBN 0-8284-0175-6.

Some 4,500 entries arranged chronologically covering fourth century B.C. to 1911. 125 post-1911 items added in the reprint. **(a)**

Bibliography of Recreational Mathematics. Fourth Edition. William L. Schaaf. Reston, VA: National Council of Teachers of Mathematics, 1970-78. 4v. V. 1: 148pp. ISBN 0-87353-021-7. V. 2: 191pp. ISBN 0-87353-022-5. V. 3: 187pp. ISBN 0-87353-023-3. V. 4: 178pp. ISBN 0-87353-128-0. $10.00/volume. $40.00 (set). ISBN 0-87351-20-9 (set).

Classified arrangement covers books and articles. Volumes overlap chronologically. Glossary in volume 3 with supplement in Volume 4. **(a,p)**

Bibliography of Russian Mathematics Books. George Elmer Forsythe. New York: Chelsea, 1956. 106pp. o.p.

Listing of 600 plus books in pure and applied mathematics published since 1930. Classified subject index and transliteration table. **(a)**

Bibliography of Statistical Bibliographies. Henry Oliver Lancaster. Edinburgh: Oliver and Boyd, 1968. ix + 103pp. $18.00. ISBN 0-934454-12-4.

University level materials in statistics, mathematical statistics, and probability. Arranged in two classes: "Persons" for contributors to statistical theory and "Subject" containing methodology and applications. Author index includes authors, editors, compilers, and biographees. **(a)**

Bibliography of Statistical Literature. V. 1: 1950-1958. V. 2 1940-1949. V. 3: Pre-1940. Maurice G. Kendall, Alison G. Doig. Salem, NH: Ayer, 1981. 3v. V. 1: xii + 297pp. V. 2: 190pp. V. 3: vi + 356pp. $75.00 (set). ISBN 0-405-13881-4 (set); 0-405-13882-2 (V. 1).

Comprehensive listing of major articles written on statistics since sixteenth century. For current information consult *Statistical Theory and Method Abstracts* (See p. 125). **(a)**

Bibliography on Chaos. Ed: Zhang Shu-Yu. Directions in Chaos; 5. Teaneck, NJ: World Scientific, 1991. vi + 514pp. $68.00 (hbk); $32.00 (pbk). ISBN 981-02-0580-5 (hbk); 981-02-0581-3.

A list of books and articles related to chaotic dynamics which had appeared by the end of 1990. **(a)**

Bibliography on Time Series and Stochastic Processes: An International Team Project. International Statistical Institute. Ed: Herman Wold. Cambridge, MA: MIT Press, 1965. xv + 516pp. o.p.

Covers research on time series and stochastic processes up to 1965. For current information consult *Statistical Theory and Method Abstracts* (see p. 125). **(a)**

Bulletin de Bibliographie, d'Histoire et de Biographie Mathématiques. Ed: M. Terquem. v.1-8, 1855-1862. Paris: Mallet-Bachelier. o.p.

Useful for referencing the older literature. Note that titles of articles have been translated into French. Reviews and biographies in French. Each volume has subject and author indexes. **(a)**

Bullettino di Bibliografia e di Storia della Scienze Matematiche e Fisiche. Ed: B. Boncompagni. v.1-20, 1868-1887. New York: Johnson Reprint Corp., 1964. o.p. ISBN 03-84-06306-3.

Useful for referencing the older literature. Contributions in Italian, French, or Latin. Index in v.20. **(a)**

Children's Mathematics Books: A Critical Bibliography. Margaret Matthias, Diane Thiessen. Chicago: American Library Association, 1979. vi + 61pp. o.p. ISBN 0-8389-0285-5.

170 titles in mathematics for early grades, preschool to grade 7. Classified, critically annotated, and rated as to quality. Covers counting, geometry, measurement, number concepts, time, and miscellaneous mathematical topics. **(p)**

Chronological Annotated Bibliography of Order Statistics. H. Leon Harter. Columbus, OH: American Sciences Press, 1983-1992. *Volume I: Pre-1950.* iv + 515pp. ISBN 0-9235950-21-4. *Volume II: 1950-1959.* vi + 993pp. ISBN 0-935950-05-2. *Volume III: 1960-1961.* vi + 214pp. ISBN 0-935950-21-4. *Volume IV: 1962-1963.* viii + 173. ISBN 0-935950-22-2. *Volume V: 1964-1965.* vi + 180pp. ISBN 0-935950-23-0. *Volume VI: 1966-1967.* vi + 232pp. ISBN 0-935950-24-9. *Volume VII: 1968-1969.* vi + 292pp. ISBN 0-935950-25-7. $95.00/Vol.

An eight volume set, of which only the citations and indexes (1960-69) volume needs to be published. Books, articles, dissertations, theses, technical reports, and notes on order statistics, excluding rank statistics, for periods indicated. Each succeeding volume, in addition to covering its time period, includes publications identified as relevant since the publishing of the previous volumes. Some simple citation indexing plus author and subject indexes. **(a)**

Combinatorial Optimization: Annotated Bibliographies. Ed: M. O'Heigeartaigh, J. K. Lenstra, A. H. G. Rinnooy Kan. New York: Wiley, 1985. vii + 204pp. $56.95. ISBN 0-471-90490-2.

Collection of annotated bibliographies covering several major areas within field of optimization. **(a)**

Counting Books Are More than Numbers: An Annotated Action Bibliography. Patricia L. Roberts. Hamden, CT: Library Professional Publications, 1990. viii + 270pp. $32.50. ISBN 0-208-02216-3.

More than 350 books are annotated in this bibliography which seeks to show teachers, parents, and librarians how to introduce the concepts and skills of mathematics to young children with the aid of picture books. Includes an artist-author and title index. **(p)**

Dictionary and Bibliography of Discrete Distributions. Ganapati Parashuram Patil, Sharadchandrea W. Joshi. New York: Hafner, 1968. xii + 268pp. o.p.

Bibliography section contains over 2,900 entries arranged by author with a classified subject index. Dictionary section is classified and indexed. **(a)**

French Mathematical Seminars: A Union List. Second Edition. Nancy D. Anderson. Providence, RI: American Mathematical Society, 1989. xv + 178pp. $36.00. ISBN 0-8218-0129-5.

Union list for North American universities' holdings providing accurate bibliographic entries, cross references, sponsoring and issuing body, dates, normal form of entry, name and address of publisher, and other useful information. National Union Catalog symbols used for holdings. **(a)**

Handbooks and Tables in Science and Technology. Second Edition. Ed: Russell H. Powell. Phoenix, AZ: Oryx, 1983. vi + 297pp. $35.50. ISBN 0-89774-039-4.

Covers over 3,400 handbooks and tables in science and technology with thorough indexing. Very useful tool. **(p,s)**

History of Mathematics from Antiquity to the Present: A Selective Bibliography. Joseph W. Dauben. New York: Garland, 1985. xxxix + 508pp. o.p. ISBN: 0-8240-9284-8.

Annotated bibliography of 2,000 sources in history of mathematics. Arranged by broad categories. **(a,p)**

Integer Programming and Related Areas: A Classified Bibliography. Ed: Dirk Hausmann. New York: Springer-Verlag, 1978. Lecture Notes in Economics and Mathematical Systems;160. xiv + 314pp. o.p. ISBN 0-387-08939-*.

Comprehensive bibliography on discrete optimization containing 3,162 titles up to March, 1978. Some reach back to 1950s. Arranged alphabetically, the bibliography also contains a subject-classified list and a cross reference list of co-authors. **(a)**

Integer Programming and Related Areas: A Classified Bibliography. Ed: R. von Randow. New York: Springer-Verlag, 1985. Lecture Notes in Economics and Mathematical Systems;197. xiv + 348pp. $36.00. ISBN 0-387-11203-0.

A computer printout, this book provides useful information on applications of integer programming, complexity, dynamic programming, graph theoretic results, packing, and shortest paths. **(a)**

Integer Programming and Related Areas: A Classified Bibliography, 1981-1984. Ed: R. von Randow. New York: Springer-Verlag, 1985. Lecture Notes in Economics and Mathematical Systems;243. xx + 386pp. $41.00. ISBN 0-387-15226-1.

Covers mid-1981 to autumn 1984 as well as any earlier publications not included in previous volumes. Topics included are theory and methods of general integer programming, combinatorial and graph-theoretical optimization problems related to integer programming, and applications of integer programming. **(a)**

Integer Programming and Related Areas: A Classified Bibliography, 1984-1987. Ed: R. von Randow. Berlin: Springer-Verlag, 1990. Lecture Notes in Economics and Mathematical Systems;346. xvi + 514pp. $56.00. ISBN 0-387-52687-0.

Lists works on integer programming and combinatorial optimization. Intended as a reference source for working mathematicians and researchers as well as industrial operations research departments. **(a)**

International Bibliography of Journals in Mathematical Education. Gert Schubring, Jutta Richter. Bielefeld, Germany: Universität Bielefeld, Institut für Didaktik der Mathematik, 1980. 108pp. Price not available.

International coverage of some 300 mathematical education journals. Entries include name, publisher, editor, subscription address, language, and annotation on contents. **(a)**

Library Recommendations for Undergraduate Mathematics. Ed: Lynn Arthur Steen. MAA Reports;4. Washington, DC: Mathematical Association of America, 1992. xi + 194pp. $15.00. ISBN 0-88385-076-1.

Revision of MAA's long out-of-date *Basic Library List for Four-Year Colleges* (1976). Companion volume to *Two-Year College Mathematics Library Recommendations* (see below). 3000 titles arranged in 25 chapters and coded with asterisks into 4 levels of priority (200 essential, 400 highly recommended, 800 recommended, and 1600 listed). One of the purposes of this book is "to ensure that college libraries have important reference works and classic sources in the mathematical sciences." Both in-print and out-of-print titles have been selected. Also includes a section on periodicals and journals and author index. **(a)**

Mathematics Education in Secondary Schools and Two-Year Colleges: A Sourcebook. Paul J. Campbell, Louise S. Grinstein. New York: Garland, 1988. xvii + 439pp. $71.00. ISBN 0-8240-8522-1.

An annotated bibliography of books and articles with brief essays by specialists at the beginning of each of 20 sections. **(p)**

Omega–Bibliography of Mathematical Logic. Ed: Gert H. Müller, Wolfgang Lenski. New York: Springer-Verlag, 1987. 6v. V. 1: xxxix + 483pp. $189.00. ISBN 0-387-17321-8. V. 2: xxxvii + 468pp. $189.00. ISBN 0-387-15521-X. V. 3: xiv + 615pp. $219.00. ISBN 0-387-15522-8. V. 4: xlv + 696pp. $219.00. ISBN 0-387-15523-6. V. 5: li + 790pp. $249.00. ISBN 0-387-15525-2. V. 6: xli + 404pp. $179.00. ISBN 0-387-15524-4. Set: $1,178.00. ISBN 0-387-17457-5.

Classified, comprehensive bibliography of works in mathematical logic published between 1879 and 1985 following *Mathematical Reviews* section 03 categories with some modifications. Each volume includes an introduction, subject index, author index, and source index. **(a)**

Pre-Meiji Works in the Library of Congress. Compiled by Shojo Honda. Washington, DC: Library of Congress, 1982. vii + 64pp. o.p. ISBN 0-8444-0392-X.

Bibliography of printed books and manuscripts from the Asian Division of the Library of Congress covering pre-Meiji Japanese mathematics from the Seventeenth Century to 1867. Entries arranged alphabetically. Titles Romanized using Hepburn system. **(a)**

Pythagoras: An Annotated Bibliography. Luis E. Navia. New York: Garland, 1990. (Garland Reference Library of the Humanities; 1,128). xviii + 381pp. $50.00. ISBN 0-8240-4380-4.

A full list of Pythagorean secondary literature. The 1,197 entries are broadly grouped but accessible through narrow internal divisions and author and name indexes. Despite having some flaws (e.g., no subject index), the source should be useful to researchers. **(a,p)**

Rara Arithmetica; A Catalogue of the Arithmetics Written before the Year MDCI... David Eugene Smith. New York: Chelsea, 1970. xvii + 725pp. $45.00. ISBN 0-8284-0192-6.

Includes addenda originally published in 1939. Catalog of Plimpton collection now at Columbia University Library. (**a**)

Tomber's Bibliography and Index in Nonassociative Algebras: Combined Volumes I, II, III. Marvin Tomber et al. Nonantum, MA: Hadronic Press, 1984. 535pp. o.p. ISBN 0-911767-17-7.

Contains early works on associative algebra compiled from a variety of sources, with *Mathematical Reviews* being the primary one. "Little effort was made to verify sources." (**a**)

Two-Year College Mathematics Library Recommendations. Ed: Lynn Arthur Steen. MAA Reports; 5. Washington, DC: Mathematical Association of America, 1992. xi + 76pp. $11.00. ISBN 0-88385-077-X.

The long-awaited revision of MAA's *Basic Library List for Two-Year Colleges* (1980). Organization and coverage is similar to its companion volume, *Library Recommendations for Undergraduate Mathematics* (see above). This list contains 1200 titles specially selected from the 3000 titles of the larger work. (**a,p**)

Women and Mathematics, Science and Engineering: A Partially Annotated Bibliography with Emphasis on Mathematics and with References on Related Topics. Else Hoyrup. Skriftserie fra Roskilde Universitesbibliotek;4. Roskilde: Roskilde University Library, 1978. vi + 62pp. Not in trade.

Partially annotated bibliography of mainly American and Danish papers and books primarily from the sixties and seventies. Some of the topics include sex roles in textbooks, women mathematicians, and sex differences in spatial visualization, creativity, cognition, and problem solving. (**a**)

DICTIONARIES

A. J. Lohwater's Russian-English Dictionary of the Mathematical Sciences. Second Edition, Revised and Expanded. Ed: R. P. Boas. Providence, RI: American Mathematical Society, 1990. xi + 343pp. $50.00 (hbk); $35.00 (pbk). ISBN 0-8218-0160-0 (hbk); 0-8218-0133-3 (pbk).

The opening grammar synopsis of this important reference book has been rewritten. Stress marks have been added to Russian words. The vocabulary has been "extensively enlarged" to reflect contemporary Russian writing in the mathematical sciences. Intended only as an aid in translating Russian to English, not vice versa. (**a,s**)

Angliĭskiĭ Tolkovyĭ Slovar' po Kibernetike i Prikladnoĭ Matematike. Ed: M. M. Glushko. Moscow: Moskov. Gos. Univ., 1986. 184pp. o.p.

Title translates to "English explanatory dictionary in cybernetics and applied mathematics." Contains 400 keywords and 5,000 dictionary items. (**a**)

Anglo-Russkiĭ Slovar' Matematicheskikh Terminov. Ed: P. S. Aleksandrov et al. Moscow: Izdatel'stvo Inostrannoĭ Literatury, 1962. Price not available.

Alphabetical listings provide in-context translations. One appendix describes units of measure and another is a rather detailed bilingual guide to English grammar and usage. **(a)**

The Artificial Intelligence Dictionary. Ellen Thro. San Marcos, CA: Slawson Communications/Microtrend Books, 1991. xiii + 407pp. $24.95. ISBN 0-915391-36-8.

Contains more than 1,300 terms from artificial intelligence and related subjects like expert systems, data structures, computer languages, robotics, probability, and software. Takes a general, jargon-free approach. Among the topics of the 12 appendixes are lists of symbols, British versus American terms, abbreviations, and sources of additional information. **(p,s)**

Cambridge Italian Dictionary. Ed: Barbara Reynolds. New York: Cambridge University Press, 1985. 2276pp. o.p.

An updated version of a classic comprehensive Italian-English, English-Italian dictionary. **(a)**

Cassell's German-English English-German Dictionary. Rev: Harold T. Betteridge. New York: Macmillan, 1978. xx + 1580pp. $25.00;$26.00. ISBN 0-02-522920-6 (standard);0-02-522920-3 (thumb indexed).

A revision of the earlier works, this new dictionary includes new words and usages. The English-German section contains many helpful examples of idiomatic usage. **(a)**

Cassell's Latin Dictionary: Latin-English, English-Latin. Ed: D. P. Simpson. New York: Macmillan, 1977. xviii + 833 pp. $24.95. ISBN 0-02-522-580-4.

This remains a useful reference tool. **(a)**

Cheshsko-Russkiĭ Matematicheskiĭ Slovar'. Ed: Ernest Kol'man. Moscow: Sovetskaia͡ Ent͡siklopedii͡a, 1971. 237pp. o.p.

This Czech-Russian mathematical dictionary contains approximately 8,000 terms with the words listed alphabetically and the phrases listed by keyword. No cross-referencing. **(a)**

Chinese-English Glossary of the Mathematical Sciences. John DeFrancis. Providence, RI: American Mathematical Society, 1964. o.p. 275pp.

Contains 16,540 terms from a number of sources, with the majority coming from mathematics dictionaries published in Peking (1956), Taipei (1958), Tokyo (1961), and Moscow (1959). **(a)**

Computer Dictionary. Third Edition. Donald D. Spencer. Ormond Beach, FL: Camelot Publishing, 1991. ix + 457pp. $24.95. ISBN 0-89218-209-1.

More than 5,800 computer terms, phrases, acronyms, and abbreviations alphabetically-arranged and defined for readers at all levels. Includes terms from allied fields such as mathematics, robotics, data communications, and electronics.

Also covers newer areas of computer science like object-oriented programming, computer graphics, desktop publishing, and artificial intelligence. **(a,p,s)**

A Concise Dictionary of Mathematics. Christopher Clapham. Oxford: Oxford University Press, 1990. 203pp. $19.50. ISBN 0-19-866156-8.

Intended as a general undergraduate reference work for mathematics, this book does not include statistics, computing, or applied mathematics terms. It does include entries for names of mathematicians. **(a)**

Crescent Dictionary of Mathematics. William Karush. Palo Alto: Seymour Publications, 1987. 313pp. $14.95. ISBN 0-86651-352-3.

Reprint of book originally published by Macmillan in 1962. Basic dictionary for high school and college with 1,422 entries covering field of mathematics. References in appendix A. **(a,p)**

Diccionario Matemático; Español-Ingles, Ingles-Español. Mathematics Dictionary; Spanish-English, English-Spanish. Mariano Garcia Rodriguez. New York: Hobbs, Dorman, 1965. vii + 78pp. Price not available.

Only word and phrase equivalents in Spanish and English are presented. **(a)**

Dictionar Poliglot de Matematica, Mechanica si Astronomie. Stefan Gheorghita. Bucharest: Editura Tehnica, 1978. xvi + 664pp. Price not available.

This is a polyglot dictionary with listings arranged alphabetically in English. The equivalents in Romanian, German, French, and Russian follow. Indices in each of the latter four languages direct the user to the appropriate English language entry. **(a)**

Dictionary of Engineering and Technology, with Extensive Treatment of the Most Modern Techniques and Processes. V.1: German-English. V.2: English-German. Fifth Edition. Richard Ernst. New York: Oxford University Press, 1989 (V. 1); 1985 (V. 2). V.1: 1,258pp. $125.00. ISBN 0-19-520820-X. V.2: 1,422pp. $95.00. ISBN 0-19-520485-9.

Contains more than 197,000 entries covering all major industries and including new terms from electronics and computer technology. Recommended for larger-sized reference collections. **(a)**

Dictionary of Gaming, Modelling and Simulation. G. Ian Gibbs. Ann Arbor, MI: Books on Demand UMI, 1978. xi + 161pp. $49.50 (hbk.); $43.50 (pbk.). ISBN 0-317-11019-5. Order #: 2021906.

Well illustrated. Covers statistical tests, mathematical formulas and related terms. Dated. **(a)**

Dictionary of Logical Terms and Symbols. Carol Greenstein, Carol Horn. New York: Van Nostrand Reinhold, 1978. xiii + 188pp. o.p. ISBN 0-442-22834-1 (hbk.); 0-442-22836-8 (pbk.).

Presents notation used in logic, engineering, and computer science with tables for translation from one system to another. Less detailed than *Dictionary of Symbols of Mathematical Logic* (see below). **(a)**

Dictionary of Mathematical Games, Puzzles, and Amusements. Harry Edwin Eiss. Westport, CT: Greenwood, 1988. xvi + 279pp. $49.95. ISBN 0-313-24714-5.

Collection of essays on games and puzzles and related mathematical topics such as Euclid's theory of parallels. Includes many historical references. **(p)**

Dictionary of Mathematical Sciences. Second Edition. Leo Joseph Herland. New York: Ungar, 1965. 2v. V. 1: xii + 323pp. V. 2: 349pp. o.p. ISBN 0-8044-4393-9 (V. 1); 0-8044-4394-7 (V. 2).

German to English and English to German with equivalents given under each term. **(a)**

Dictionary of Mathematics. C. C. T. Baker. New York: Hart, 1965; (c1961). v + 338pp. Price not available.

Basic dictionary for college and high school students. Includes biographical entries, definitions, processes, formulae, and tables. Illustrated. **(a,p)**

Dictionary of Mathematics. Ed: E. J. Borowski, J. M. Borwein. London: Collins, 1989. o.p. ISBN: 0-00-434347-6.

Containing over 4,000 entries and 400 diagrams, this can be used by people with a wide range of ability levels. It includes worked examples and capsule biographies of major mathematicians. **(p)**

Dictionary of Mathematics. Ed: J. A. Glenn, G. H. Littler. Totowa, NJ: Barnes & Noble, 1984. x + 240pp. o.p. ISBN: 0-389-20451-X.

Definitions of terms gathered from indexes of British college texts are of uneven quality. All technical terms used are also defined. **(a)**

Dictionary of Mathematics. T. Alaric Millington, William Millington. New York: Barnes and Noble, 1971 (c. 1966). ix + 259pp. $8.95. ISBN 0-06-463311-X.

Covers all areas of mathematics with brief definitions suitable for beginning students. Dated. **(a,p)**

Dictionary of Mathematics in Four Languages: English, German, French, Russian. Günther Eisenreich, Ralf Sube. Amsterdam: Elsevier North-Holland, 1982. 2v. (1,459pp.). $218.50 set. ISBN 0-444-99706-7.

This is a polyglot mathematical dictionary containing some 35,000 terms. Very thorough coverage. An alphanumeric code and the columnar format help simplify the use of this comprehensive dictionary. Explanatory notes help users with the problems of synonyms and homonyms. *Wörterbuch der Mathematik*, also by Eisenreich and Sube, is the same book, organized in German. **(a,p,s)**

Dictionary of Mathematics Terms. Douglas Downing. Woodbury, NY: Barron's, 1987. xiv + 241pp. $9.95. ISBN: 0-8120-2641-1.

Definitions of uneven quality for 600 terms. Includes cross references. Symbols list, brief tables, etc. Reference source best suited to public library. **(p)**

Dictionary of Named Effects and Laws in Chemistry, Physics and Mathematics. Fourth Edition. D. W. G. Ballentyne, D. R. Lovett. London; New York: Chapman & Hall, 1980. viii + 346pp. o.p. ISBN 0-412-22380-5 (hbk.); 0-412-22390-2 (pbk.).

Brief explanations of 1,400 effects, laws, formulas, units, definitions, and theorems from mathematics and physical sciences named after individuals. **(a,p)**

A Dictionary of Real Numbers. Jonathan Borwein, Peter Borwein. Pacific Grove, CA: Wadsworth, 1990. viii + 424pp. $79.95. ISBN 0-534-12840-8.

Given a number to be understood, this book tells which classical computations have approximately the same outcome. Even with 100,000 entries the book is not complete, but it is the first of its kind. Intended for undergraduate audiences. **(a,p)**

Dictionary of Scientific Units Including Dimensionless Numbers and Scales. Fifth Edition. H. G. Jerrard, D. B. McNeil. London; New York: Chapman & Hall, 1986. ix + 222pp. $29.95. ISBN 0-412-28090-6.

Definitions for more than 850 named units and dimensionless numbers. Appendices for tables of weights and measures, physical constants and conversion factors for SI (Systeme International) and CGS units. Fifth Edition provides more precise values for the fundamental physical constants and more exact definitions of other units (e.g., candela) which have now been approved by the International Committees. **(p,s)**

Dictionary of Statistical, Scientific and Technical Terms. Hardeo Sahai, Jose Berrios. New York: French and European Publications, 1981. ix + 143pp. $35.00. ISBN: 0-8288-2352-9.

English to Spanish and Spanish to English terms in statistics and related fields. About 3,000 terms per section including phrases and proper names. **(a)**

A Dictionary of Statistical Terms. Fifth Edition. F. H. C. Marriott. New York: Wiley, 1990. viii + 223pp. $62.95. ISBN 0-470-21349-3.

Marriott's updating of the late Maurice G. Kendall's and the retired William R. Buckland's basic and important work. Contains 400 new entries and omits a number of Italian terms. Most references indicate the author and date of publication where the term was first used. **(a,p,s)**

Dictionary of Symbols of Mathematical Logic. Second Corrected Printing. Ed: R. Feys, F. B. Fitch. Amsterdam: North Holland Publishing, 1973. xiv + 171pp. o.p. ISBN 0-7204-2250-7.

Presents concepts and notations of formal deductive systems of symbolic logic employed by most mathematical and philosophical texts. More detailed than *Dictionary of Logical Terms and Symbols* (see above). **(a,p)**

Dictionary/Outline of Basic Statistics. John E. Freund. New York: Dover, 1991. vii + 195pp. $6.95. ISBN 0-486-66796-0.

Reprint of book published by McGraw-Hill in 1966. Dictionary of statistical terms plus outline of statistical formulas. **(a)**

Dictionnaire de Mathématiques. Lucien Chambadal. New York: French and European Publications, 1982. 302 + [10]pp. $69.95. ISBN 0-8288-1896-7.

Covers terms in algebra, geometry, and differential and integral calculus. Definitions include examples. Index to notations. **(a)**

Dictionnaire des Mathématiques. Alain Bouvier, Michel George. Paris: Presses Universitaires de France, 1979. xiv + 832pp. o.p. ISBN 2-13-035427-0.

Illustrated French language mathematical dictionary. **(a)**

Encyclopaedic Dictionary of Mathematics for Engineers and Applied Scientists. I. N. Sneddon. (See p. 100).

English-French-Spanish-Russian Systematic Glossary of the Terminology of Statistical Methods. Isaac Paenson. Oxford: Pergamon, 1970. xxxviii + 517pp. $190.00. ISBN 0-08-012285-X.

Arranged by topical chapters to place terms in context. Individual terms are defined in all four languages. Separate index for each language. **(a)**

English-Greek Dictionary of Pure and Applied Mathematics with Greek and English Appendices. Memas Kolaitis. Athens: Technical Chamber of Greece, 1976. 2v. (cvi, 1516pp.). Price not available.

Exhaustive, now somewhat dated dictionary of Greek equivalents to English language mathematical terms. Indexes. **(a)**

English-Greek Mathematical Dictionary. C. P. Tzelekis. Athens: Athens Pub. Center, 1973. ix + 234pp. Price not available.

7,800 entries of English mathematical terms with their Greek equivalents for Greek speaking students. Appendices. **(a)**

English-Russian Dictionary of the Mathematical Sciences. Ed: Pavel S. Aleksandrov. Moscow, 1962. 369pp. o.p.

English to Russian equivalents for some 6,000 mathematical terms plus outline of English grammar, table of measurements. No index. **(a)**

Facts on File Dictionary of Mathematics. Revised Edition. Ed: Carol Gibson. New York: Facts on File, 1988. 235pp. $24.95 (hbk.); $12.95 (pbk.). ISBN 0-8160-1867-7 (hbk.); 0-8160-2365-4 (pbk.).

Basic definitions for non-specialists clearly presented and cross-indexed. **(a,p)**

German-English Mathematical Vocabulary. Second Edition. Sheila Macintyre, Edith Witte. New York: Interscience, 1966. x + 95pp. o.p.

Covers German-English equivalents for pure mathematics and basics of German grammar in one slim volume. **(a)**

German-English Mathematics Dictionary. Compiled by Charles Hyman. New York: Interlanguage Dictionaries, 1960. 131pp. Price not available.

Based on terms appearing in text books and similar works. 8,500 terms specific to mathematics. German with English equivalents. No definitions. **(a)**

German-English Science Dictionary. Louis De Vries, Leon Jacolev. New York: McGraw-Hill, 1978. xlii + 592 pp. o.p.

Dated reference work which does contain valuable suggestions for translators. **(a)**

Glossary of the Mathematical and Computing Sciences. Compiled by Loh Shiu-chang et al. Hong Kong: The Chinese University of Hong Kong, 1976. 2v. V.1: iv + 494. V.2: iv + 551. Price not available.

Contains 30,000 terms used in connection with mathematics and computing science. English with Chinese equivalents in one volume, Chinese to English in other. **(a)**

Han-Ying Shuxue Cihui. Ed: Yu Xia Qi. Beijing: Kexue Chubanshe, 1986. iv + 1,130pp. Price not available.

Chinese-English mathematical dictionary of 34,000 entries giving Chinese characters, pinyin romanization, and English translation. Arranged alphabetically by pinyin form. There is an index by number of strokes as well. **(a)**

The Harper Collins Dictionary of Computer Terms. Ian R. Sinclair. New York: Harper Perennial, 1991. 261pp. $25.00 (hbk.); $10.95 (pbk). ISBN 0-06-271505-4 (hbk.); 0-06-461016-0 (pbk.).

Contains 1,700 entries including selected trade names but not biographical entries. Well written, inexpensive, and current. Coverage compares favorably to other similar dictionaries. **(p)**

The Harper Collins Dictionary of Mathematics. E. J. Borowski. New York, NY: Harper Perennial, 1991. x + 659pp. $14.95. ISBN 0-06-461019-5.

Originally published as *Collins Dictionary of Mathematics*. Dictionary of some 6,000 terms at undergraduate level and above. Appendix of symbols and conventions is useful. **(a,p)**

The Harper Collins Dictionary of Statistics. Roger Porkess. New York: Harper Perennial, 1991. xi + 267pp. $25.00. ISBN 0-06-271527-5.

Intended for use by mathematics or statistics students this book includes definitions and worked examples. **(a)**

Harrap's New Standard French and English Dictionary. Ed: J. E. Mansion. New York: Prentice Hall, 1980. $125.00 for each Part. ISBN 0-317-45641-5 (Pt. 1: French to English); ISBN: 0-317-45642-3 (Pt. 2: English to French).

This two-part, four-volume set is a comprehensive bilingual dictionary. **(a)**

Inglisur-Kartuli Matematikuri Leksikoni. C. Gabeskiria. Tbilisi: Tbilis. Gos. Univ., 1983. 168pp. o.p.

8,000 English to Georgian equivalent terms in vocabulary adequate for lower graduate level mathematics and mathematical sciences. Includes lists of mathematicians, abbreviations, and foreign words used in English. **(a)**

International Dictionary of Applied Mathematics. Ed: W. F. Friedberger. Princeton, NJ: Van Nostrand Reinhold, 1960. 1,173pp. o.p.

Excellent encyclopedic dictionary defining 8,000 terms and methods for advanced scientists and engineers. Has French, German, Spanish and Russian indexes. Something of a classic. **(a,s)**

Kleine Duden-Mathematik. Ed: Hermann Engesser. Mannheim: Bibliographisches Institut, 1986. 480pp. $9.77. ISBN 3-411-02180-2.

Intended to aid with practical mathematical problems of daily life including data processing and information science terms in this context. Explains abstract mathematical concepts in understandable language with many illustrations and examples.

Langenscheidt New Muret-Sanders Encyclopedic Dictionary of the English and German Languages. Ed: Otto Springer. Berlin: Langenscheidt, 1974. (883pp. $125.00. Part I. Vol. 1. ISBN: 0-88729-001-9). (960 pp. $125.00. Part I, Vol. 2. ISBN: 0-88729-002-7). (973pp. $125.00. Part II, Vol. 1. ISBN: 0-685-06934-6). (1048pp. $125.00. Part II, Vol. 2. ISBN: 0-88729-004-3).

Two parts, with each part in two volumes. Part I, Vol. 1 contains A-M, English-German; Part I, Vol 2 contains N-Z, English-German. Part II, Vol. 1 contains A-K, German-English, and Part II. Vol. 2 includes L-Z, German-English. This is periodically revised. **(a)**

Lexikon der Algebra. Günther Eisenreich. Berlin: Akademie-Verlag, 1989. 677pp. $49.46. ISBN 3-05-500231-8.

Comprehensive, one volume German dictionary of algebra from recognized authority in mathematical lexicography. Contains many "see" references to longer key entries of up to six pages, most with several references for additional study. **(a)**

Matematicheskie Terminy: Spravochnik. N. V. Aleksandrova. Moscow: Izdat. Vysshaia͡ Shkola, 1978. 189pp. o.p.

Brief entries for origin of various mathematical terms, each with sources. Index including English language entries to mathematicians mentioned. **(a)**

Matematicheskiĭ Slovar'. Ulo Kaazik. Tallinn: Valgus, 1985. 293pp. o.p.

A Russian language dictionary of mathematical words and concepts. Intended as a ready-reference work, it covers some 5,000 high school and general college level mathematical terms. Indexes to names and notation. **(a)**

Matematicheskiĭ Slovar' Vyssheĭ Shkoly. V. T. Vodnev et al. Minsk: Vysheĭshaia͡ Shkola, 1984. 526pp. o.p.

This dictionary is characterized by rather brief entries and is intended primarily for university students. **(a)**

Matematika Terminderinin Orusa-Kyrgyza Sozdugu. Ch. Dzhanybekov, R. Usubakunov. Frunze: Ilim, 1978. 500 + [2]pp. o.p.

Contains 10,000 Russian mathematical words with Kirghiz translations. Intended for Kirghiz students reading Russian texts. **(a)**

Matematikai Kislexikon. Ed: Farkas Miklós. Budapest: Műszaki Könyvkiadó, 1972. 443pp. Price not available.

Hungarian mathematical dictionary. **(a)**

Matematikatorteneti ABC. Fifth Edition. Marton Sain. Budapest: Tankonyvkiadó Vállalat, 1987. 328pp. Price not available. ISBN 96-31-79716-3.

Contains dates, facts, and miscellany in mathematics for students and teachers. Some biographical entries and mathematical symbols plus a bibliography. **(a)**

Mathematics Dictionary. Fifth Edition. Ed: Robert C. James. New York: Van Nostrand Reinhold, 1992. 548pp. $42.95. ISBN 0-442-00741-8.

A high quality general mathematics dictionary of some 8,000 terms, covering arithmetic through calculus into statistics plus sketches of major contributors. Multilingual index. **(a,p)**

Mathematics Dictionary: English-Arabic (with an Arabic Index): Covering the Terms (and Definitions) of Traditional and Modern Mathematics, Mechanics and Computers. Jordanian Committee for Arabisation, Ministry of Education, Amman. Beirut: Librairie du Liban, 1975. 308pp. Price not available.

Terms given in English; explanation in Arabic. **(a)**

Mathematics Illustrated Dictionary: Facts, Figures, and People. Revised Edition. Jeanne Bendick. New York: Franklin Watts, 1989. 247pp. $14.90. ISBN 0-531-10664-0.

Well-designed, easy-to-use dictionary of mathematical terms, concepts, processes, and people. Author organizes formulas in tables at back of book. Other tables cover mathematical symbols, logs, square roots, and the metric system. Revisions include more basic computer terms, additional entries for mathematicians, and improved illustrations. **(a,p)**

La Mathématique Parlée par Ceux Qui L'Enseignement. Paris: Commission du Dictionnaire de l 'A. P. M. E. P., 1967. Price not available.

This is issued on cards rather than in bound pages and is for primary use by pre-college students.

Mathematischer Fachwörtschatz: Englisch-Deutsch, Deutsch-Englisch. Traute Pfeil. Jena: Friederich-Schiller-Universität Jena, 1980. 100pp. Price not available.

Hierarchical vocabulary of mathematical terms giving English-German and German-English equivalents. **(a)**

Mathematisches Begriffswörterbuch. Herbert Meschkowski. Mannheim: Bibliographisches Institut, 1976. 315pp. Price not available. ISBN: 3-411-05099-3.

This illustrated German language dictionary includes formulas, examples, graphs, useful "see" and "see also" references, and a bibliography. **(a)**

Mathematisches Vokabular. E. B. Klaften. Munich: Wilhelm Lampl. 1961. x + 186pp. $10.39. ISBN 3-87910-151-5.

English-German and German-English vocabulary of elementary mathematics topically arranged. Covers general terms, arithmetics, algebra, plane geometry, solid geometry, trigonometry, calculus, and coordinates. **(a)**

Mathematisches Wörterbuch: Alphabetische Zusammenstellung sammtlicher in die mathematischen Wissenschaften gehorender Gegenstande in erlkarenden und beweisenden synthetisch und analitisch bearbeiteten Abhandlungen. Ludwig Hoffmann. Berlin: G. Bosselmann; Wiegandt & Hemple, 1858-1867. 7v. in 5. o.p.

First edition of "the fullest mathematical dictionary ever published." -Sotheran, Second Supplement, 1700. This is still a valuable work for its historical interest. **(a)**

McGraw-Hill Dictionary of Physics and Mathematics. Ed: Daniel Lapedes. New York: McGraw-Hill, 1978. xiv + 1120pp. $48.50. ISBN 0-07-045480-9.

Includes some 20,000 definitions of terms in physics, mathematics, and related fields. Some illustrations. 46 pages of appendices. **(a,p,s)**

Mehrsprachenwörterbuch Mathematischer Begriffe. Herbert Meschkowski. Mannheim: Bibliographisches Institut, 1972. 131pp. $23.49. ISBN 3-411-01409-1.

Multilingual. First part is list of German terms with corresponding terms in English, French, Russian, and Italian. Second part is lists of mathematical terms in these four languages with their respective equivalents. **(a)**

Nemetsko-Russkiĭ Matematicheskiĭ Slovar'. Third Edition. L. A. Kaluzhnin et al. Moscow: Russkiĭ I͡Azyk, 1980. 559pp. o.p.

This German-Russian mathematical dictionary contains some 30,000 entries and includes a listing of names as well as glossaries of terms. **(a)**

New Mathematics Dictionary and Handbook (See p. 117).

The Oxford Russian-English Dictionary. Marcus Wheeler. Oxford: Clarendon Press, 1972. xiii + 918pp. Price not available. ISBN 0-19-864111-7.

This useful general-purpose dictionary of written and spoken Russian is intended for users whose native language is English. **(a)**

Penguin Dictionary of Curious and Interesting Geometry. David Wells. London: Penguin Books, 1991. xiv + 285pp. $20.00. ISBN 0-14-011813-6.

Brocard points, dragon curves, and Islamic tessellations are just a few of the hundreds of shapes, famous and obscure, ancient and modern, that are illustrated and briefly described. Good for browsing and can provide excellent stimulation for math club projects. (**p**)

Penguin Dictionary of Mathematics. Ed: John Daintith, R. D. Nelson. New York: Penguin, 1989. 350pp. $8.95. ISBN 0-14-051119-9.

2,800 definitions in pure and applied mathematics for college students includes 200 biographical entries. Cross references. No computer terminology. (**a,p**)

Polsko-Russkii Matematicheskii Slovar'. Second Edition. N. E. Miklaszewsli. Moscow: Russkii Iazyk, 1981. 368 pp. o.p.

This revised edition of the Polish-Russian mathematical dictionary contains approximately 24,000 terms. (**a**)

Quantities and Units of Measurement: A Dictionary and Handbook. J. V. Drazil. New York: French and European Publications, 1983. 313pp. $110.00. ISBN 0-8288-0009-X.

Originally published as *Dictionary of Quantities and Units* (1971). Two sections: first is a dictionary of units of measurement, symbols and abbreviations; second part a handbook of quantities and selected constants. (**a**)

Riiaziiiat Terminlari Lughati: Ingilisja-Rusja-Azarbaijanja. Ed: H. N. Aghaiev, F. G. Magsudov, G. A. Mustafaiev. Baku: Elm, 1979. 344pp. o.p. Added title: *Slovar' Matematicheskikh Terminov (na Angliiskom, Russkom, Azerbaidzhanskom Jazykah)*.

Over 4,000 English mathematical terms with Russian and Azerbaijani equivalents. (**a**)

Romanian-English Dictionary and Grammar for the Mathematical Sciences. S. H. Gould, P. E. Obreanu. Providence, RI: American Mathematical Society, 1967; 1979. v + 51pp. $17.00. ISBN 0-8218-0038-8.

Basic information for readers of mathematical Romanian. It covers basic construction of the language and defines Romanian words commonly appearing in mathematical works. (**a**)

Rosiis'ko-Ukrains'kyĭ Matematychnyĭ Slovnyk. Fedir Sydorovych Hudymenko. Kiev: Akademii͡a Nauk Ukrainskoĭ S. S. R., 1960. 162pp. o.p.

This is a Russian-Ukranian mathematical dictionary of some 12,000 terms. (**a**)

Russian-English Dictionary of Mathematics. Oleg P. Efimov. Boca Raton, FL: CRC Press, 1993. viii + 419pp. $89.95. ISBN 0-8493-4456-5.

Comprehensive vocabulary, containing more than 27,000 entries, for readers and writers of Russian mathematical literature. It embraces all the major branches of mathematics, from elementary to advanced topics in toology and higer algebra. (**a**).

Russian-English/English-Russian Glossary of Statistical Terms. Samuel Kotz. Edinburgh: Oliver & Boyd, 1971. vii + 87pp. Price not available. ISBN 0-05-002446-9.

Based on and supplemental to Kendall and Buckland's *Dictionary of Statistical Terms*, Third Edition, it covers 2,500 terms giving Russian equivalents plus somewhat more in Russian to English. **(a)**

Russian-English Mathematical Dictionary: Words and Phrases in Pure and Applied Mathematics. L. M. Milne-Thomson. Ann Arbor, MI: Books on Demand UMI, 1962. xiv + 191pp. $51.30. ISBN 0-317-28131-3. Order #: 2055742.

Developed as aid for translating Russian works in applied mathematics, dictionary provides English equivalents to some 8,500 Russian terms supplemented by 28 page outline of Russian grammar. **(a)**

Russian-English Mathematical Vocabulary. J. Burlak. Edinburgh: Oliver & Boyd, 1963. 305pp. Price not available.

Includes "Short Guide to Reading Russian" by K. Brooke which emphasizes grammar. Vocabulary gives English equivalents to some 13,000 Russian (in Cyrillic) mathematical terms. **(a,s)**

Russian-English Vocabulary with Grammatical Sketch: To Be Used in Reading Mathematical Papers. Gabrielle Rainich. Providence, RI: American Mathematical Society, 1950, 1980. 66pp. $15.00. ISBN 0-8218-0037-X.

This title will aid mathematicians in reading Russian mathematical papers. Vocabulary is based on materials supplied by Society members. Short grammatical sketch by A. H. Kuipers. **(a,s)**

Russko-Angliiskiĭ Matematicheskiĭ Slovar'. Russian-English Mathematical Dictionary. A. Kramer. Trenton, NJ: Published by Author, 1961. iii + 123pp. o.p.

The author writes, "this dictionary more than adequately covers all mathematical terminology likely to be encountered in Russian elementary and advanced scientific and technical literature containing more than 5,000 entries." It does not provide definitions, merely Russian-English equivalents. **(a)**

Russko-Angliiskiĭ Matematicheskiĭ Slovar'-Minimum. M. M. Glushko. Moscow: Moskov. Gos. Univ., 1988. 142pp. o.p. ISBN 5-211-00180-X.

Dictionary of 5,000 plus entries based on Lohwater's *Russian-English Dictionary of the Mathematical Sciences* with some updating. **(a,s)**

Russko-Anglo-Nemetsko-Frant͡suzskiĭ Matematicheskiĭ Slovar'. V. B. Orlov, N. S. Skorokhod, A. B. Sosinskiĭ. Moscow: Russkiĭ I͡Azyk, 1987. 298pp. o.p.

3,000 basic terms for those utilizing mathematics in their professions, university students, etc. Alphabetical lists of English, German, and French equivalents. **(a)**

Short Dictionary of Mathematics. C. H. McDowell. New York: Philosophical Library, 1957. xiii + 61pp. Price not available.

In two parts. First covers arithmetic and algebra. Second has plane trigonometry and geometry. Provides definitions (some with illustrations), list of signs and symbols, and very simple tables. **(a)**

Simon and Schuster's International Dictionary. English/Spanish Spanish/English. Ed: Tana de Gamez. New York: Simon and Schuster, 1973. xviii + 1605 pp. o.p. ISBN 0671212672 (thumb-indexed).

This single-volume work is a comprehensive bilingual dictionary. **(a)**

Slovar' Matematicheskikh Terminov Na Angliĭskom, Russkom, Armi͡anskom, Nemet͡skom, Frant͡suzksom I͡Azykakh. A. O. Tonian. Erevan: Akademii͡a Nauk Armianskoĭ S. S. R., 1965. 237pp. o.p.

The entries in this polyglot mathematical dictionary are listed in English, with equivalents given in Russian, Armenian, German, and French. Indices in the latter four languages direct the user to the desired English entry. The format, languages in columns, entries in rows, is easy to use. **(a)**

Statistical Dictionary of Terms and Symbols. Albert K. Kurtz, Harold A. Edgerton. New York: Wiley, 1939; reprint 1967. xiii + 191pp. o.p.

A excellent dictionary with detailed definitions and explanations of symbols. **(a,s)**

Statistical Vocabulary. Second Edition. Inter-American Statistical Institute. Washington, DC: Pan American Union, 1960. xi + 83pp. Price not available.

1,300 English language terms with equivalents in Spanish, Portugese, and French. Separate word indexes for non-English languages. **(a,s)**

Statisztikai Szótár; 1700 Statisztikai Kifejezés hét Nyelven. Kozponti Statisztikai Hivatal. Budapest: Statisztikai Kiadó Vállalat, 1960. vii + 171pp. Price not available.

Statistical dictionary of 1,700 terms in seven languages. Major listing in Russian with equivalents in Hungarian, Bulgarian, Czech, Polish, German, and English.

Sugaku Ei-Wa Wa-Ei Jitten. Mathematics: English-Japanese and Japanese-English Dictionary. Yusaku Komatsu. Tokyo: Kyoritsu Shuppan, 1979. vi + 358pp. Price not available.

Presents general mathematical terms in Japanese characters with English language equivalents in one section and mathematical terms in English with their Japanese equivalents in the second part. **(a)**

Tolkovyĭ Matematicheskiĭ Slovar'. A. M. Mikisha, V. B. Orlov. Moscow: Russkiĭ I͡Azyk, 1988. 240pp. o.p. ISBN 5-2000-0246-X.

Russian language dictionary of basic mathematical terms. **(a)**

Topical Dictionary of Statistics. Gary L. Tietjen. New York: Chapman and Hall, 1986. xii + 171pp. $24.50. ISBN 0-412-01201-4.

Topically arranged, it attempts to provide context, applications, and related terms in addition to basic definitions. Definitions of key words or phrases are located through index. **(a,s)**

Transliterated Dictionary of the Russian Language. Ed: Eugene Garfield. Philadelphia: ISI Press, 1979. xi + 382pp. $25.00. ISBN: 0-89-495003-7.

Another general-purpose dictionary for users whose native language is English.

Vieweg Mathematik Lexikon. Second Edition. Otto Kerner et al. Braunschweig: Friedr. Vieweg, 1988. xii + 377pp. $23.49. ISBN 3-528-06308-4.

Reference tool for basic mathematics courses at West German universities. Covers analysis including differential equations and function theory, algebras, topology, probability and statistics plus some basic geometry and number theory.

Vocabularie Mathematic in Interlingue Contraductionin Angles (English), Frances (Français) e German (Deutsch). C. E. Sjöstedt. Stockholm: Bokförlaget Natur och Kultur, 1970. 87pp. Price not available.

This is a dictionary with equivalent mathematical terms in Interlingua, English, French, and German.

Webster's New Biographical Dictionary. Springfield, MA: Merriam-Webster, 1988. xvii + 1130pp. $21.95. ISBN 0-87779-543-6.

This is the most current update of this old standard.

Webster's New World Dictionary of Mathematics. William Karush. New York: Webster's New World; Distributed by Prentice Hall Trade, 1989. 317pp. $11.95. ISBN 0-13-192667-5.

A revised edition of the 1962 *Crescent Dictionary of Mathematics* (see p. 88). Important mathematical advances since then have been added, especially terms related to the field of computers. The list of outside references has also been updated and expanded. Contains more than 1,400 entries and is aimed at the general reader with a curiosity about mathematics as well as students and teachers of high school and college mathematics. **(a,p)**

Webster's Third New International Dictionary of the English Language Unabridged. Ed: Philip Babcock Gove. Springfield, MA: Merriam-Webster, 1986. 2662pp. $99.95; $109.95. ISBN 0-87779-201-1;0-87779-206-2.

The recommended English language dictionary. **(a,p)**

ENCYCLOPEDIAS

Encyclopaedia of Mathematical Sciences. Ed: R. V. Gamkrelidze. New York: Springer-Verlag, 1988-. (Series not yet complete.)

Individual volumes have been published out of order with various prices. Contact publisher for more information). Comprised of survey articles that include basic definitions, history and motivating examples, and descriptions of the frontiers of research. Authors are mostly from the former USSR and are often among the most influential in their respective fields. Highly recommended. **(a,p,s)**

Encyclopaedia of Mathematics. Ed: M. Hazewinkel. Dordrecht; Norwell, MA: Kluwer Academic, 1987-1993. 10v. (1-9 published so far; tenth volume will be index volume). $199/Volume. ISBN 1-556-08010-7 (set).

First nine volumes of a ten volume English language translation (with editorial updates and annotations) of Soviet *Matematicheskaia Entsiklopediia* (1977-85). Including surveys for non-specialists, articles on results or problems, and definitions, this aims to be a reference work for all parts of mathematics. Most articles signed by Russian authors and include AMS classification numbers and bibliography with added current Western sources. **(a)**

Encyclopaedic Dictionary of Mathematics for Engineers and Applied Scientists. I. N. Sneddon. Elkins Park, PA: Franklin, 1976. viii + 800pp. $324.00. ISBN 0-08-016767-5.

Comprehensive one-volume dictionary. Longer entries are signed and include references. Relates mathematical concepts and techniques most commonly used in engineering to basics of major branches of mathematics through cross referencing. **(a)**

Encyclopedia of Artificial Intelligence. Volumes 1 & 2. Ed: Stuart C. Shapiro. Reprint of 1987 original. New York: John Wiley & Sons, 1990. Vol 1: xxiv + 679pp. Vol 2: pp. 680-1219. $95.00. ISBN 0-471-52079-9.

Brings together core knowledge from all fields and related disciplines of artificial intelligence. Articles written by recognized research experts on the topic and contain bibliographies and extensive cross references. More than 450 tables and figures. **(s)**

Encyclopedia of Mathematics and Its Applications. Reading, MA: Addison-Wesley, 1976-.

41 volumes of this series have been published so far. Individual volumes are extremely well done and are recommended for any research collection. **(a,p,s)**

Encyclopedia of Physics. Ed: Rita G. Lerner, George L. Trigg. Reading, MA: Addison-Wesley, 1981. xvi + 1157pp. o.p. ISBN 0-201-04313-0.

Five hundred brief articles covering all of physics written to be useful to non-specialists as well as mathematicians and scientists. Some articles deal with mathematical subjects such as Lie groups, matrices, and vector and tensor analysis. **(a,p,s)**

Encyclopedia of Statistical Sciences. Ed: Samuel Kotz, Norman L. Johnson, Campbell B. Read. New York: Wiley, 1982-88. 9v. + 1 supp. $138.00/Volume (V. 1-4, 6-7, 9); $150.00 (V. 5); $165.00 (V. 8); $81.95 (supp.). $1,110.00 (set, V. 1-9). ISBN 0-471-05546-8 (V. 1); 0-471-81274-9 (supp.); 0-471-05544-1 (set).

Quality encyclopedia with many longer signed articles. Short articles are frequently unsigned, but nearly all articles have references. Intended for knowledgeable readers but not the specialist. Includes articles on organizations, journals, contributors to the field, formulas, etc. Volume 9 includes index to entire work. **(a,p,s)**

Encyclopedic Dictionary of Mathematics. Second Edition. Ed: Kiyosi Ito. Cambridge MA: MIT Press, 1987. 4v. (xvii + 2148pp.). $365.00 (set). ISBN 0-262-09026-0.

Translation of the respected *Sugaku Jiten*, published by Mathematical Society of Japan (see p. 102). Comprehensive with most entries specific rather than general. Includes biographical information. Volume 4 contains appendices of formula and mathematical tables, list of mathematical journals, and name and subject indexes. **(a,s)**

Enzyklopädie der mathematischen Wissenschaften, mit Einschluss ihrer Anwendungen. Second Edition. Leipzig: Teubner, 1939-.

Thirteen books have been published (out of order) in this unfinished series. Most recent volume was published in 1967. Comprehensive and detailed coverage of mathematics and its various applied fields. Authoritative work noted for accuracy. **(a,s)**

Fachlexikon ABC Mathematik. Ed: Walter Gellert et al. Thun: Harri Deutsch, 1978. 624pp. Price not available. ISBN: 3-87144336-0.

Contains some 700 articles. Not indexed.

International Encyclopedia of Statistics. Ed: William H. Kruskal, Judith M. Tanur. New York: Free Press, 1978. 2v. V.1: xxi + 666pp. $90.00. ISBN 0-02-917970-X. V. 2: pp. 667-1350. $90.00. ISBN 0-02-917980-7. $225.00 (set). ISBN 0-02-917960-2 (set).

Updating and expansion of statistics articles from *International Encyclopedia of the Social Sciences* (Macmillan, 1968). Original 70 articles increased to 75 and biographies from 45 to 57. New or revised material is indicated. **(a)**

Kuwait Science Encyclopedia: Mathematics. Fozi Mustafa Dannan et al. Kuwait: Kuwait Foundation for the Advancement of Sciences, 1984. 4v. Price not available.

Text is in modern Arabic with signs, etc. in Latin characters. Covers more than 2,100 mathematical words, terms, and names including biographies and major works from different periods. Can be used as Arabic to English dictionary of mathematics as well. **(a)**

Lexikon der Optimierung. Berlin: Akademie-Verlag, 1986. vi + 388pp. $29.68. ISBN 3-05-500031-5.

Lexicon of optimization containing definitions, descriptions, numerical procedures from theory of optimization, and optimal control. Has English-French-German-Russian lexicon of equivalent terms. **(a)**

Malai͡a Matematicheskai͡a Ent͡siklopedii͡a. E. Frid et al. Ed: I͡U. N. Blagoveshchenskiĭ, B. Kocsis. Budapest: Akadémiai Kiadó, 1976. 693pp. Price not available. ISBN 963-05-0844-3.

Russian translation of Hungarian original. Classified arrangement with eight major headings. Index and references. **(a)**

Matematicheskai͡a Ent͡siklopedii͡a. I. M. Vinogradov. Moscow: Sovetskai͡a Entsiklopedii͡a, 1977-85. 5v. V. 1: 1,151pp. V. 2: 1,104pp. V. 3: 1,184pp. V. 4: 1,216pp. V. 5: 1,246pp. o.p.

The original of *Encylopaedia of Mathematics* (see p. 99), this approaches being an alphabetical list of phrases explained in the terminology of its discipline supplemented by longer pieces on key topics. No biographical or historical entries. **(a)**

Matematicheskiĭ Entsiklopedicheskiĭ Slovar'. ÎU. V. Prokhorov. Moscow: Sovet. Entsiklopediia, 1988. 845pp. o.p.

Encyclopedic dictionary containing some 3,500 articles based on *Matematicheskaia Entsiklopediia* (ME) (see above). Includes a biographical dictionary with 900 entries and many other features. Inexpensive substitute for five volume ME and a useful source of information on terms found in Russian-English dictionaries. **(a)**

Mathematics Encyclopedia. Ed: Max S. Shapiro. Garden City, NY: Doubleday, 1977. 289pp. $5.95. ISBN 0-385-12427-9.

Useful general work for schools and public libraries, covers all branches of mathematics. Contains tables, formulas, and symbols. **(a,p)**

Mathematisches Wörterbuch: mit Einbeziehung der theoretischen Physik. Third Edition. Ed: Josef Naas, Hermann Ludwig Schmid. Berlin: Akademie-Verlag; New York: Pergamon, 1979. 2v. V. 1: xv + 1043. V. 2: viii + 952. $395.67 (set). ISBN 3-519-02400-4 (set).

Comprehensive, containing definitions, articles on trends and 400 biographies. Some articles include reference notes. Appendixes. **(a)**

Prentice-Hall Encyclopedia of Mathematics. Beverly Henderson West. Englewood Cliffs, NJ: Prentice-Hall, 1982. xv + 683pp. $39.00. ISBN 0-13-696013-8.

Eighty articles on various aspects of mathematics plus tables of symbols, weights and measures, and logarithms. Includes biographical entries, cross-references and references for additional reading. This is a comprehensive resource, perhaps most appropriate for undergraduates or non-professionals. **(p)**

Sugaku Jiten. Third Edition. Toyko: Iwanami Shoten, 1985. 1,609pp. + four leaves of plates. Price not available. ISBN 4-00-080016-7.

Highly respected Japanese encyclopedia of mathematics covering 450 topics arranged under 21 divisions. Translated into English as *Encyclopedic Dictionary of Mathematics*, edited by Kiysio Ito (see 100). **(a)**

Universal Encyclopedia of Mathematics. New York: Simon and Schuster, 1964. 715pp. o.p.

General tool covering arithmetic through calculus with individual sections on formulas and functions tables. English edition by Allen & Unwin (London, 1964) includes mathematical tables. **(a,p)**

VNR Concise Encyclopedia of Mathematics. Second Edition. Ed: W. Gottwald. New York: Van Nostrand Reinhold, 1989. 776pp. + 56pp. of plates. $34.95. ISBN 0-442-20590-2.

Based on *Kleine Enzyklopädie der Mathematik*, English language version offers comprehensive coverage in three parts (elementary, higher, and contemporary mathematics) with photographs and biographies. General work for educated readers, not just mathematics subject specialists. Excellent source to consult for answering difficult questions posed by non-specialists. **(a,p)**

GUIDES

Abstracts and Indexes in Science and Technology: A Descriptive Guide. Second Edition. Dolores B. Owen, Marguerite M. Hanchey. Metuchen, NJ: Scarecrow Press, 1985. xv + 235pp. $20.00. ISBN 0-8108-1712-8.

Mathematics and statistics is one of several scientific and technical subject areas whose indexes and abstracts are examined. Each title is described in outline form with information on arrangement, coverage, scope, locating material, abstracts, indexes, other material, and periodicals scanned. **(a)**

Complete Categorized Guide to Statistical Selection and Ranking Procedures. Edward J. Dudewicz, Joo Ok Koo. Columbus, OH: American Sciences Press, 1982. v + 627pp. $89.75. ISBN 0-935950-03-6.

Classified bibliography including reports, theses and published items. Index by author and chronological indexes by subjects and author. Reprints of reviews when available are included. **(a)**

Current Information Sources in Mathematics: An Annotated Guide to Books and Periodicals, 1960-72. Elie M. Dick. Littleton, CO: Libraries Unlimited, 1973. x + 281pp. o.p. ISBN 0-87287-047-2.

Serves as supplement to Parke's *Guide to the Literature of Mathematics and Physics* (New York: Dover, 1958) (See p. 104). Covers 1,600 English language mathematics monographs published 1960-1972 in form of a classified, annotated bibliography. Sections on reference books, periodicals, publishers, and professional organizations. **(a,p,s)**

Führer durch die mathematische Literatur. Hermann Felix Müller. Nendeln: Kraus Reprint, 1979. x + 252pp. Price not available. ISBN 3-262-01421-4.

Guide to mathematical literature with attention to historically important works. Reprint of original 1909 edition. **(a)**

Guida allo Studio della Storia delle Matematiche: Generalit'a Didattica, Bibliografia. Appendice: Questioni Storiche Concernenti le Scienze Esatte. Second Edition, Revised and Augmented. Gino Loria. Milano: U. Hoepli, 1946. xix + 385pp. Price not available.

Guide to the literature of the history of mathematics, covering all periods and countries. Many typographical errors. **(a)**

Guide to Statistical Methods and to the Pertinent Literature. Lothar Sachs. Berlin; New York: Springer-Verlag, 1986. xii + 212pp. $32.00. ISBN 0-387-16835-4.

English/German bilingual bibliography of 1,449 items with keyword and subject heading index of 5,500 terms. Intended for students and researchers using intermediate and multivariate methods. **(a)**

Guide to the Literature of Engineering, Mathematics, and the Physical Sciences. Sylvia Goldman. Baltimore, MD: Johns Hopkins University. Applied Physics Laboratory, 1959. 52 leaves. o.p.

Classified arrangement similar to others (e.g., Parke below), but less thorough coverage. Selected items of importance to Applied Physics Laboratory. No annotations.

Guide to the Literature of Mathematics and Physics Including Related Works on Engineering Science. Second Revised Edition. Nathan Grier Parke. New York: Dover, 1958. xviii + 436pp. o.p.

Introduced by discussions of research and library use, main body is a bibliography of 5,000 titles in classified arrangement with occasional notes. Emphasis on applied mathematics. More current is *Current Information Sources in Mathematics* by Elie Dick (See p. 103). **(a,p,s)**

Guide to the Literature of Mathematics Today. Joong Fang. Hauppauge, NY: Paideia, 1972. 267pp. Price not available. ISBN 0-912490-10-1.

Classified arrangement based on author's particular philosophical approach to mathematics. Discussion and lists of major terms used in mathematics. Specific entries are very brief. **(a)**

History of Statistics, Their Development and Progress in Many Countries. John Koren. American Statistical Association. New York: Macmillan, 1918. xii + 773pp. $43.00. ISBN 0-8337-1950-5. Reprinted New York: B. Franklin, 1970.

Issued to commemorate 75th anniversary of American Statistical Association. Primarily useful for history of official statistical publications.

How to Find Out in Mathematics: Guide to Sources of Information. Second Revised Edition. John E. Pemberton. Elmsford, NY: Pergamon, 1969. xiv + 193pp. o.p. ISBN 0-08-006824-3 (hbk.); 0-08-006823-5 (pbk.).

Classic manual of sources for librarians and students. Chapters cover topics relevant to mathematics: careers, types of research materials, math education; sources for tables, history, biography; and bibliographies. Separate part for mathematics applications. Discussions of collection development, Russian materials, and the government. **(a,p)**

Information Sources in Science and Technology. C. D. Hurt. Englewood, CO: Libraries Unlimited, 1988. xiv + 362pp. $28.50. ISBN 0-87287-582-2.

Mathematics is one of 16 disciplines covered in this useful and relatively up-to-date book. Brief reviews given for guides to the literature, abstracts, indexes, bibliographies, encyclopedias, dictionaries, handbooks and tables, and directories. **(a,p,s)**

Russian for the Mathematician. S. H. Gould. New York; Berlin: Springer-Verlag, 1972. xi + 211pp. $42.00. ISBN 0-387-05811-7.

Textbook of Russian prepared for mathematicians by the translations editor of the American Mathematical Society. Covers alphabet, pronunciation, inflection, aspect, and vocabulary. Supported by a generous selection of readings. **(a,s)**

Scientific and Engineering Literature: A Guide to Reference Sources. Third Edition. H. R. Malinowsky. Littleton, CO: Libraries Unlimited, 1980. 342pp. o.p. ISBN 0-87287-230-0 (hbk.); 0-87287-245-9 (pbk.).

Classified arrangement by field, then by type of reference tool. Bibliography and author/title/subject index. Typical, somewhat limited coverage of mathematics. **(a,p)**

Short Guide Through the Literature on Semirings. Kazimierz Glazek. Wroclaw: Univ. of Wroclaw, Mathematical Institute, 1985. 147pp. Price not available.

Includes 1,270 item bibliography of articles to 1984 on semirings. Treatment of concepts varies from theoretical to purely algebraic. **(a)**

Special Functions for Engineers and Applied Mathematicians. Second Edition. Larry C. Andrews. New York: McGraw-Hill, 1992. xix + 479pp. $59.50. ISBN: 0-07-001848-0.

This revised edition of the 1985 original contains more applications from a number of fields, such as statics, dynamics, and fiber optics. It can be used as a textbook in courses that include higher mathematics or as a reference book for practitioners. Many worked examples. **(a)**

Use of Mathematical Literature. Ed: A. R. Dorling. London: Butterworths, 1977. xii + 260pp. $57.44. ISBN 0-408-70913-8.

Graduate level guide covering the nature of mathematical literature with reference to organization of mathematics, key organizations and journals, and reference materials, Covers history of mathematics. Specialized subjects such as rings and algebras or group theory are covered as separate topics. **(a,p,s)**.

A User's Guide to Principal Components. J. Edward Jackson. New York: John Wiley & Sons, Inc., 1991. xvii + 569 pp. $69.95. ISBN: 0-471-62267-2.

Intended primarily for practitioners of multivariate analysis, this is a practical rather than theoretical text. It can serve as a sourcebook for principal components and might be used as a supplementary text for multivariate courses. The book includes tables, graphs, and worked examples. **(a)**

Using the Mathematical Literature: A Practical Guide. Barbara Kirsch Schaefer. Books in Library & Information Science;25. Ann Arbor, MI: Books on Demand UMI, 1979. ix + 141pp. $38.80. ISBN 0-685-23675-7. Order #: 2029014.

A brief work that helps provide insight into the mathematical literature and its effectice use. Useful starting point for literature searches. **(a,p)**

HANDBOOKS AND MANUALS

Analysis of Variance. Ed: P. R. Krishnaiah. Handbook of Statistics;1. New York: North-Holland, 1980. xvii + 1002pp. $105.25. ISBN 0-444-85335-9.

First volume of a series of reference books in statistical methodology and applications, each volume to be devoted to a particular topic. Contains twenty-five chapters written by prominent statisticians and covers most of the useful techniques in univariate and multivariate ANOVA. Material is primarily expository, but technical, and well-documented. **(a,p,s)**

Anleitung zum praktischen Gebrauch der LaPlace-Transformation und der Z-Transformation. Sixth Edition. Gustav Doetsch. Munich: R. Oldenbourg Verlag, 1989. 256pp. $42.04. ISBN 3-486-21310-5.

Quality survey of Laplace transformation methods in applied fields. Techniques fully explained. Includes examples (See p. 109 for English translation). **(a)**

Atlas of Functions. Jerome Spanier, Keith B. Oldham. Washington: Hemisphere, 1987. ix + 700pp. $165.95. ISBN: 0-89116-573-8.

Vital rethinking of old fashioned handbook of mathematical tables, it presents information on families of functions incorporating definitions, expansions, computer generated graphs, etc. Basic new reference tool. **(a,p)**

Beta Mathematics Handbook. Second Edition. Lennart Rade, Bertil Westergren. Lund, Sweden: Studentlitteratur; Boca Raton, FL: CRC Press, 1990. 494pp. $44.50. ISBN 91-44-25052-5.

General purpose university level handbook. Covers definitions, theorems, methods, algorithms, formulas, tables, etc. with a glossary and detailed index. Second edition contains corections and additional treatment of certain subjects including, but not limited to, basic concepts in discrete mathematics, real numbers in different bases, complex matrices, and linear differential equations. **(a,p)**

Boundary Element Reference Book. Ed: J. Mackerle, C. A. Brebbia. Berlin; New York: Springer-Verlag, 1988. viii + 382pp. $99.00. ISBN 0-387-18584-4.

Covers topic for last 10 years providing a review of key works with 300 references, classified and annotated listing of all boundary element (BE) works published to 1988, BE software available, and a who's who in BE. **(a)**

Collected Algorithms from ACM. New York: Association for Computing Machinery, 1978-. 4v. + Quarterly supplements. Looseleaf. $425.00 for four back volumes. $70.00 for each year's quarterly supplements. ISBN 0-89791-017-6 (V. 1); 0-89791-026-5 (V. 2); 0-89791-103-2 (V. 3); 0-89791-297-7 (V. 4). ISSN 0149-1989.

A compilation of algorithms published in *Communications of the ACM* (1958-) and *Transactions on Mathematical Software* (1975-). Entries list identifying title of algorithm, source, author, institutional address, certification status of the algorithm, and remarks. Highly recommended for mathematical research collections. **(a,s)**

Computer Graphics Handbook: Geometry and Mathematics. Michael E. Mortenson. New York: Industrial Press, 1990. xii + 259pp. $24.95. ISBN 0-8311-1002-3.

There is no discussion of computer hardware and software or even the central issues relating to computer displays, so the title is deceptive. It is a good reference book on geometry and mathematics, but potential buyers should review it first to determine its usefulness. **(a,p)**

A Computer Laboratory Manual for Number Theory. Donald G. Maim. Wentworth, NH: COMPress, 1980. viii + 256pp. o.p. ISBN 0-933694-13-X.

Among the topics covered arc unique factorization and the Euclidean algorithm, congruences, and the distribution of primes. Includes lengthy list of references. Should be a standard reference text for every number theory enthusiast. **(a)**

Correspondence Analysis Handbook. J.-P. Benzecri. New York: Dekker, 1992. xii + 665pp. $135.00. ISBN: 0-8247-8437-5.

Users of data analysis are the intended audience for this handbook. It contains many illustrations and examples as well as a Fortran program for correspondence analysis suitable for implementation on an IBM PC or compatible using Microsoft FORTRAN 77. The index doubles as a glossary. **(a)**

CRC Handbook of Chemistry and Physics: A Ready-Reference Book of Chemical and Physical Data. Seventy-First Edition. Ed: David R. Lide. Boca Raton, FL: CRC Press, 1990. 1v. (various pagings). $99.50. ISBN 0-8493-0471-1.

The essential reference tool for physical data. Tables in this edition have been reconfigured to facilitate browsing. More tables on health and the environment have been added. Some tables have been eliminated because of the power of pocket calculators and personal computers to quickly generate numeric values. **(a,p,s)**

CRC Handbook of Mathematical Curves and Surfaces. David H. von Seggern. Boca Raton, FL: CRC Press, 1990. 286pp. $49.95. ISBN 0-8493-0155-6.

Pictures of families of functions on right-hand pages, facing displays of formulas with relevant parameters: algebraic curves, transcendental functions, orthogonal and non-orthogonal polynomials, special functions (e.g., Legendre Bessel), probability densities, catastrophe curves, spirals and helices, algebraic and transcendental surfaces, and nondifferentiable curves and surfaces (including polygons and polyhedra). A handy reference for those without a computer. **(a,p,s)**

CRC Handbook of Mathematical Sciences. Sixth Edition. Ed: William H. Beyer. Boca Raton, FL: CRC Press, 1987. 860pp. $84.95. ISBN: 0-8493-0656-6.

Standard desk reference for integral tables, special functions, probability and statistics, differential equations, analytic geometry, and more. Continues *Handbook of Tables for Mathematics* and *Handbook of Mathematical Tables.* **(a,p,s)**

CRC Handbook of Percentage Points of the Inverse Gaussian Distribution. Boca Raton, FL: CRC Press, 1989. 300pp. $110.00. ISBN: 0-8493-3626-0.

Detailed introduction with references to support the tables which take up the remainder of the book. (**a**)

Deskbook of Math Formulas and Tables. Joanne Buhl Auth. New York: Van Nostrand Reinhold, 1985. viii + 219pp. o.p. ISBN 0-442-21106-6.

A handy volume with helpful formulas, techniques, and tables of types used daily. Refresher, with clear, simple explanations and examples. (**p**)

DTV-Atlas zur Mathematik. Volume 1. Fourth Edition. Volume 2. Third Edition. Fritz Reinhardt, Heinrich Söder. Munich: Deutscher Taschenbuch Verlag, 1978. 2v. *V. 1*: 266pp. ISBN 3-423-03007-0. *V. 2*: pp. 274-498. ISBN 3-423-03008-9. $12.24/Volume.

Summary of comtemporary mathematics arranged by major fields with several subdivisions for each field. Many colored illustrations with concise text. (**a**)

Engineering Formulas. Sixth Edition. Kurt Gieck. New York: McGraw-Hill, 1990. 1v. (various pagings). $29.95. ISBN 0-07-023455-8.

Engineer's desk reference that is also good for work in the field because of its small size. Covers most mathematical formulas and tables necessary to the practice of engineering. Highly recommended for any technically oriented collection. (**p**)

Engineering Mathematics Handbook: Definitions, Theorems, Formulas, Tables. Third Edition. Jan H. Tuma. New York: McGraw-Hill, 1987. xiii + 498pp. $59.50. ISBN 0-07-065443-3.

Standard desk reference for engineering and mathematics arranged in parts by type of mathematics covered. Individual parts include definitions, basic concepts and more advanced concepts. (**a,p**)

Exact Solutions of Relativistic Wave Equations. V. G. Bagrov, D. M. Gitman. Dordecht: Kluwer Academic, 1990. Mathematics and Its Applications (Soviet Series);39. x + 323pp. $113.00. ISBN 0-7923-0215-X.

Contains a systematic study of all known exact solutions of the Dirac and Klein-Gordon equations for a charged particle in an external electromagnetic field. Gives exact solutions of the generalized Dirac-Pauli equation. (**a**)

Field Theory Handbook. Second Edition, Corrected Third Printing. P. Moon, D. E. Spencer. New York: Springer-Verlag, 1988. viii + 236pp. $49.50. ISBN 0-387-18430-9.

Compilation of equations for those working in field. Tabulates properties of 40 coordinate systems with Laplace and Helmholtz equations and separation equations and their solutions. Also includes tabulation of related ordinary differential equations with solutions plus bibliography. (**a**)

Finite Element Handbook. Ed: H. Kardestuncer, D. H. Norrie. New York: McGraw-Hill, 1987. xxiv + 1,381pp. $105.00. ISBN 0-07-033305-X.

Wide-ranging, it covers all aspects of the finite element method. Almost 100 contributing authors. Contains extensive, up-to-date bibliographies. (**a,s**)

Formulaire de Mathématiques. Third Edition. L. Chambadal. Paris: Dunod, 1985. 188pp. $11.38. ISBN 2-04-015759-X.

For students preparing for university scientific education, it covers definitions, theorems, and formulas. **(a,p)**

The Geometrical Foundation of Natural Structure: A Source Book of Design. Robert Williams. New York: Dover, 1979. xv + 265pp. $9.95. ISBN 0-486-23729-X.

Corrected reprint of 1972 *Natural Structure*, this book provides a detailed resource of theoretical information for tessellations, packings and arrangements for designers, scientists, and engineers.

Green's Functions and Transfer Functions Handbook. A. G. Butkovskiĭ. Chichester: Ellis Horwood, 1982. 236pp. $62.85. ISBN 0-85312-447-7.

Collection of tables for groups r,m,n for Green functions of different initial and boundary conditions. Some discussion of basic concepts of system theory and feedback effects. Includes references. **(a)**

Guide to Medical Mathematics. D. A. Franklin, G. B. Newman. New York: Wiley, 1973. ix + 453pp. o.p. ISBN 0-470-27520-0.

Written for medical researchers, it assumes familarity with only arithmetic and geometry. It presents graph construction, trigonometric functions, calculus, series, and differential equations. Useful examples. **(p)**

Guide to the Applications of the Laplace and Z-transforms. Second Edition. Gustav Doetsch. New York: Van Nostrand Reinhold, 1971. xii + 240pp. o.p. ISBN 0-442-02163-1.

English translation of *Anleitung*... (see p. 106). **(a)**

Handbook for Linear Regression. Mary Sue Younger. North Scituate, MA: Duxbury Press, 1979. xiv + 569pp. $23.00. ISBN 0-87872-187-8.

Covers simple and multiple linear regression methods. Subject index and four useful appendices. **(a,p)**

Handbook for Matrix Computations. Thomas F. Coleman, Charles Van Loan. Philadelphia, PA: SIAM, 1988. vii + 264pp. $24.00. ISBN 0-89871-227-0.

Technical work for those doing scientific computations using matrices. After reviewing Fortran 77, covers BLAS (basic linear algebra subprograms), LINPACK (Fortran codes for linear equations and least squares calculations), and MATLAB (interactive system for matrix computation). **(a,s)**

Handbook of Algorithms and Data Structures In Pascal and C. Second Edition. G. H. Gonnet, R. Baeza-Yates. New York: Addison-Wesley, 1991. xiv + 424pp. $25.56. ISBN 0-201-41607-7.

Covers searching, sorting, selection, and arithmetic algorithms. Implementations in both C and Pascal are given for important algorithms. New edition incorporates many recent results and a new chapter on text searching. Bibliography contains references to thirty-six textbooks and 1,350 published research papers. **(p,s)**

Handbook of Applicable Mathematics. Ed: Walter Ledermann, R. F. Churchhouse. Chichester, England: Wiley, 1980-1991. 8v. V. 1-4: $185.00 each. V. 5 & 6: $375.00 each. V. 7: $110.00. V.8: $49.95. $1,500.00 (set). ISBN 0-471-90804-5 (set); 0-471-27704-5 (V. 1); 0-471-27821-1 (V. 2); 0-471-27947-1 (V. 3); 0-471-10141-9 (V. 4); 0-471-90023-0 (V. 5, set of two parts); 0-471-90567-4 (V. 5A); 0-471-90568-2 (V. 5B); 0-471-90024-9 (V. 6, set of two parts); 0-471-90274-8 (V. 6A); 0-471-90272-1 (V. 6B); 0-471-91825-3 (V. 7); 0-471-92792-9 (V. 8).

Quality reference set focuses on practical and applied mathematics. Covers algebra, numerical methods, geometry, combinatorics, probability, analysis, and statistics in authoritative, readable articles that include bibliographies. Volume 8 contains contents and general index. **(a,s)**

Handbook of Applied Mathematics. Fourth Edition. Edward E. Grazda, Martin Ernest Jansson. New York: Van Nostrand Reinhold, 1977, c1966. vi + 1,119pp. o.p. ISBN 0-88275-615-X.

Covers useful applications of mathematics including arithmetic, algebra, geometry, trigonometry, and differential calculus to basic trades. First chapters review mathematical operations, balance apply to trades. **(a,p)**

Handbook of Applied Mathematics: Selected Results and Methods. Second Edition. Ed: Carl E. Pearson. New York: Van Nostrand Reinhold, 1983. xiii + 1,307pp. o.p. ISBN 0-442-23866-5.

Covers mathematical procedures and applications for scientific and engineering problems with formulas for vector analysis, tensors, and complex variables. **(a,p)**

Handbook of Boolean Algebras. Ed: J. Donald Monk, Robert Bonnet. Amsterdam; New York: North-Holland, 1989. V. 1: xix + 323pp. $72.00. ISBN 0-444-70261-X. V. 2: xix + 415pp. $118.00. ISBN 0-444-87152-7. V. 3: xix + 650pp. $161.00. ISBN 0-444-87153-5. $334.25 (set). ISBN 0-444-87291-4 (set).

Treatment of those areas of the theory of Boolean algebras of most interest to pure mathematicians. Also includes detailed discussions of the major developments in the theory of Boolean algebras that have occurred over the last quarter century since R. Sikorski's *Boolean Algebras* in 1964. **(a)**

Handbook of Differential Equations. Second Edition. Daniel Zwillinger. San Diego, CA: Academic Press, 1992. xx + 787pp. $56.00. ISBN 0-12-784391-4.

A compilation of the most important and widely applicable methods for solving and approximating differential equations. **(a,p,s)**

Handbook of Econometrics. Ed: Zvi Griliches, Michael D. Intriligator. New York: Elsevier Science, 1983-84. Volume I: xxvii + 771pp. $75.00. ISBN 0-444-86185-8. Volume II: xxvi + 686pp. $75.00. ISBN 0-444-86186-6.

First two of three volumes of a planned 35 chapter survey of econometrics intended to serve as a reference and teaching supplement. Topics covered include mathematical and statistical methods, time series, etc. **(a)**

Handbook of Estimates in the Theory of Numbers. Blair Spearman, Kenneth S. Williams. Carleton Math Series;14. Ottawa, ON: Department of Mathematics, Carleton University, 1975. 191pp. o.p.

Classified (17 headings) collection of asymptotic estimates including pertinent references. **(a)**

Handbook of Fourier Theorems. D. C. Champeney. Cambridge; New York: Cambridge University Press, 1987 (hbk.); 1989 (pbk.). xi + 185pp. $44.50 (hbk.); $22.95 (pbk.). ISBN: 0-521-26503-7 (hbk.); 0-521-36688-7 (pbk.).

For professionals familiar with Fourier theory and applications who want more detailed knowledge. Includes an introduction to Lebesque integration. No proofs. **(a,s)**

Handbook of Hypergeometric Integrals: Theory, Applications, Tables, Computer Programs. Harold Exton. New York: Halsted, 1978. 316pp. o.p. ISBN 0-470-26342-3.

Useful reference for mathematicians, statisticans, physicists, and computer analysts. Presents a structured list of analytical formulas. **(a)**

A Handbook of Integer Sequences. N. J. A. Sloane. New York: Academic Press, 1973. xiii + 206pp. $59.95. ISBN 0-12-648550-X.

Bulk of this book is a table of 2,300 sequences, arranged in lexicographic order, each accompanied by a brief description and a reference to the literature. **(a)**

Handbook of Integral Transforms of Higher Transcendental Functions. O. I. Marichev. Chichester: Ellis Horwood; New York: John Wiley & Sons, 1983. 336pp. o.p. ISBN 0-85312-528-7.

Translation of 1978 Russian original. Presents method of calculating integrals from special functions including theory and tables of formulas. **(a,s)**

Handbook of Introductory Statistical Methods. C. Philip Cox. New York: Wiley, 1987. xxi + 272pp. $44.95. ISBN 0-471-81971-9.

Summary of many statistical techniques for estimation and hypothesis testing. A convenient reference source of statistical procedures for practicing scientists in quantitative research disciplines. Can also complement instruction in first course for graduate students who intend to use statistical methods in their own research. **(a,p)**

Handbook of Mathematical Economics. Ed: Kenneth J. Arrow, Michael D. Intriligator. Amsterdam; New York: North-Holland Publishing, Elsevier, 1981. 3v. (xvii + 2264pp.). $80.00/Volume. ISBN 0-444-86126-2 (V. 1); 0-444-86127-0 (V. 2); 0-444-86128-9 (V. 3).

Compilation of survey papers written by widely recognized experts for students and teachers of mathematical economics. **(a)**

Handbook of Mathematical Formulas. Hans Jochen Bartsch. New York: Academic Press, 1974. 528pp. $64.00. ISBN 0-12-080050-0.

Collection of formulas covering entire field. Translated from German. **(a,p)**

Handbook of Mathematical Functions with Formulas, Graphs, and Mathematical Tables. Tenth Printing. Milton Abramowitz, Irene A. Stegun. New York: Wiley, 1984. xvi + 1046pp. $69.95. ISBN 0-471-80007-4.

Standard work, containing special functions formulas, numerical tables, and mathematical properties of tabulated fuctions. **(a,p)**

Handbook of Mathematical Logic. Ed: Jon Barwise, et al. Amsterdam; New York: North-Holland, 1977. xi + 1,165pp. o.p. ISBN 0-7204-2285-X.

Reviews contemporary developments in logic for four key areas: model, set, recursion, and proof theories. Each part has several chapters all written by recognized authorities. **(a)**

Handbook of Mathematical Tables and Formulas. Fifth Edition. Richard S. Burington. New York: McGraw-Hill, 1973. x + 500pp. $38.68. ISBN 0-07-009015-7.

In two parts. First covers formulas and theorems. Second part includes tables of logarithms, several functions, powers & roots, probability distributions, annuity, etc. **(a,p)**

Handbook of Mathematics. Ed: L. Kuipers and R. Timman. Trans: I. N. Sneddon. International Series of Monographs in Pure and Applied Mathematics;99. New York: Pergamon, 1969. xi + 782pp. o.p. ISBN 0-08-011857-7.

Translated from Dutch, this classic handbook of general mathematics includes comprehensive coverage of all major divisions plus history of mathematics. **(p)**

Handbook of Mathematics. Third Revised Edition. I. N. Bronshtein, K. A. Semendi͡aev. Transl: K. A. Hirsch. New York: Van Nostrand Reinhold, 1985. xv + 973pp. $52.95. ISBN 0-442-21171-6.

English translation based on 19th German edition, *Taschenbuch der Mathematik* (See p. 119). Originally in Russian, there have been several East German editions. **(a)**

A Handbook of Methods of Approximate Fourier Transformation and Inversion of the Laplace Transformation. V. I. Krylov, N. S. Skobli͡a. Moscow: Mir, 1985. 271pp. o.p. Reprint of 1977 edition.

Contains valuable information but suffers from having no numerical examples and a short bibliography which references only Russian books. **(a)**

Handbook of Nonlinear Regression Models. David A. Ratkowsky. New York: Dekker, 1990. ix + 241pp. $79.75. ISBN 0-8247-8189-9.

Background and philosophy of nonlinear regression modeling followed by most commonly used models with up to six parameters. Close-to-linear models are emphasized. **(a,s)**

Handbook of Numerical Analysis. Ed: P. G. Ciarlet, J. L. Lions. Amsterdam: North-Holland, 1990-1991. *Volume 1: Finite Difference Methods: Solutions of Equations in*

RN. viii + 652pp. $125.75. ISBN 0-444-70366-7. *Volume 2: Finite Element Methods, Pt. I.* x + 928pp. $157.25. ISBN 0-444-70365-9.

Intended to provide mathematical rigor in the analysis of numerical schemes. First volume contains three papers which present the basic concepts and principles of numerical analysis in three subjects: finite difference schemes for parabolic equations, splitting methods and ADI, and least squares. The first article in the second volume is an introduction to finite elements and is followed by articles covering basic error estimates for elliptic problems, local behavior in finite element methods, eigenvalue problems, and evolution problems. **(a,s)**

Handbook of Numerical Analysis Applications. Jaroslav Pachner. New York: McGraw-Hill, 1984. xvii + 429pp. o.p. ISBN 0-07-048057-5.

In two main sections: review of numerical methods in science and engineering, and list of computer programs for implementation. Useful access to microcomputer programs for those familiar with numerical analysis.

Handbook of Numerical and Statistical Techniques with Examples Mainly from the Life Sciences. J. H. Pollard. Cambridge; New York: Cambridge University Press, 1981. xvi + 349pp. $69.50 (hbk.); $32.50 (pbk.). ISBN 0-521-21440-8 (hbk.); 0-521-29750-8 (pbk.).

Easily used handbook of statistics and statistical techniques with examples from life sciences. Divided into three parts: numerical techniques, statistical techniques, and both techniques in discussing least squares regression. **(a)**

Handbook of Numerical Matrix Inversions and Solution of Linear Equations. Joan R. Westlake. New York: Wiley, 1968; reprinted Huntington, NY: Krieger, 1975. viii + 171pp. o.p. ISBN 0-8827-5225-1.

Covers direct methods, iterative methods, error measures, and scaling focusing on computer solutions. Includes glossary, listing of matrix algebra theorems, etc. plus some test matrices. **(a)**

Handbook of Operations Research. Joseph J. Moder, Salah E. Elmaghraby. New York: Van Nostrand Reinhold, 1978. 2v. V. 1: xxiv + 622. V. 2: xxiv + 659. o.p. ISBN 0-442-24595-5 (V. 1); 0-442-24596-3 (V. 2).

Titles of two volumes are *Foundations and Fundamentals* and *Models and Applications* and reflect contents. **(a)**

Handbook of Probability and Statistics with Tables. Second Edition. Richard Stevens Burington, Donald Curtis May, Jr. New York: McGraw-Hill, 1970. xiv + 462pp. o.p. ISBN 0-07-009030-0.

Ready reference source for basic formulas, definitions, specialized tables and more. Includes coverage of nonparametric methods, regression theory, variance analysis, and sampling techniques. **(a,p,s)**

Handbook of Random Number Generation and Testing with TESTRAND Computer Code. Edward J. Dudewicz, Thomas G. Ralley. Columbus, OH: American Sciences Press, 1981. xi + 634pp. $95.00. ISBN 0-935950-01-X.

Designed as quality test of random number generators. Discusses generator's role in Monte Carlo and simulation techniques. Lists twenty generators, recommendations, sample programs, and references. **(a)**

Handbook of Sequential Analysis. Ed: B. K. Ghosh, P. K. Sen. Statistics, Textbooks and Monographs;118. New York: Dekker, 1991. xv + 637pp. $150.00. ISBN 0-8247-8408-1.

Over 30 statisticians describe all major developments in sequential analysis through 1990, introducing basic elements and covering theory, methods, and applications in a wide range of disciplines. **(a)**

Handbook of Set-Theoretical Topology. Ed: Kenneth Kunen, Jerry E. Vaughan. Amsterdam; New York: North-Holland, 1984. vii + 1,273pp. $148.75. ISBN 0-444-86580-2.

Collection of 24 papers reflecting contemporary thought on the subject. **(a)**

Handbook of Statistical Distributions. Jagdish K. Patel, C. H. Kapadia, D. B. Owen. Ann Arbor, MI: Books on Demand UMI, 1976. xiv + 302pp. $84.30. ISBN 0-318-35010-6. Order #: 2030870.

Source for most frequently used distributions by statisticians. References to most available but not necessarily original sources. Emphasis on practical results. **(a,p)**

Handbook of Statistical Methods for Engineers and Scientists. Ed: Harrison M. Wadsworth, Jr. New York: McGraw-Hill, 1990. x + 706pp. $79.50. ISBN 0-07-067674-7.

Twenty sections by 25 authors beginning with basic statistics. Emphasis on quality control. Also includes survey sampling, Bayesian analysis, experimental design, regression, simulation, robust and nonparametric methods, and time series. Good first reference for engineers. Includes section references. **(a,p,s)**

Handbook of Stochastic Methods for Physics, Chemistry, and the Natural Sciences. Second Edition. C. W. Gardiner. Berlin; New York: Springer-Verlag, 1985. xx + 442pp. $55.00. ISBN 0-387-15607-0.

Covers stochastic methods frequently used by theoretical physicists and chemists. Restricted to Markov processes with diffusions as focus. Three parts: introduction, basic mathematical methods, and applications and generalizations. Useful examples. **(a,s)**

Handbook of Terms Used in Algebra and Analysis. A. G. Howson. Cambridge: Cambridge University Press, 1972. ix + 238pp. $22.89. ISBN 0-521-09695-2.

Divided into 38 chapters containing brief articles on major subjects in algebra and analysis. Key terms bold face in context. Appendix of named theorems and properties. Indexes of symbols and subjects. **(a)**

Handbook of Terms Used in Chaos and Quantum Chaos. Willi-Hans Steeb. Mannheim: B. I. Wissenschaftsverlag, 1991. 181pp. DM 39.00. ISBN: 3-411-15061-0.

Includes bibliographical references and an index. **(a)**

Handbook of the Logistic Distribution. Ed: N. Balakrishnan. New York: Dekker, 1991. xv + 601pp. $115.00. ISBN: 0-8247-8587-8.

This book contains both theory and practical applications, detailed descriptions of univariate and multivariate generalizations of the distribution, worked examples, and a comprehensive bibliography. Useful to students and practitioners alike. **(a,p)**

Handbook of the Normal Distribution. Jagdish K. Patel, Campbell B. Read. New York: Marcel Dekker, 1982. ix + 337pp. $79.95. ISBN 0-8247-1541-1.

Basic source of information and references on normal distribution including historical development. Relates normal distribution to other distributions, as well. Intended for research or applications of statistical methodology. References. **(a)**

Handbuch der Bedienungstheorie. I: Grundlagen und Methoden. II: Formeln und andere Ergebnisse. B. V. Gnedenko, D. König. Berlin: Akademie-Verlag, 1983-84. 2v. *V. 1*: 519pp. *V. 2*: 607pp. Price not available.

Most comprehensive collection of formulas (no proofs) and results related to queueing theory including Markov processes and insensitivity theory. Volume 1 covers background while Volume 2 is on standard queueing models. Vocabulary list in English, German, Russian, and French plus references and list of symbols. **(a)**

Integrals and Series. A. P. Prudnikov, I͡U. A. Brychkov, O. I. Marichev. Transl: N. M. Queen, G. G. Gould. New York: Gordon and Breach Science, 1986-1990. *Volume 1: Elementary Functions*. 798pp. *Volume 2: Special Functions*. 750pp. $175.00/Volume. $350.00 (set). ISBN 2-88124-097-6 (set). *Volume. 3: More Special Functions*. 800pp. $175.00. ISBN 2-88124-682-6. English language translation of *Integraly i Ri͡ady* (see below).

Volume 1 covers elementary functions, Volume 2 is on special functions, and Volume 3 contains tables of indefinite and definite integrals, finite sums and series, and includes the functions of Struve, Weber, Anger, Lommel, Kelvin, and several others. Also contains tables of representations of generalized hypergeometric functions, and tables of Mellin transforms of a wide class of elementary and special functions, combined with tables of special cases of the Meijer G-function. Intended for specialists and students. Basically a photo offset of original with small amount of translation added. **(a,s)**

Integraly i Ri͡ady. A. P. Prudnikov, I͡U. A. Brychkov, O. I. Marichev. Moscow: Nauka, 1983. 748pp. o.p.

In Russian, it covers indefinite and definite integrals, and finite sums and series with special functions. Supplement titled *Dopolnitelnye Glavy* published in 1986. For English translation, *Integrals and Series* (see above).

Kompendium zur Elementärmathematik. A. G. Zypkin. Transl: Stefan Ivanov. Thun: Harri Deutsch, 1986. 567pp. $23.31. ISBN 3-87144-776-5.

Compendium of elementary mathematics. German translation of Russian original. For use as a reference work and textbook. Many examples. **(a)**

Manual for Translators of Mathematical Russian. Sydney H. Gould. Providence, RI: American Mathematical Society, 1966; 1986. iv + 44pp. $15.00. ISBN 0-8218-0028-0.

Suggestions on translating Russian mathematics. Applicable to related sciences (e.g., physics) as well. **(a)**

Manual of Mathematics. Granino Arthur Korn, Theresa M. Korn. New York: McGraw-Hill, 1967. 391pp. o.p.

Chapters are outlines of basic mathematical subjects and closely parallel the Korns' *Mathematical Handbook for Scientists and Engineers* (see below). Subject index. **(p)**

The Mathematica Handbook. Martha L. Abell, James P. Braselton. Boston: Academic Press, 1992. xvi + 789pp. $32.50. ISBN: 0-12-041535-6.

This book is intended to provide a convenient reference of all built-in **Mathematica** Version 2.0 objects to both beginning and advanced users. **(a,p)**

Mathematica Quick Reference. Nancy Blachman. Berkeley, CA: Variable Symbols, 1990. $14.95.

A spiral-bound, slim (4 x 8.5 in.) alphabetic list of all **Mathematica** built-in objects (functions, variables, options) and packages, giving exact syntax and brief description. Handy, easily portable, and very useful. **(a,p)**

Mathematical Handbook for Scientists and Engineers: Definitions, Theorems, and Formulas for Reference and Review. Second Edition, Enlarged & Revised. Granino Arthur Korn, Theresa M. Korn. New York: McGraw-Hill, 1968. xvii + 1,130pp. $112.00. ISBN 0-07-035370-0.

Covers mathematical definitions, theorems and formulas for professionals and students. Efficient format. Important formulas and definitions are boxed, topical reviews are in large print, and discussion and advanced topics in smaller print. Symbols and notations glosary. **(a)**

Mathematical Handbook, Higher Mathematics. Third Printing. M. Vygodskiǐ. Transl: George Yankovsky. Moscow: MIR Pub., 1978. 872pp. Price not available.

Russian work covering plane and solid analytical geometry, calculus, ordinary differential equations, "remarkable" curves. Many examples, no problems, and few proofs. Name and subject indexes. **(a)**

Mathematics Manual: Methods and Principles of the Various Branches of Mathematics for Reference, Problem Solving, and Review. Frederick S. Merritt. New York: McGraw-Hill, 1962. 378pp. o.p.

From simple arithmetic through higher mathematics including matrices, tensors, probabilities, and statistics, it presents important definitions, principles, theorems, corollaries, relationships, and methods of various branches of mathematics. References and examples. **(a)**

MathSci User Guide. Second Edition. Providence, RI: American Mathematical Society, 1990. various pagings (loose-leaf). $85.00. ISBN 0-8218-0233-X.

This manual assists users of the MathSci database. In addition to the usual searching instructions and aids, there are chapters listing serial titles covered, classifications schemes, an index to subject terms, keywords, institution codes and addresses, and information on TeX and MathSci. **(a,s)**

Multigrid Techniques: 1984 Guide with Applications to Fluid Dynamics. Achi Brandt. Bonn: St. Augustin, 1984. viii + 183pp. (GMD-Studien; 85). $23.57. ISBN 3-88457-081-1.

Intended for people who have some knowledge of multigrid techniques. Divided into three parts: "Stages in developing fast solvers," "Advanced techniques and insights," and "Applications to fluid dynamics." Serves as a detailed guide. **(a)**

Network Optimisation Practice: A Computational Guide. David K. Smith. New York: Halsted Press, 1982. 237pp. $68.95. ISBN 0-470-27347-X.

Contains standard algorithms for the basic network optimization problems. Format for each problem is statement of problem, development of algorithm, numerical example, and implementation of algorithm by a BASIC program and a PASCAL program. Can be used as a reference or a text for students of network analysis. **(a)**

New Mathematics Dictionary and Handbook. Robert W. Marks. New York: Bantam, 1964. 186pp. o.p.

Nontechnical, holistic work covering historical items, biography, basic procedures and operations, and tabular data for average reader. **(a,p)**

On Mathematics and Mathematicians. Robert Edouard Moritz. New York: Dover, 1958. vii + 410pp. o.p.

Reprint of *Memorabilia Mathematica; or, The Philomath's Quotation Book*, Macmillan, 1914. Some 2,100 quotations on mathematics in classified arrangement with thorough indexing. **(a,p)**

Operations Research Handbook: Standard Algorithms and Methods with Examples. H. A. Eiselt, H. von Frajar. New York: Walter de Gruyter, 1977. 398pp. o.p. ISBN 3-11-007055-3.

Algorithms for computer programming are presented in form of principles, description, and examples. **(a)**

Outliers in Statistical Data. Second Edition. V. Barnett, T. Lewis. Chichester, England: Wiley, 1984. xiv + 463p. $123.00. ISBN 0-471-90507-0.

Standard reference work on outliers. Provides relevant illustration and tabulation as well as suggestions for future research. Definite must for any researcher involved in correlation or regression. **(a,s)**

Pocketbook of Integrals and Mathematical Formulas. 2nd Edition. Ronald J. Tallarida. Boca Raton, FL: CRC Press, 1992. 225pp. $14.95. ISBN 0-8493-0142-4.

This small book contains the information on integrals, derivatives, series, statistics, etc. which is not readily provided by calculators plus a comprehensive table of integrals. Appendices and index. **(a,p,s)**

Pocketbook of Mathematical Functions. Milton Abramowitz et al. Thun: Harri Deutsch, 1984. 468pp. $24.61. ISBN 3-87144-818-4.

Abridged version of *Handbook of Mathematical Functions* (see p. 111) by Abramowitz and Stegun. Popular work with many printings. **(a,p,s)**

Poradnik Matematyczny. R. Bartoszynski et al. Warsaw: Polish Scientific Publishers, 1978. 908pp. Price not available.

Reference book of mathematical concepts includes proofs in some cases. Covers elementary logic, set theory, algebra, geometry, analysis, differential calculus, linear algebra, graphs, nomography and numerical approximation. **(a)**

Practical Guide to Splines. Carl de Boor. New York; Heidelberg: Springer-Verlag, 1978. xxiv + 392pp. $35.00. ISBN 0-387-90356-9.

Topical arrangement in 17 chapters covering theory to application with 120 plus example problems and 106 references. **(a)**

Practical Math Handbook for the Building Trades. Paul Calter. New York: Prentice-Hall, 1983. x + 294pp. $17.25. ISBN 0-13-692228-7.

Precalculus level with general mathematical formulas and tables, section on pocket calculator use, tables and formulas specific to building trades such as plumbing, heating, and estimating. **(p)**

The Practical Statistician: Simplified Handbook of Statistics. Monterey, CA: Brooks/Cole, 1975. vii + 384pp. $22.95 (pbk.). ISBN: 0-8185-0127-8.

A useful guide to using statistics as a research tool, this book enables its users to determine what analysis is appropriate and also shows how to do the computations. It can serve as the text for a research-methods course or as a resource in the practical application of inferential statistics. **(a,p)**

Praktikum po Matematicheskomu Programmirovaniu͡. I. L. Akulich. Riga: Latviisk. Gos. Univ., 1975. 140pp. Price not available.

Intended primarily for economics students, book presents an introduction to linear programming. Divided into four sections covering the general problem of linear programming and its geometrization, the simplex method, dual problems, and the transportation problem.

Quantities and Units of Measurement: A Dictionary and Handbook. (See p. 96)

Reading the Numbers: A Survival Guide to the Measurements, Numbers, and Sizes Encountered in Everyday Life. Mary Blocksma. New York: Viking Penguin, 1989. xv + 224pp. $7.95. ISBN 0-14-010654-5.

Can be used to answer a variety of questions dealing with numeric data. Also useful as a tool to help readers interpret tables and statistics in another source. **(a,p)**

Shuxue Shouce. Beijing: Renmin Jiaoyu Chubanshe, 1979. Pagination and price not available.

Mathematics handbook prepared by Chinese mathematicians which includes formulas, numerical tables, and sections on mathematical subjects such as differential geometry, analytic functions, etc. **(a)**

SI Metric Handbook. John L. Feirer. New York: Schribner, 1977. ca. 400pp. o.s.i. ISBN 0-87002-908-8.

Standard reference to SI (Systeme International) measurement system for those unfamiliar with the system. **(p)**

Spravochnik po Obyknovennym Different͡sial'nym Uravnenii͡am. Fifth Edition. E. Kamke. Moscow: Izdat. Nauka, 1976. 576pp. o.p.

Manual of ordinary differential equations translated and revised from German *Differentialgleichungen*, 6th edition. Two parts: methods, theorems, etc. useful to solving/studying differential equations and tables. **(a)**

Spravochnik po Teorii Veroi͡atnosteĭ i Matematicheskoĭ Statistike. Second Edition. V. S. Koroli͡uk et al. Moscow: Nauka, 1985. 640pp. o.p.

Manual on probability theory and mathematical statistics in three parts: probability theory, stochastic processes, and mathematical statistics. Some items overly brief, others absent. **(a)**

Survey of Applicable Mathematics. Ed: Karel Rektorys. Transl: Rudolf Vyborny et al. Cambridge, MA: MIT Press, 1969. 1,369pp. o.p.

Survey for engineers and related professionals. Formulas and theorems are explained in text and examples. Indexes. **(a)**

Taschenbuch der Mathematik. Twentieth Edition. I. N. Bronshtein, K. A. Semendi͡aev. Moscow: Nauka; Leipzig: Teubner. 1989. xi + 860pp. $27.20. ISBN 3-8714-4492-8.

The English language translation of same work is entitled *Handbook of Mathematics* (see p. 112).

Taschenbuch der Mathematik, Ergänzende Kapitel. Sixth Edition. I. N. Bronshtein, K. A. Semendi͡aev. Leipzig: Teubner, 1990. v + 234pp. $11.48. ISBN 3-322-00782-0.

Supplementary volume discusses analysis, mathematical methods of operations research, and mathematical information processing. Intended as supplement to Bronshtein's and Semendi͡aev's work of same title (see above).

Time Series in the Frequency Domain. Ed: D. R. Brillinger, P. R. Krishnaiah. Handbook of Statistics;3. New York: North-Holland, 1983. xiv + 485pp. $95.00. ISBN 0-444-86726-0.

Book contains 20 articles and has something to offer to both novices and experts. Should undoubtedly become a standard reference work for students and researchers in the time series field. **(a,p,s)**

Times Series in the Time Domain. Ed: E. J. Hannan, P. R. Krishnaiah, M. M. Rao. Handbook of Statistics;5. New York: Elsevier, 1985. xiv + 490pp. $95.00 ISBN 0-444-87629-4.

One of a series of works on statistical methodology and applications, this book is concerned with time domain methods, which involve parametric representations of the temporal relationships. **(a,s**

User's Guide to Spectral Sequences. John McCleary. Wilmington, DE: Publish or Perish, 1985. xiv + 423pp. (Mathematics Lecture Series; 12). $40.00. ISBN 0-914098-21-7.

Divided into three parts: Algebra, Topology, and Sins of Omission. Intended for graduate students in algebraic topology who have already had a first course and who need to learn how to compute with spectral sequences. Also can be used by an algebraist without a topological background who needs spectral sequences. **(a)**

INDEXES

Abbreviations of Names of Serials Reviewed in Mathematical Reviews. Providence, RI: American Mathematical Society. Annual. $9.00.

Reprinted from the December index of Mathematical Reviews, this listing by abbreviations provides complete title, publisher, and place. **(a)**

Author Institution Codes and Addresses Used in Current Mathematical Publications and Mathematical Reviews. Providence, RI: American Mathematical Association. Annual. $9.00.

Reprinted from the December issue of *Mathematical Reviews*, it provides addresses for a large number of institutions where an author's address clearly coincides. **(a)**

CMCI CompuMath Citation Index. Philadelphia, PA: Institute for Scientific Information, 1981-. 3/year. $1495.00. ISSN 0730-6199.

In three volume parts: citation index, corporate and source index, and permuterm subject index. Covers mathematics, mathematical statistics, computers, electronic data processing, operations, statistics, and related research. **(a)**

Computing Reviews. New York: Association for Computing Machinery, 1961-. Monthly. $115.00. ISSN 0010-4884.

Short, signed reviews with an emphasis on hardware aspects. Author rebuttals appear occasionally. **(s)**

Cumulative Index to Mathematics of Computation 1970-1984, Volumes 24-43. Providence, RI: American Mathematical Society, 1987. 503pp. $59.00. ISBN 0-8218-0093-0.

Cumulative index a primary journal in mathematical literature. Divided into author, subject, and key areas, it is a useful tool for collections holding Mathematics of Computation. **(a)**

Current Index to Statistics. Alexandria, VA: American Statistical Association, 1975-. Annual. $60.00. ISSN 0364-1228.

Annual author and keyword index to statistical journal and monographic literature. Included in MathSci online database. **(a)**

Current Mathematical Publications. Providence, RI: American Mathematical Society, 1969-. 17/year. $314.00. ISSN 0361-4794.

Biweekly current awareness tool published by *Mathematical Reviews.* Its issues contain citations to mathematical items received. Many, but not all, will later appear in Mathematical Reviews (see below). Items are arranged by subject and then alphabetically by author. Formed by merger of *Contents of Contemporary Mathematical Journals and New Publications.* Includes tables of contents of selected mathematical journal as well as special sections for books, videocassettes, and serials listed in each particular issue, institution codes and address. Author and key indexes. **(a,s)**

Directory of Published Proceedings, Series SEMT–Science/Engineering/Medicine/Technology. White Plains, NY: InterDok Corp., 1965-. Annual. $150.00. ISSN 0093-5816.

"A bibliographic directory of preprints and published proceedings of congresses, conferences, symposia, meetings, seminars, and summer schools which have been held worldwide from 1964 to date." Entries are arranged chronologically and provide place, name of meeting, principal sponsors, publisher or distributor, address, order numbers, price, and, if necessary, title and series or journal of publication, editor, and pagination. Includes sponsor and subject indexes. **(s)**

Index of Mathematical Papers. Providence, RI: American Mathematical Society, 1970-1973. 4v. Volume 1. $85.00. Volume 2-5. $166.00. ISSN 0019-3917.

For four years, these were the annual author and subject indexes to *Contents of Contemporary Mathematical Journals* and its successor, *Current Mathematical Publications.* These issues (yellow paperbound volumes) have dual numbering: whole number 1-7 and volume and number, from v.1:1 to v.5. These issues can be safely sent to remote storage as they essentially duplicate the *Mathematical Reviews* indexes for this period. From 1974-to present (orange apperbound volumes) the *Index of Mathematical Papers* continues as secondary numbering on the annual indexes of *Mathematical Reviews.* **(a)**

Index to Statistics and Probability: The Citation Index. John W. Tukey et al. (V. 7 by Frederick G. Kilgour). Providence, RI: American Mathematical Society, 1973-. 7v. $52.00 (V. 1); $118.00 (V. 2); $134.00 (V. 3 & 4 combined); $106.00 (V. 5); $58.00 (V. 7). $303.00 (set, V. 1-5). ISBN 0-8827-4000-8 (V. 1); 0-8827-4001-6 (V. 2); 0-8827-4002-4 (V. 3 & 4); 0-8827-4004-0 (V. 5); 0-8218-4006-7 (V. 7).

Authoritative work that provides access to 25,000 items on statistics for period 1902 and 1968. Includes indexes to 113 major works, a citation index, index by words or phrases, lists of authors and titles arranged by journal and date, and by author giving the journals in which papers were published. Sixth volume (yet to be published) provides a permuted index to minimum abbreviations on all journals and other works appearing in earlier volumes. Seventh volume is a cumulation of the "back-of-the-book" index entries from 96 English-language books in the field of librarianship and information science. The "Tukey Index" is also part of the online MathSci database. **(a)**

Index to the Mathematics of Computation, 1943-69. Providence, RI: American Mathematical Society, 1972. 461pp. $43.00. ISBN 0-8218-4000-2.

Author and subject access to approximatley 6,000 papers appearing in *Mathematical Reviews* from 1943 through 1969. Subject entries lead to appropriate author entries.

Index to Translations Selected by the American Mathematical Society. Providence, RI: American Mathematical Society, 1966 (V. 1); 1973 (V. 2). 2v. V. 1: iii + 90pp. $20.00. ISBN 0-8281-0042-6. V. 2: iii + 93pp. $32.00. ISBN 0-8218-0059-0.

Volume 1 covers *American Mathematical Society Translations–Series 1, American Mathematical Society Translations–Series 2* (Volumes 1-50), and *Selected Translations in Mathematical Statistics and Probability* (Volumes 1-5). Volume 2 covers *American Mathematical Society Translations–Series 2* (Volumes 51-100) and *Selected Translations in Mathematical Statistics and Probability* (Volumes 6-13). Both volumes contain author and subject indexes. **(a)**

Jahrbuch über die Fortschritte der Mathematik. Berlin: Walter de Gruyter, 1868-1942. 68v. ISSN 0179-2849.

The major abstracting tool for mathematics prior to *Mathematical Reviews* appearance in 1940. Published in German, it provides international coverage using a classified subject arrangement. **(a)**

Lecture Notes in Mathematics 1-500: An Index and Other Useful Information; Lecture Notes in Mathematics 501-1000. Ed: A. Dold, B. Eckmann. New York: Springer-Verlag, 1975; 1984. 78pp.; 108pp. Free.

Contents of *Lecture Notes* numbers 1 through 500 and 501 through 1,000. Title, author, and subject indexes, rationale for series, and instructions for manuscript preparation. **(a,s)**

MAST: Minimum Abbreviations of Serial Titles Mathematics. Mary L. Tompkins. North Hollywood, CA: Western Periodicals, 1969. viii + 427pp. Price not available.

Attempts to standardize serials titles abbreviations and provide approach to deciphering unusual title abbreviations. In two parts, index to titles and detailed master list of 2,000 plus titles. Also useful as listing of mathematics serials. **(a)**

Mathematical Reviews. Providence, RI: American Mathematical Society, 1940-. Monthly. $4,010.00. ISSN 0025-5629.

Monthly abstracting service provides international reviews of mathematical and mathematics related works and includes French, German, and Russian abstracts. Complete citations given, monographs identified by star preceding author's name. Critical abstracts/reviews arranged by subject and are signed. Reviews generally appear 18 months after work is listed in *Current Mathematical Publications* (*CMP*) (see above). Percentage of *CMP* citations which are not reviewed in *MR* has risen from 12% to nearly 20%. Author and key indexes appear monthly, annually, and cumulatively (see below). AMS uses a transliteration system for Russian and Chinese which differs from that used by libraries; comparative tables are provided. Use with *MathSci User Guide* (see p. 116). **(a,p,s)**

Mathematical Reviews Index. Providence, RI: American Mathematical Society, 1940- . Annual and cumulations. Divided into two parts, author/key index and subject index. Author cumulations are 1940-59 [$427.00], 1960-64 [$337], 1965-72 [$670], 1973-79 [$1,702 (Author and Subject combined price)], 1980-84 [$2,060 (Author and Subject combined price)]. Subject cumulations are 1940-58 [$368], 1959-72 [$1,331], 1973-79, 1980-84. No further cumulations are planned. Yearly author and subject indexes available for 1989 [$744 (Author and Subject combined price)], 1990 [$803 (Author and Subject comined price)], and 1991 [$802 (Author and Subject combined price)]. Discounts available for some other combinations.

Because classification schemes changed over the years, care must be taken in selecting subject headings to be searched with each cumulation. **(a,s)**

Mathematics of Computation. Providence, RI: American Mathematical Society, 1943-. Quarterly. $205. ISSN 0025-5718.

Publication devoted to reporting advances in numerical analysis, application of computer methods, mathematical tables, and aids to computation in general. "Reviews and Descriptions of Tables and Books" section lists unpublished tables available by request from the editorial office. Volumes 1 through 13 were called *Mathematical Tables and Other Aids to Computation.* **(a)**

Probability Theory Subject Indexes from Mathematical Reviews. Providence, RI: American Mathematical Society, 1987. iii + 469pp. $75.00. ISBN 0-8218-0108-2.

Provides complete list of all reviews published in *Mathematical Reviews* from 1940 through 1984. Classified arrangement using AMS subject headings with entries consisting of author, title, (except 1940-1958) and MR reference number. Similar to "Reviews in" series except there are no reviews/abstracts. **(a)**

Quality Control and Applied Statistics. New York: Executive Sciences Institute, 1956-. Monthly. $149.00. ISSN 0033-5207.

International digest of statistical literature. Classified arrangement of signed abstracts with annual author/subject indexes. **(a)**

Referativnyĭ Zhurnal. Matematika. Moscow: Vsesoi͡uznyĭ Institut Nauchnoĭ i Tekhnicheskoĭ Informat͡sii, 1953-. Monthly. $756.00. ISSN 0034-2467.

In Russian. Indexing and abstracting source providing international coverage of mathematical articles and books. Especially valuable for finding Russian publications not indexed by *Mathematical Reviews* (see p. 122) and *Zentralblatt für Mathematik und ihre Grenzgebiete* (see 126). Reviews in each issue are grouped by broad subject areas and indexed by author and subject. Annual cumulative author and subject indexes. **(a)**

Reviews in Complex Analysis, 1980-1986. Providence RI: American Mathematical Society, 1989. 4v. (xiv + 3064pp.). $307.00. ISBN 0-8218-0127-9.

Collection of some 13,000 reviews on the subject appearing in *Mathematical Reviews* from 1980 through 1986. Author and key indexes. **(a)**

Reviews in Functional Analysis, 1980-1986. Providence, RI: American Mathematical Society, 1989. 4v. (xii + 2461pp.). $265.00. ISBN 0-8218-0134-1.

Compendium of all reviews on functional analysis appearing in *Mathematical Reviews* from 1980 through 1986. Author and key indexes. **(a)**

Reviews in Global Analysis, 1980-1986. Providence, RI: American Mathematical Society, 1988. 5v. V. 1: ix + 786pp. V. 2: v + pp. 787-1620. V. 3: v + pp. 1621-2418. V. 4: v + pp. 2419-3112. V. 5: v + pp. 3113-3920. $325.00 (set). ISBN 0-8218-0104-X (set).

Reprints of all reviews on the subject appearing in *Mathematical Reviews* from 1980 through 1986. Author and key indexes provided. **(a)**

Reviews in Graph Theory 1940-78. Ed: William G. Brown. Providence, RI: American Mathematical Society, 1980. 4v. V. 1: xvi + 588pp. $124.00. ISBN 0-8218-0210-0. V. 2: xvi + 546pp. $124.00. ISBN 0-8218-0211-9. V. 3: xvi + 529pp. $124.00. ISBN 0-8218-0212-7. V. 4: x + 327pp. $75.00. ISBN 0-8218-0213-5. $368.00 (set). ISBN 0-8218-0214-3 (set).

Approximately 10,000 reviews as printed in *Mathematical Reviews*, 1940-1978, volumes 1-56 inclusive. Volume 4 contains author, key and subject indexes. **(a)**

Reviews in K-Theory, 1940-1984. Ed: Bruce A. Magurn. Providence, RI: American Mathematical Society, 1985. viii + 811pp. $294.00. ISBN 0-8218-0088-4.

Collection of all reviews appearing in *Mathematical Reviews* from 1940 through 1984. Classified arrangement based on a special classification scheme, subject, author and key indexes, and MR number list are included. **(a)**

Reviews in Number Theory, 1940-72. Ed: William J. LeVeque. Providence, RI: American Mathematical Society, 1974. 6v. V. 1: v + 420pp. V. 2: v + 672pp. V. 3: vi + 377pp. V. 4: vi + 582pp. V. 5: v + 470pp. V. 6: ix + 410pp. $369.00 (set). ISBN 0-8218-0226-7 (set); 0-8218-0203-0 (V. 1).

Covers 14,426 reviews on number theory appearing in *Mathematical Reviews* from 1940 thorough 1972. Author index. **(a)**

Reviews in Number Theory, 1973-1983. Ed: Richard K. Guy. Providence, RI: American Mathematical Society, 1984. 6v. V. 1: viii + 460pp. V. 2: ix + 765pp. V. 3: viii + 344pp. V. 4: ix + 477pp. V. 5: ix + 699pp. V. 6: xi + 828pp. $823.00 (set). ISBN 0-8218-0218-6 (set).

Contains 17,119 reviews on number theory appearing in *Mathematical Reviews* from 1973 through 1983 with author and key indexes. **(a)**

Reviews in Numerical Analysis, 1980-1986. Providence, RI: American Mathematical

Society, 1987. 5v. V. 1: ix + 649pp. V. 2: v + pp. 650-1378. V. 3: v + pp. 1379-2006. V. 4: v + pp. 2007-2701. V. 5: v + pp. 2702-3558. $265.00 (set). ISBN 0-8218-0102-3 (set).

Contains all reviews appearing in *Mathematical Reviews* from 1980 through 1986. Individual reviews are followed by references to later *MR* reviews citing the earlier review. Volume 5 is author and key indexes. **(a)**

Reviews in Operator Theory, 1980-86. Providence, RI: American Mathematical Society, 1989. 4v. (x + 2328pp.). $265.00. ISBN 0-8218-0135-X.

Includes all *Mathematical Reviews* entries for operator theory as primary or secondary classification for the period 1980 through 1986. Provides cross-references with each review plus author and key indexes. **(a)**

Reviews in Partial Differential Equations, 1980-1986. Providence, RI: American Mathematical Association, 1988. 5v. (ix + 3998pp.). $325.00. ISBN 0-8218-0103-1.

Reprints of all reviews on the subject appearing in *Mathematical Reviews* from 1980 through 1986. Reviews are in a classified arrangement following *MR*'s system and reprinted at the end of volume 4. Volume 5 contains author and key indexes. **(a)**

Reviews in Ring Theory. Ed: Lance W. Small. Providence, RI: American Mathematical Society, 1981. 2v. V. 1: xi + 585pp. V. 2: xi + pp. 586-1114. $306.00 (set). ISBN 0-8218-0215-1 (set).

Covers all reviews printed in *Mathematical Reviews* from 1960 thorough 1979, volumes 21-58. Provides coverage of key papers from volumes 1-20, 1940-59 as well. Volume 2 includes author and key indexes. **(a)**

Reviews in Ring Theory, 1980-1984. Ed: Lance W. Small. Providence, RI: American Mathematical Society, 1986. xiv + 685pp. $99.00. ISBN 0-8218-0097-3.

Presents all reviews appearing in *Mathematical Reviews* from 1980 through 1984. Includes author and key indexes. **(a)**

Reviews of Papers in Algebraic and Differential Topology, Topological Groups, and Homological Algebra, 1940-67. Classified by Norman Earl Steenrod. Providence, RI: American Mathematical Society, 1968. 2v. (xii + 1448pp.). $87.00. ISBN 0-8218-0046-9.

Covers some 6,000 reviews appearing in *Mathematical Reviews* from 1940 through 1967. Classified arrangement of 290 headings. Author index. **(a)**

Reviews on Finite Groups, 1940-70. Ed: Daniel Gorenstein. Providence, RI: American Mathematical Society, 1974. xi + 706pp. $117.00. ISBN 0-8218-0202-X.

Reviews of 3,052 papers on finite groups appearing in *Mathematical Reviews* from 1940 through 1970. Author index. **(a)**

Reviews on Infinite Groups; as Printed in Mathematical Reviews, 1940 through 1970, Volumes 1-40 Inclusive. Ed: Gilbert Baumslag. Providence RI: American Mathematical Society, 1974. 2v. (xii + 1062pp.). $184.00. ISBN 0-8218-0201-0.

Contains reprints of reviews on infinite groups appearing in *Mathematical Reviews* from 1940 through 1970. Classified arrangement with cross references to related articles and author index with citations. **(a)**

Statistical Theory and Method Abstracts. Edinburgh, Scotland: Oliver & Boyd, 1959-1970; London: Longman, 1971-. Quarterly. $95.00. ISSN 0039-0518.

International abstracting journal with classified arrangement under broad subject headings such as probability, distributions, time series analysis, etc. Author index. **(a,p)**

Statistics: Subject Indexes from Mathematical Reviews, 1940-84. Providence, RI: American Mathematical Society, 1987. iii + 507pp. $75.00. ISBN 0-8218-0107-4.

Lists all reviews on statistics appearing in *Mathematical Reviews* from 1940-1958. Classified arrangement using AMS subject headings. Entries include author, title (except for 1940-1958) and *MR* reference number. Includes the various classification schemes used throughout the period. Similar to "Reviews in" series sans reviews. **(a)**

Zentralblatt für Mathematik und ihre Grenzgebiete. Berlin: Springer, 1931-. 30 volumes/year. $2,689.00. ISSN 0044-4235.

An alternative to *Mathematical Reviews*, this international, classified review/abstract journal has signed author reviews that are written in English, French, or German. Reviews often appear quicker in *Zbl* than *MR*. In addition to mathematics proper, it covers related subjects. Three volumes issued per year. Author and subject indexes issued annually and cumulatively every tenth volume. **(a)**

TABLES

Applied Mathematics Series. U.S. National Bureau of Standards. Washington, DC: Government Printing Office, 1948-69. 63 v. ISSN 1049-4685.

Contains mathematical tables for those engaged in scientific and technical work. Some volumes are reissues of those produced by WPA project for computation of mathematical tables. **(a,s)**

Attribute Sampling Plans, Tables of Tests and Confidence Limits for Proportions. Robert E. Odeh, D. B. Owen. New York: Dekker, 1983. xi + 368pp. $99.75. ISBN 0-8247-7136-2.

Tables for binomial, Poisson, or hypergeometric distributions. For those in quality control or acceptance sampling. Companion volume to his *Tables for Normal Tolerance Limits* (see below). **(a)**

Barlow's Tables of Squares, Cubes, Square Roots, Cube Roots and Reciprocals of All Integers Up to 12,500. Fourth Edition. Ed: Peter Barlow, L. J. Comrie. New York: Chemical Publications, 1960. xii + 258pp. o.p.

British equivalent of the *CRC Handbook of Mathematical Sciences* (see *Handbooks and Manuals*, above). **(a)**

Biometrika Tables for Statisticians. Third Edition. Ed: Egon Sharpe Pearson, H. O. Hartley. London: Biometrika Trust, 1976. 2v. V. 1: xvi + 270pp. $19.35. ISBN 0-904653-10-2. V. 2: xviii + 385pp. $22.98. ISBN 0-904653-11-0.

Handbook of choice for statisticians and frequently referenced by authors of statistical works. **(a,p,s)**

Charaktertafeln von Sylownormalisatoren Sporadischer Einfacher Gruppen. Th. Ostermann. Essen: Universität Essen, Fachbereich Mathematik, 1986. x + 187pp. Price not available.

Compilation of character tables for normalisers of noncyclic Sylow subgroups for nineteen sporadic simple groups generated by using CAYLEY and CAS systems. **(a)**

CRC Handbook of Tables for Probability and Statistics. Second Edition. William H. Beyer. Cleveland, OH: Chemical Rubber Co., 1968. xiv + 642pp. $69.95. ISBN 0-8493-0692-2.

Compilation of standard statistical tables organized in 13 topical sections such as probability and statistics, f-distribution, correlation coefficient, etc., and some miscellaneous mathematical tables. Includes index. **(a,p,s)**

CRC Standard Mathematical Tables. Twenty-Ninth Edition. Ed: William H. Beyer. Boca Raton, FL: CRC Press, 1991. 609pp. $45.00. ISBN 0-8493-0629-9.

Mathematical tables for scientific, engineering, industrial, and educational personnel. Includes numerical, logarithmic, and trigonometry tables, plus binomial functions, and differential equations. Widely used ready reference tool. **(a,p,s)**

Eight-Place Tables of Trigonometric Functions for Every Second of Arc, with an Appendix on the Computation to Twenty Places. Jean Peters. Bronx, NY: Chelsea, 1968. xi + 954pp. o.p. ISBN 0-8284-0174-8.

Title reflects content. Eight place values for sine, cosine, tangent, and cotangent are given for every second of arc 0-90 degrees. Widely accepted as authoritative tables for the field. Classic. **(a,p)**

Five Place Tables. Sixth Edition. P. Wijdenes. Leyden: Noordhoff International, 1975. iv + 168pp. Price not available.

Logarithms of integers. Logarithms and natural values of trigonometric functions in the decimal system for each grade from 0 to 100 grades with interpolation tables. An unusual feature is the right angle being divided into 100 grades (and further divided into 100 centigrades) rather than 90 degrees.

Factorizations of b^n+1, $b = 2,3,5,6,7,10,11,12$ Up to Higher Powers. Revised Edition. John Brillhart et al. Providence, RI: American Mathematical Society, 1988. xcv + 236pp. $36.00. ISBN 0-8218-5078-4.

Presents known factorizations of numbers given in title plus historical account of hardware and development. References. **(a)**

Group Tables. A. D. Thomas, G. V. Wood. Nantwich: Shiva Publishing; Distributed by Imprint Editions, Fort Collins, CO, 1980. 174pp. (not consecutively paged). (Shiva Mathematics Series; 2). $13.95. ISBN 0-906812-02-X.

Book contains, in tabular form, information on groups of order at most 32. Character table and subgroup lattice given for groups of order less than 32. Multiplication tables for the 51 nonisomorphic groups of order 32. (**a**)

Guide to Mathematical Tables. Aleksandr Vasil'evich Lebedev, R.M.A. Fedorova. New York: Pergamon, 1960. 586pp. o.p.

Translation of Russian *Spravochnik po Matematicheskikh Tablit͡s*. Similar to Fletcher's *Index to Mathematical Tables* less the more obscure tables. Some tables are difficult to find elsewhere. Also 190 page supplement (Pergamon, 1960) with author index and other aids. (**a**)

Guide to Tables in Mathematical Statistics. Joseph Arthur Greenwood, H.O. Hartley. Ann Arbor, MI: Books on Demand UMI, 1962. lxii + 1,014pp. $160.00. ISBN 0-317-13006-4. Order #: 2021559.

Sponsored by National Academy of Science, Mathematics Division, with classified arrangement supplemented by a list of contents of several books of tables. Author and subject indexes. Remains useful despite age. (**a,p,s**)

Guide to Tables in the Theory of Numbers. Derrick Henry Lehmer. Washington, DC: National Research Council, National Academy of Sciences, 1941. xiv + 177pp. o.p. (*Bulletin of the National Research Council,* 105 (Feb. 1941).

In three parts: description of tables, bibliography arranged by author with references to descriptions, and errata in tables. (**a**)

Handbook of Statistical Tables. Donald Bruce Owen. Reading, MA: Addison-Wesley, 1962. 580pp. o.p.

Twenty categories of tables with brief notes. Includes a subject and name index. Extensive bibliography. (**a,p**)

Index mathematischer Tafelwerke und Tabellen aus allen Gebieten der Naturwissenschaften. Karl Schütte. Munich: Oldenbourg, 1966. 239pp. Price not available.

Preface in English, headings in English and German. Classified arrangement of 2,800 plus tables with indexes by author and institute. Not as extensive as Fletcher (See below). (**a**)

An Index of Mathematical Tables. Second Edition. ED: Alan Fletcher et al. Reading, MA: Addison-Wesley, 1962. 2v. (xi + 994pp.). o.p.

Classic working index to mathematical tables, international in scope and covering sixteenth century to 1961. Includes working tables and excludes those of purely historical importance. Volume 1 has historical introduction and an index by functions. Volume 2 has sources of tables and index to Part 1. (**a,p**)

Index to Mathematical Problems, 1980-1984. Ed: Stanley Rabinowitz. Indexes to Mathematical Problems;1. Westford, MA: MathPro Press, 1991. xii + 532pp. $39.95. ISBN: 0-9626401-1-5.

Highly regarded work. **(a)**

Jahnke-Emde-Lösch: Tables of Higher Functions. Seventh Edition. Friedrich Lösch. Stuttgart: Teubner, 1966. xii + 322pp. Price not available.

Dual English-German text with references included. No elementary mathematical functions. **(a)**

Life Table and Its Applications. Chin Long Chiang. Melbourne, FL: Robert E. Krieger Publishing, 1984. xix + 316pp. $39.50. ISBN 0-89874-570-5.

Brief discussion of probability and statistics followed by treatment of age specific death rates, adjustment of rates, maximum likelihood estimation, and statistics of life tables. Complete and abridged life tables given with applications. Extensive references. **(a)**

Logarithmetica Britannica, Being a Standard Table of Logarithms to 20 Decimal Places. Alexander John Thompson. Cambridge: Cambridge University Press, 1967; 1952. 2v. o.p.

Two volume reprint of the classic work done in 1952 which collected all nine volumes of the *Logarithmetica Britannica* series published between 1924 and 1952. **(a)**

New Cambridge Elementary Statistical Tables. D. V. Lindley, W. F. Scott. Cambridge; New York: Cambridge University Press, 1984. 80pp. $9.50. ISBN 0-521-26922-9. Abridged and updated version of 1953 work.

Distributions and their descriptions arranged for study and teaching of statistics. Has tables for most common parametric and nonparametric tests. **(a,p)**

New Table of Indefinite Integrals, Computer Processed. Melvin Klerer, Fred Grossman. New York: Dover, 1971. xiii + 198pp. o.p. ISBN 0-486-62714-4.

Based on traditional tables as primary sources, this table includes over 2,000 indefinite integrals. **(a)**

The Number of Mutually Orthogonal Latin Squares–A Table up to Order 10000. A. E. Brouwer. Afdeling Zuivre Wiskunde [Department of Pure Mathematics];123. Amsterdam: Mathematisch Centrum, 1979. ii + 31pp. Price not available.

A table of lower bounds for N(v), the maximum number of mutually orthogonal Latin squares of order v, for $v < 10000$. **(a)**

Numerical Tables of the Queueing Systems I. Hirotaka Sakasegawa. Computer Science Monographs;10. Toyko: Institute of Statistical Mathematics, 1978. 116pp. Price not available.

Gives mean queue length, variance of numbers in system, mean waiting time, etc. **(a)**

Parts per Million Values for Estimating Quality Levels. Robert E. Odeh. New York: Marcel Dekker, 1988. xiv + 347pp. $110.00. ISBN 0-8247-7950-9.

Carefully computed tables for estimating quality levels from 20,000 ppm to 1 ppm. Supplements author's *Tables for Normal Tolerance Limits, Sampling Plans, and Screening* (see below). **(a,p)**

Pocket Book of Statistical Tables. Comp: Robert E. Odeh et al. New York: Dekker, 1977. x + 166pp. $49.75. ISBN 0-8247-6515-X.

Collection of statistical tables not easily calculated in convenient form. Each table has brief description and some references. **(a,p)**

Quaternary Quadratic Forms: Computer Generated Tables. Gordon L. Nipp. New York: Springer-Verlag, 1991. vi + 155pp. $59.00. ISBN 0-387-97601-9.

The book comes with a 3.5 inch disk in which the tables of the book have been condensed, allowing mathematicians to investigate and answer questions about quaternary forms by means of computer searches. Forms in the tables are separated into classes and genera. Included are the number of automorphs for each class and the mass of each genus as well as much other complex and computer-intensive information. **(a)**

Royal Society Mathematical Tables. Royal Society of London. Cambridge: Cambridge University Press, 1950-64. 11v. o.p.

Continues 10 volume series, *Mathematical Tables*, of British Association for Advancement of Science (1931-1952, 10v.). A well respected collection of tables published for Royal Society's Mathematical Tables Committee by University Press, Cambridge. Includes Farley series, Riemann zeta function, Bessel functions, logarithms, etc. **(a)**

Selected Tables in Mathematical Statistics. Providence, RI: American Mathematical Society, 1970-. Irregular. Price varies. ISSN 0094-8837.

Currently at volume 11, series collects selected tables and accompanying discussion of underlying distributions. Volumes are scholarly works in mathematical statistics. Some volumes are collections of lengthy articles, others are monographs. No cumulative indexing. **(a)**

SI and Metrication Conversion Tables. G. Socrates, L. J. Sapper. London: Newnes Butterworths, 1969. 28pp. o.p.

Recognized, established units are arranged aphabetically in table that gives comparable SI (Systeme International) unit. **(a)**

Statistical Tables and Formulae. Stephen Kokoska, Christopher Nevison. New York; Berlin: Springer-Verlag, 1989. 88pp. $11.00. ISBN 0-387-96873-3.

Collection of probability and statistics formulas plus standard tables, including tables for nonparametric tests. No index. **(a,p)**

Statistical Tables for Biological, Agricultural and Medical Research. Sixth Edition, Revised and Enlarged. Ronald Aylmer Fisher, Frank Yates. New York: Hafner, 1963. x + 146pp. o.p. ISBN 0-05-000872-2.

Authoritative collection of tables plus bibliography of sources on statistical method. (**a**)

Statistical Tables For Multivariate Analysis. Heinz Kres. Transl: Peter Wadsack. New York; Berlin: Springer-Verlag, 1983. xxii + 504pp. $79.00. ISBN 0-387-90909-5.

English language translation of German (1975) collection of 30 tables from various sources for multivariate hypothesis testing, multivariate normal distribution, and multivariate problems. (**a**)

Statistical Tables for Science, Engineering, Management and Business Studies. Second Edition. J. Murdoch, J. A. Barnes. New York: Wiley, 1977. 46pp. o.p. ISBN 0-470-62516-3.

Thirty-five tables divided into 7 groups: basic distributions and significance, statistical quality control, significance tables for runs, attribute single sampling plans, random numbers, and various mathematical and accounting tables. (**a,p**)

Statistical Tables. Second Edition. F. James Rohlf, Robert R. Sokal. San Francisco: Freeman, 1981. xii + 219pp. $14.95. ISBN 0-7167-1258-X.

Originally written to accompany authors' work *Biometry*, the volume proved to be a useful collection of tables and is still in print. (**a,p**)

Statistics Tables: For Mathematicians, Engineers, Economists and the Behavioral and Management Sciences. H. R. Neave, London: Allen & Unwin, 1978. 87pp. $6.98. ISBN 0-04-001001-5.

Collection of statistical tables helpful to many users. (**a,p**)

Supplement to Table of Sines and Cosines to Ten Decimal Places at Thousandths of a Degree. Herbert E. Salzer, Norman Levine. New York: Applied Science Publications, 1986. 68pp. $3.50. ISBN 0-915061-02-3.

First part contains a proof that the error due to the computational operations in any of the formulas for linear inverse interpolation does not exceed the tabular uncertainty error. The second part gives sin x from x = 0 degrees to x = 5.740 degrees in steps of 0.001 degrees to ten significant figures. Also supplies some corrections to original 1962 Pergamon book. (**a**)

Tabelle per L'uso del Test di Kolmogorov nel Caso Discreto. Roberto Marvulli. Turin: G. Giappichelli, 1980. 81 + [3]pp. Price not available.

Tables for using Kolmogorov test in discrete cases. Extends utility of Kolmogorov test from continuous case to discrete cases using series of operative tables. (**a**)

Table of Definite and Infinite Integrals. Alexander Apelblat. New York: Elsevier Scientific, 1983. x + 457pp. o.p. ISBN 0-444-42151-3.

Intended as reference tool to be used in conjuction with works on definite and infinite integrals. Includes references. (**a**)

Table of Series and Products. Eldon R. Hansen. Englewood Cliffs, NJ: Prentice-Hall, 1975. xviii + 523pp. o.p. ISBN 0-13-881938-6.

List of series that allow expression of mathematical functions in terms of more basic functions. Also provides closed form for given series. **(a)**

Table of the Relative Class Numbers h (K) of Imaginary Cyclic Quartic Fields K with h*(K) = 2 (MOD4) and Conductor f < 416,000*. K. Hardy et al. Carleton-Ottawa Mathematical Lecture Note Series;8. Ottawa, ON: Carleton University, Department of Mathematics and Statistics, 1987. 282pp. o.p. ISSN: 0827-3669.

Contains extensive tables. **(a)**

Tables and Graphs of Oscillator Spheroidal Functions. I͡U. A. Kulchitskiĭ et al. Dubna: Joint Institute Nuclear Research, 1986. 76pp.

Tabulation of prolate and oblate spheroidal functions of oscillator for some of first values of quantum numbers of functions. Applications to radio physics, quantum mechanics, and optics. **(a)**

Tables, Data and Formulae for Mathematicians. A. Greer, D. J. Hancox. Cheltenham: Stanley Thornes, 1978. [5] + 118pp. $5.21. ISBN 0-85950-037-3.

Topically arranged formulae plus basic tables and data. Indexed. **(a,p)**

Tables for Converting Polynomials and Power Series into Chebyshev Series. Herbert E. Salzer, Norman Levine. New York: Applied Science Publications, 1984. x + 65pp. $12.00. ISBN 0-915061-01-5.

For expressing polynomials as linear combinations of Chebyschev polynomials for various degrees in handy arrangement. Discussion of Chebyshev series and references. **(a)**

Tables for Lagrangian Interpolation Using Chebyshev Points. Herbert E. Salzer, Norman Levine, Saul Serben. New York: Applied Science Publications, 1984. i + 323pp. $28.00. ISBN 0-915061-00-7.

Covers advantages of taking zeroes of Chebyshev polynomials as nodes of Lagrangian interpolation. Includes description of computation of tables, use in economization of polynomials, quadrature and interpolation. Bibliography. **(a)**

Tables for Normal Tolerance Limits, Sampling Plans, and Screening. Robert E. Odeh, Donald B. Owen. New York; Basel: Marcel Dekker, 1980. ix + 316pp. $89.75. ISBN 0-8247-6944-9.

Tables based on univariate and bivariate normal distributions for industrial statisticians and engineers. Includes mathematical derivations, bibliography, index of symbols and general index. Supplement to *Parts per Million Values for Estimating Quality Levels* (see 129). **(a,p)**

Tables for Numerical Integration. H. V. Smith. New York: Oxford, 1982. v + 26pp. $7.50. ISBN 0-19-520586-3.

Compilation of tables for global error bounds of quadrature formulae, eg.: Gauss-Legendre, Lobatto, etc. **(a)**

Tables Minorant la Racine n-ième du Discriminant d'un Corps de Degré n. Francisco Diaz y Diaz. Orsay: Université de Paris-Sud, Département de Mathématique, 1980. 59pp. (not consecutively paged). Price not available.

Tables of lower bounds for the nth root of the discriminant of a field of degree *n*. **(a)**

Tables of Dimensions, Indices, and Branching Rules for Representations of Simple Lie Algebras. W. G. McKay, J. Patera. New York: Marcel Dekker, 1981. v + 317pp. $105.00. ISBN 0-8247-1227-7.

This book includes reductions, or branching rules of complex simple Lie algebras to their maximal semisimple subalgebras, table of dimensions, and of 2nd and 4th indices of the representations. Useful to those in pure mathematics, particle physics, and quantum chemistry. **(a)**

Tables of Dominant Weight Multiplicities for Representations of Simple Lie Algebras. M. R. Bremner, R. V. Moody, J. Patera. New York: Dekker, 1985. v + 340pp. $99.75. ISBN: 0-8247-7270-9.

Presents computer generated tables for Lie algebra representations. Primarily positive roots tables and multiplicity tables. Bibliography. **(a)**

Tables of Fourier Transforms and Fourier Transforms of Distributions. F. Oberhettinger. New York: Springer-Verlag, 1990. viii + 259pp. $35.00. ISBN 0-387-50630-6.

"New, revised, and enlarged version of the previously published book by the author entitled *Tabellen zur Fourier Transformation* (Springer-Verlag, 1957)." Impressive collection of functions tabulated. **(a,s)**

Tables of Integral Transforms. Ed: A. Erdelyi. Bateman Manuscript Project. California Institute of Technology. New York: McGraw-Hill, 1954. 2v. V. 1: xx + 391pp. V. 2: xvi + 451pp. o.p.

Extensive collection of integral transforms useful in many fields. Indexes of notations. **(a)**

Tables of Integrals and Other Mathematical Data. Fourth Edition. Herbert Bristol Dwight. New York: Macmillan, 1961. 336pp. $39.75. ISBN 0-02-331170-3.

Organized by type of function, includes formulas, series, integrals, and derivatives. Tables of numerical values for several functions in appendix. Some references, mostly secondary. **(a,p)**

Tables of Integrals, Series, and Products. Corrected and Enlarged Edition. I. S. Gradshtein, I. M. Ryzhik. Transl: Alan Jeffrey. New York: Academic, 1980. xlv + 1160pp. $49.95. ISBN: 0-12-294760-6.

Translated from Russian, useful work for collections in need of tables of integrals. Classified arrangement. **(a,p)**

Tables of Physical and Chemical Constants and Some Mathematical Functions. Fifteenth Edition. Ed: G. W. C. Kayne, T. H. Laby. New York: Wiley, 1986. 477pp. $52.95. ISBN 0-582-46354-8.

Contains important tables in three areas: general physics, chemistry, mathematical functions. All tabulated values are in SI (Systeme International) units. References. **(a,p)**

Tables of Representations of Simple Lie Algebras. Vol. 1. W. G. McKay, J. Patera, D. W. Rand. Montreal: Université de Montréal, Centre de Recherches Mathématiques, 1990. vi + 328pp. $46.00. ISBN 2-921120-06-2.

These tables will meet the needs of those in physics, mathematics, and chemistry. Subsequent volumes are planned. **(a)**

Tables of the SU(MN) Contains SU(M) X SU(N) Coefficients of Fractional Parentage. J. Q. Chen et al. Teaneck, NJ: World Scientific Press, 1991. viii + 447pp. $58.00. ISBN 981-02-0113-3.

Contains tables of the Clebsch-Gordan coefficients that are primarily of interest to mathematical and theoretical physicists. **(a)**

Tables of the Distribution of the Correlation Coefficient. Dolun Oksoy, Leo A. Aroian. Schenectady, NY: Union College and University, Institute of Administration and Management, 1982. xxxiii + 132 leaves. o.p.

Tables are constructed: areas found using revised Garwood formula, Hotelling's formula for ordinates, and Newton-Gregory for numerical integration. Percentage points calculated with Aitken's interpolation formula. **(a)**

Tables of the F- and Related Distributions With Algorithms. K. V. Mardia, P.J. Zemroch. London; New York: Academic Press, Harcourt Brace Jovanovich, 1978. x + 256pp. $103.00. ISBN 0-12-471140-5.

Tables of percentage points of *F, t* and *X* squared distributions given with both degrees of freedom parameters able to have fractional values. Interpolation explained. Extensive list of Algol 60 algorithms. No treatment of applications. **(a)**

Tables of the First 15000 Zeros of the Riemann Zeta Function to 28 Significant Digits, and Related Quantities. H. J. J. te Riele. Afdeling Numerieke Wiskunde [Department of Numerical Mathematics]; 67. Amsterdam: Mathematisch Centrum, 1979. 155pp. (not consecutively paged). Price not available.

Contents reflect title. **(a)**

Tablit͡sy Modifit͡sirovannykh Funkt͡siĭ Besseli͡a. I͡U. M. Rappoport. Moscow: Nauka, 1979. 337pp. o.p.

Introduction describes method of computation of tables. Tables contain real and imaginary halves of modified Bessel functions. **(a)**

Tablit͡sy Sferoidal'nyh Volnovyh Funkt͡siĭ i ikh Pervyh Proizvodnykh. Ed: V. I. Krylov et al. Minsk: Izdat. Nauka i Tekhnika, 1973-1976. *V. 1.* 443pp. *V. 2.* 418pp. o.p.

Tables of spheroidal wave functions and their first derivatives. **(a)**

Tablitsy Funkt͡siĭ Struve i Integralov ot Nikh. M. M. Agrest et al. Moscow: Nauka, 1982. 332pp. o.p.

Introduction contains tables for asymptotic expansions, infinite series formulas for Struve and modified Struve functions, and identities about finite sums of gamma functions. Appendix has computational methods for integrals of Struve functions involving exponential factors. **(a)**

Tablit͡sy Matematicheskoĭ Statistiki. Third Edition. L. N. Bol'shev, N. V. Smirnov. Moscow: Nauka, 1983. 415pp. o.p.

Bibliography and commentary by D. S. Shmerling. Collection of tables of mathematical statistics in first update since 1968. Subject index. **(a)**

Tablit͡sy Neopredelennykh Integralov ot Vysshikh Transt͡sendentnykh Funkt͡siĭ. I͡U. F. Filippov. Khar'kov: Vishcha Shkola, 1983. 108pp. o.p.

Tables of indefinite integrals of higher transcendental functions in Russian. Presents new method of calculating indefinite integrals of transcendental functions. **(a)**

Tablit͡sy Nepolnykh Integralov Lipshit͡sa-Khankeli͡a. M. M. Agrest, I͡U. M. Rappoport, M. M. Rikenglaz. Moscow: Nauka, 1982. 393pp. (loose errata). o.p.

Tables of incomplete Lipschitz-Hankel integrals. **(a)**

Tafeln der mathematischen Statistik. Third Edition. Paul Heinz Müller, Peter Neumann, Regina Storm. Leipzig: VEB Fachbuchverlag, 1977. 276pp. $11.14.

Tables useful for solution of common statistical problems. Explanations and applications of each table given explicitly. Summaries in Russian, English, French, Spanish, and Arabic. **(a)**

WRITING GUIDES

The Chicago Manual of Style. 13th Edition. University of Chicago, 1982. ix + 737pp. $37.50. ISBN 0-226-10390-0.

Long considered the standard reference tool for authors, editors, and copywriters. This edition has been revised and updated to reflect "the impact of the new technology on the entire editing and publishing process." **(a,p,s)**

Consultant's Guide to Proposal Writing: How To Satisfy Your Client and Double Your Income. Second Edition. Herman Holtz. New York: Wiley, 1990. xiv + 303pp. $24.95. ISBN 0-471-51569-8.

Proposal writing and proposal using (for marketing purposes). Subtitle indicates general orientation. Covers fundamentals of proposals, their preparation, and sale. Basics applicable to any field (e.g., design, outline, format, etc.). **(a,p)**

Handbook for Academic Authors. Revised Edition. Beth Luey. Cambridge; New York: Cambridge University Press, 1990. xiv + 273pp. $32.50 (hbk.); $12.95 (pbk.). ISBN 0-521-39494-5 (hbk.); 0-521-39646-8 (pbk.).

Primarily for book authors. Covers publisher's contracts, copy editing, editors, etc. Has good annotated bibliography on dictionaries, style manuals, grammar guides, and related works. Revised edition includes a chapter on electronic manuscripts and electronic publishing. **(a)**

Handbook of Technical Writing Practices. Ed: Stello Jordan. Ann Arbor: Books on Demand UMI, 1971. 2v. (ix + 1374pp.). $160.00/Volume. ISBN 0-317-10865-4 (V. 1); 0-317-10866-2 (V. 2). Order #: 2055514.

Part of Wiley Series on Human Communication. Comprehensive reference handbook covering products of technical writers, related functions (editing, illustrating, data processing), and management of technical writing. Also guides and references including style manuals and data sources. **(p)**

How to Write and Publish a Scientific Paper. Third Edition. Robert A. Day. New York: Oryx Press, 1988. xi + 211pp. $25.00 (hbk.); $16.95 (pbk.). ISBN 0-89774-472-1 (hbk.); 0-89774-456-X (pbk.).

General cookbook approach based on author's experience includes several short how to chapters (e.g., how to write a conference report, thesis, review paper or how to present papers orally). Useful appendices plus references and index. **(a,p)**

How to Write Mathematics. Norman E. Steenrod et al. Providence, RI: American Mathematical Society, 1973. 64pp. $15.00. ISBN 0-8218-0055-0.

Covers approaches to exposition, specifically for research monographs and textbooks. Three short essays by Steenrod, Paul Halmos, and Menchen M. Schiffer on how they individually approach writing a work. **(a)**

A Manual for Authors of Mathematical Papers. Eighth Revised Edition. Providence, RI: American Mathematical Society, 1990. 20pp. Free. 0-8218-0022-1.

Covers mechanical aspects, not intellectual content. Useful in preparation of papers in form acceptable to the AMS. **(a,s)**

Mathematics into Type; Copy Editing and Proofreading of Mathematics for Editorial Assistants and Authors. Revised Edition. Ellen Swanson. Providence, RI: American Mathematical Society, 1979; 1987. x + 90pp. $20.00. ISBN 0-8218-0053-1.

Technical details of publishing mathematics manuscripts from copy editors' perspective. Glossary, bibliography, and index. Reprinted in 1986 with corrections. **(a,p)**

The Printing of Mathematics: Aids for Authors and Editors and Rules for Compositors and Readers at the University Press, Oxford. T. W. Chaundy, P. R. Barrett, Charles Batey. Ann Arbor, MI: Books on Demand UMI, 1954. $29.80. ISBN 0-317-10261-3. Order #: 2051896.

Reprint of 1954 book published by Oxford University Press. Includes sections on mechanics of mathematical printing, recommendations to mathematical authors, and rules for the composition of mathematics at the University Press. Discusses preparation of manuscripts and how they can best be set in type. **(a)**

Technical Editor's and Secretary's Desk Guide. George Freedman, Deborah A. Freedman. New York: McGraw-Hill, 1985. xxxviii + 538pp. $39.95. ISBN: 0-07-021918-4.

General guide on preparing manuscripts that contain college level mathematics including mathematics, physics, electronics, and chemistry. Explains use and meaning of various symbols used by mathematicians. **(a,p)**

Writing Mathematics Well: A Manual for Authors. Leonard Gillman. Washington, DC: Mathematical Association of America, 1987. ix + 49pp. $6.00. ISBN 0-88385-443-0.

Compact manual covers basics of organization, presentation, and usage in preparation of manuscripts. Good outline of fundamentals. Bibliography. **(a,p)**

BIOGRAPHIES

American Men & Women of Science 1989-90: A Biographical Directory of Today's Leaders in Physical, Biological and Related Sciences. Seventeeth Edition. Ed: Basil Daley et al. New York: R. R. Bowker, 1989. 8 volumes $650.00/set. ISBN 0-8352-2568-2.

Almost 4,000 of the 125,000 entries are new with this edition. The selection criteria for inclusion are: distinguished achievement, high quality of research, or holding a highly responsible position. The information in this valuable reference work is also available online via DIALOG. **(a)**

Biograficheskiĭ Slovar' Deiateleĭ v Oblasti Matematiki. Ed: A. I. Borodin, A. S. Bugaĭ. Kiev: Radians'ka Shkola, 1979. 607pp. o.p.

Over 2,000 short biographies in Russian including illustrations, index, and bibliography. **(a)**

Biographical Dictionary of Mathematicians: Reference Biographies from the Dictionary of Scientific Biography. New York: Scribner, 1991. 4v. (vii + 2696pp.). $175.00. ISBN 0-684-19282-9 (set); 0-684-19288-8 (V. 1); 0-684-19289-6 (V. 2); 0-684-19290-X (V. 3); 0-684-19291-8 (V. 4).

The four-volume set includes the full text (including drawings and bibliographies) of entries of mathematicians in the parent set, *Dictionary of Scientific Biography*, including Supplements I and II. 1,019 articles sketch the life and work of mathematicians from all regions of the world and from antiquity to modern times. **(a,p,s)**

Biographischliterarisches Handwörterbuch der exakten Naturwissenschaften. J. C. Poggendorff. Berlin: Akademie-Verlag, 1863-.

Well known series issued in many parts containing biographical and bibliographical material on figures from natural sciences including mathematics. Original publisher through 1985 was J. A. Barth of Leipzig. **(a)**

Breakthroughs. Claire L. Parkinson. Boston, MA: G. K. Hall, 1985. xii + 576pp. $40.00. ISBN 0-8161-8706-1.

From subtitle: "A chronology of great achievements in science and mathematics, 1200-1930." Of 2,800 events presented in telegraphic style, over 700 are in mathematics. Includes list of sources, name and subject indexes. **(p)**

Lexikon Bedeutender Mathematiker. Ed: Siegfried Gottwald, Hans Joachim Ilgauds, Karl Heinz Schlote. Leipzig: Bibliographisches Institut, 1990. 504 + [3]pp.; 8pp. of plates. $22.26. ISBN 3-8171-1164-9.

In German. Short biographies of more than 1,600 mathematicians worldwide from ancient times to the present. **(a)**

Matematiki Mekhaniki: Biograficheskiĭ Spravochnik. Alekseĭ Nikolaevich Bogoli͡ubov. Kiev: Naukova Dumka, 1983. 637pp. o.p.

In Cyrillic, it contains brief biographies of 1,500 scholars contributing to mathematics or mechanics. Photographs, chronology of major events in both fields, bibliography, and index to persons are included. **(a)**

The Mathematical Practitioners of Hanoverian England, 1714-1840. Eva Germaine Rimington Taylor. Irvine, CA: Reprint Services, 1989; 1966. xv + 503pp. $59.00. ISBN 0-7812-0636-X.

Reprint of 1966 book published by Cambridge University Press. Sequel to her earlier work (see below) with similar arrangement. All mathematicians mentioned in biographical section are in one large alphabetical index issued separately and published by Cambridge University Press in 1980. **(a)**

Math Equals: Biographies of Women Mathematicians + Related Activities. Teri Perl. Menlo Park, CA: Addison-Wesley, 1978. v. + 250pp. o.p. ISBN 0-20105709-3.

A bool writtten in order to help change sex role stereotypes and increase awareness of career choices. Short bibliography. **(a,p)**

The Mathematical Practitioners of Tudor and Stuart England. Eva Germaine Rimington Taylor. Irvine, CA: Reprint Services, 1989; 1954. xi + 442pp. $59.00. ISBN 0-685-10499-0.

Originally published in 1954 by Cambridge. Includes narrative accounts, biographies of 582 mathematicians, and annotated bibliography of period's mathematical works. List of secondary references. **(a)**

Mathematics: Who's Who. Ed: B. K. Dass. Delhi: Analytic, 1984. ii + 187pp. $45.00.

Collection of some 350 one paragraph biographies, most with a photograph, of "active" mathematicians worldwide. **(a,s)**

Men of Mathematics. Eric Temple Bell. New York: Simon & Schuster, 1986. xvii + 590pp. $13.95. ISBN 0-671-62818-6.

Accounts of 34 historical figures instrumental in evolution of mathematics. Persons were selected because of importance of their work to modern mathematics and quality of their personal character. **(a,p)**

The Newton Handbook. Derek Gjertsen. London; New York: Routledge & Kegan Paul, 1986. xiv + 665pp. $59.95. ISBN: 0-7102-0279-2.

Alphabetical mini encyclopedia of Newtonia with digests and bibliographic details of published and unpublished manuscripts. **(a)**

Thinkers of the Twentieth Century. Second Edition. Ed: Roland Turner. Chicago: St. James Press, 1987. viii + 977pp. $75.00. ISBN 0-912289-83-X.

Contains 462 entries of a page or more covering important figures who worked primarily in the twentieth century. The entries include curriculum vitae, bibliographies, lists of critical writings, and a signed review. However, there is sloppy proofreading. **(p)**

Vydaiushchiesia Matematiki: Biograficheskiĭ Slovar'-Spravochnik. A. I. Borodin, A. S. Bugaĭ Kiev: Radians'ka Shkola, 1987. 653pp. o.p.

Short biographies. International coverage, but with emphasis on Soviet mathematicians. Includes bibliographical references and indexes. **(a)**

Women in Science: Antiquity through the Nineteenth Century: A Biographical Dictionary with Annotated Bibliography. Marilyn Bailey Ogilvie. Cambridge, MA: MIT Pr., 1988. xiii + 254pp. $32.50 (hbk.); $12.50 (pbk.). ISBN 0-262-15031-X (hbk.); 0-262-65038-X (pbk.).

The introduction is a historical essay about science and women. Each biographical sketch includes lists of major works by the subject herself and works about her. A good source book for courses on women and science. **(p)**

Women of Mathematics: A Biobibliographic Sourcebook. Ed: Louise S. Grinstein, Paul J. Campbell. Westport, CT: Greenwood, 1987. xx + 292pp. $65.00. ISBN 0-313-24849-4.

A well done work profiling 43 women born before 1925 who made contributions of note to field of mathematics. Each entry includes a biography, summary of work, and bibliography of works by and about the individual. **(a,p)**

DIRECTORIES

Assistantships and Graduate Fellowships in the Mathematical Sciences. Providence, RI: American Mathematical Society, 1991. vii + 142pp. $15.00. ISBN 0-8218-0171-6. ISSN 1040-7650.

A source of information on graduate programs, this includes names and partial addresses (college or university, city, state) of department chairpersons and useful facts about graduate programs, including stipends, fees, and service required for assistantships. Some programs are described in greater detail in boxed advertisements. **(a)**

Combined Membership List of the American Mathematical Society, Mathematical Association of America, Society for Industrial and Applied Mathematics. Providence, RI: American Mathematical Society. Annual. 1956-. $36.00 (Non-members); $29.00 (Members). ISSN 0569-6461.

Provides alphabetical and geographical listings for individual members plus listings for institutions and other categories of members. **(a,s)**

Directory of Members. American Statistical Association. Alexandria: American Statistical Association, 1991. xlviii + 307pp. $125.00.

Includes constitutions, by-laws, charters, past and present officers, and membership listings for the four groups represented: the American Statistical Association, the Biometric Society: Eastern & Western North American Regions, the Institute of Mathematical Statistics, and the Statistical Society of Canada. **(a)**

Directory of Statistical Microcomputer Software, 1988 Edition. Wayne A. Woodward et al. New York: Marcel Dekker, 1988. iii + 744pp. $79.75. ISBN 0-8247-7846-4.

Noncritical descriptions of 200 plus statistical programs and packages. Entries include statistical features and configurations supported, documentation available, users, cost, etc. Tables show which programs perform which tasks and if they will run on certain systems/computers. **(a,s)**

Directory of Statisticians. Washington, DC: American Statistical Association, 1961-. Irregular. $95.00. ISSN 0278-405X. Combined memberships of ASA, Biometric Society, Institute of Mathematical Statistics. Alphabetically by name. Includes, address, position, education plus geographical listing. Formerly *Statisticians and Others in Allied Professions.* **(a,s)**

Directory of Women in the Mathematical Sciences. Providence, RI: American Mathematical Society, 1981. o.p. ISSN 0732-5967. Earlier directories, 1973/74-1976/77, called *Directory of Women Mathematicians.*

An incomplete directory of women mathematicians (i.e., many women are not listed as they did not return the questionnaire), compiled by the Committee on Women in Mathematics of the AMS, MAA, NCTM, and SIAM. Gives name, mailing address, highest academic degree, title of position, institution where employed, and brief list of field(s) of interest and publications. Historical interest. **(a,p)**

Employment Information in the Mathematical Sciences. Providence RI: American Mathematical Society, 1978-. 6/Year. $135.00. ISSN 0163-3287. Published for AMS-MAA-SIAM Committee on Employment Opportunities.

Employment register, it provides information on open positions alphabetically by state/province with index by name of institution. Preponderance of academic positions. Continues *Employment Information for Mathematicians.* **(a)**

The Gourman Report: A Rating of Graduate and Professional Programs in American and International Universities. Jack Gourman. Los Angeles: National Education Standards. iv + 119pp. o.p. ISBN 0-918192-03-X.

While out of date, this is one of the few sources for information on comparative rankings of American graduate programs. Mathematics is one of 47 American programs rated. Includes discussion of how rankings were obtained. A companion volume, *The Gourman Report: A Rating of Undergraduate Programs in American and International Universities* (1980) is not listed. **(a)**

Graduate Programs in the Physical Sciences and Mathematics. Ed: Theresa C. Moore, Raymond D. Sacchetti. Princeton, NJ: Peterson's Guides, 1991. Annual. ix + 728pp. $28.95. ISBN: 1-56079-024-5 (paper). ISSN 0894-9379.

Updated annually, the Guide provides a general overview of graduate education, a dictionary of institutions having physical science and mathematics programs, and a listing of academic and professional programs in the physical sciences and mathematics. **(a,p)**

Guidebook to Departments in the Mathematical Sciences in the United States and Canada. Sixth Edition. Washington, DC: Mathematical Association of America, 1975. 99pp. o.p.

In 3 parts, this directory provides summary information on location, size, staff, library facilities, course offerings, and special features for four year colleges and PhD-granting universities. U.S. departments arranged alphabetically by state; Canada by institution. Still useful, even though very out-of-date. **(a,p)**

Mathematical Sciences Professional Directory. Providence, RI: American Mathematical Society, 1983-. Annual. $35.00. ISSN 0737-4356.

Covers structure, boards, and committees of AMS and associated professional associations with addresses for organizational officers. Academic institutions listing is by state giving key personnel and telephone numbers. Provides addresses of publishers and editors of mathematical journals. Index of colleges and universities. **(a,p,s)**

The National Faculty Directory. Detroit, MI: Gale Research Co., 1970-. Annual. $784.00, including supplements. ISSN 0077-4472.

Alphabetical list of nearly 600,000 teaching faculty at the colleges and universities of North America. An excellent source for locating the mathematician not found in the *Combined Membership List*... (see above). **(a)**

Prizes in Mathematics. Margie L. Voelker. Minneapolis, MN: University of Minnesota, 1986. 24 leaves. (Mathematics Report (University of Minnesota, School of Mathematics) 85-139). o.p.

List of awards and their winners. Individual awards entries include sponsor, eligibility, field, type of award, frequency, information source, recipients. Separate index by winners noting award and year. **(a)**

Statistical Services Directory. Second Edition. Detroit: Gale, 1984. 461pp. $200.00. ISBN 0-8103-0668-9.

Subtitled "A guide to organizations, corporations, professional and trade associations, research centers, universities, publishers, foundations, and goverment agencies that provide statistical services." Also information on statistical serial publications. **(a,p)**

Translations & Translators. Ed: Stefan Congrat-Butlar. New York: R. R. Bowker Co., 1979. xi + 241 pp. o.p. ISBN: 0-8352-1158-4.

Contains an interesting overview of the history of translation as a profession. Of more practical use, however, is a list of translators. **(a)**

World Directory of Historians of Mathematics. Second Edition. Ed: Kenneth O. May, Laura Roebuck. Toronto, ON: Historia Mathematica, 1972-. Irregular. Price not available. ISSN 0315-1700.

Alphabetical listing of scholars working in history of mathematics. Entries include names, mailing addresses, and fields of interest. Indexed by subjects and countries. **(a)**

World Directory of Mathematicians. Ed: George Daniel Mostow. Helsinki: International Mathematical Union (imprint varies), 1958-. Quadrennial. (1990 is latest edition). Available from American Mathematical Society, Providence, RI) $40.00. ISSN 0512-2740.

Includes addresses and may give title and affiliation. Geographical listing gives name only; also list of mathematical organizations. Available through AMS. **(a,p,s)**

The World of Learning. London: Europa; distributed by Taylor and Francis, 1947-. Annual. $310.00. ISSN 0084-2117.

The most comprehensive and timely international directory of colleges and universities and other learned institutions. Latest edition has more than 26,000 institutions arranged alphabetically by country. Also listings for 150,000 key academic staff and professors and more than 400 international organizations concerned with education. **(a,p,s)**

Chapter 7

BOOKS FOR A CONTEMPORARY MATHEMATICS COLLECTION

Nancy D. Anderson

CONTENTS

CRITERIA FOR SELECTION

The monographs selected for this list were culled from the literature (primarily *The American Mathematical Monthly* and *Mathematical Reviews*), recommendations from the faculty of the Department of Mathematics, University of Illinois at Urbana-Champaign, and the author's 20 years of experience as a mathematics librarian. Place of publication is given as the current address of the publisher in order to facilitate acquisition. Titles were included only if they were in print at the time this book was completed. Appropriate titles by type of library are indicated by (**a**) academic, (**p**) public, or (**s**) special. Coding was done in collaboration with the chapter authors.

This list is made up of books that are relatively easy to acquire (i.e, they are in print and generally English-language) and, therefore, serves as a basic core collection. Each mathematics library should, however, reflect the needs of its users. Thus, with the help of this list, the librarian will be free to spend time and effort in acquiring those out-of-print or otherwise difficult to identify/obtain materials that are recommended by the mathematicians who will be using the collection.

Another excellent source for collection development is *Library Recommendations for Undergraduate Mathematics* (Mathematical Association of America, 1992). This list was developed by mathematicians for libraries with the emphasis on undergraduate mathematics. Books were selected without regard to availability.

GENERAL

Actions for Renewing U.S. Mathematical Sciences Departments. Washington, DC: National Research Council, 1990. vii + 29pp. (Available only with *Moving Beyond Myths: Revitalizing Undergraduate Mathematics* [see p. 153]).

Advice to chairs of university mathematics departments on how these can "become active participants" in the effort to renew U.S. mathematics. (**a**)

Africa Counts: Number and Pattern in African Culture. Claudia Zaslavsky. Chicago: Chicago Review Press, 1979. x + 328pp. $11.95 paperbound. ISBN 0-88208-104-7.

A beautiful and passionate book, full of pictures and figures illustrating the development of mathematics among the tribes of sub-Saharan Africa: numeration systems, cultural significance of numbers, mathematical games, geometric form and pattern in architecture and art. (**p**)

Archimedes' Revenge: The Joys and Perils of Mathematics. Paul Hoffman. New York: Fawcett, 1989. viii + 285pp. $4.95. ISBN 0-449-21750-7.

A vivid account of selected challenges in modern mathematics as told through the actions and words of individuals working on these problems. (**p**)

An Assessment of Research-Doctorate Programs in the United States: Mathematical & Physical Sciences. Ed: Lyle V. Jones, Gardner Lindzey, Porter E. Coggeshall. Washington, DC: National Academy Press, 1982. xii + 243pp. $10.50. ISBN 0-309-03299-7.

Report of a survey on characteristics and quality in Ph.D. programs. Data from 16 measures are presented in raw, standardized, and graphical forms. (**a**)

The Beauty of Doing Mathematics: Three Public Dialogues. Serge Lang. New York: Springer-Verlag, 1985. x + 127pp. $25.00. ISBN 0-387-96149-6.

English translation of three highly interactive public lectures delivered by Lang at the Palais de la Decouverte in Paris. **(p)**

Beyond Numeracy: Ruminations of a Numbers Man. John Allen Paulos. New York: Alfred A. Knopf, 1991. xiii + 285pp. $21.50. ISBN 0-394-58640-9.

Seventy brief topical essays covering mathematical vocabulary from algebra to Zeno plus a list of the "top 40" mathematicians from antiquity to the present. **(a,p)**

Boojums All the Way Through: Communicating Science in a Prosaic Age. N. David Mermin. New York: Cambridge University Press, 1990. xxi + 308pp. $16.95. ISBN 0-52138-880-5.

Collection of warm, witty, and lucid essays for lay people about mathematics and physics, mostly reprinted from *Physics Today*. **(p)**

Categories for the Working Mathematician. Fourth Corrected Printing. S. Mac Lane. Graduate Texts in Mathematics;5. New York: Springer-Verlag, 1988. ix + 262pp. $29.80. ISBN 0-387-90035-7.

Reprint of the 1971 original, an enjoyable and convincing book which shows that categories provide a convenient conceptual language that can be used to illuminate and clarify a wide variety of fields of mathematical research. **(a,p)**

Chairing the Mathematical Sciences Department of the 1990s. Washington, DC: National Research Council, 1990. ix + 97pp. Complimentary. No ISBN.

Presentations, with audience comments, on funding, curricula, dealing with administrations, and statistics from 1989 meeting of mathematics department chairs sponsored by the Board of Mathematical Sciences of the National Research Council. **(a)**

A Challenge of Numbers: People in the Mathematical Sciences. Bernard L. Madison, Therese A. Hart. Washington, DC: National Academy Press, 1990. xv + 119pp. $9.95. ISBN 0-30904-190-2.

First report of the National Academy of Sciences Project MS 2000 containing a detailed analysis of human resources in the mathematical sciences. **(a,p)**

The Complete Problem Solver. John R. Hayes. Franklin Institute Press, 1981. v + 255pp. $59.95. ISBN 0-89168-028-4.

Discussion of skills for problem solving and the psychology of problem solving. References. **(p)**

Congruence of Sets, and Other Monographs. New York: Chelsea, 1967. 115 + 74 + 148 + 104pp. $17.95. ISBN 0-8284-0209-4.

Four classic essays: W. Sierpinski's *On the Congruence of Sets and Their Equivalence by Finite Decomposition*, F. Klein's *The Mathematical Theory of the Top*, C. Runge's *Graphical Methods*, and L. E. Dickson's *Introduction to the Theory of Algebraic Equations*. **(a,p)**

Conquering Mathematics: From Arithmetic to Calculus. Lloyd Motz, Jefferson Hane Weaver. New York: Plenum Press, 1991. xiv + 305pp. $23.50. ISBN 0-306-43768-6.

An informal discursive survey of basic mathematics for the literate reader who may have forgotten but can nevertheless comprehend logically and linguistically sophisticated exposition. **(p)**

Descartes' Dream: The World According to Mathematics. Philip J. Davis, Reuben Hersh. San Diego: Harcourt Brace Jovanovich, 1986. xviii + 321pp. $12.95. ISBN 0-39-543154-9.

A collection of essays which purport to be a philosophy of applied mathematics. This is not a coherent exposition, but rather a compelling collage with a central theme of caution that the limitless application of mathematics—the fulfillment of Descartes' dream—too often quenches human spirit. **(a,p)**

Developing Mathematics in Third World Countries. Ed: M. E. A. el Tom. North-Holland Mathematics Studies;33. New York: Elsevier, 1979. xi + 207pp. $72.00. ISBN 0-444-85260-3.

Sixteen invited papers and a final report with recommendations form the proceedings of an international conference held in Khartoum, 1978. This document reports on the first opportunity third world mathematicians had to exchange ideas on the development of mathematics. **(p)**

Discovering Mathematics: The Art of Investigation. A. Gardiner. New York: Oxford University Press, 1987. xiv + 206pp. $19.95;$39.95. ISBN 0-19-853265-2;0-19-853282-2.

This unique book is designed to give students experience in mathematical discovery, using four short and two extended investigations. **(p)**

Does God Play Dice? The Mathematics of Chaos. Ian Stewart. New York: Penguin Books, 1989. 317pp. $10.95. ISBN 0-14-012501-9.

Popular, witty, yet mathematically vigorous treatment of chaos theory. **(a,p,s)**

The Emperor's New Mind: Concerning Computers, Minds, and the Laws of Physics. Roger Penrose. New York: Oxford University Press, 1989. xiii + 466pp. $30.00. ISBN 0-19-851973-7.

Innovative and unconventional work designed to show that consciousness is intrinsically nonalgorithmic. From his rich experience as a mathematician, Penrose deduces that mathematicians' judgment of mathematical truths is also non-algorithmic. **(p,s)**

Ethnomathematics: A Multicultural View of Mathematical Ideas. Marcia Ascher. Pacific Grove, CA: Brooks-Cole, 1991. ix + 203pp. $38.95. ISBN 0-534-14880-8.

A thoroughly-researched and well-documented work which intends to show that while "mathematics" is a distinctly Western concept, "mathematical ideas" are universal and culturally dependent. **(a,p)**

Fifty Per Cent Proof: An Anthology of Mathematics Humour. Keith Selkirk, William Wynne Willson. Cheltenham, England: Mathematical Association and S. Thorne, 1989. 87pp. ISBN 0-7487-0166-4.

Stories, jokes, and cartoons show that mathematics can be very funny. **(p)**

Flatland: A Romance of Many Dimensions. Edwin Abbott Abbott. Princeton, NJ: Princeton University Press, 1991. 103pp. $7.95. ISBN 0-691-02525-8.

A new edition of the 1884 classic with Tom Banchoff's introduction highlighting issues of interest today. **(a,p)**

The Fractal Geometry of Nature. Benoit B. Mandelbrot. New York: W. H. Freeman, 1982. 460pp. $39.95. ISBN 0-7167-1186-9.

An extensive revision of his earlier work, *Fractals, Form, Chance and Dimension*, with many color graphics and applications to a wide range of scientific subjects. **(a,p,s)**

Fractals, Chaos, Power Laws: Minutes from an Infinite Paradise. Manfred Schroeder. New York: W. H. Freeman, 1990. xviii + 429pp. $49.00. ISBN 0-7167-2136-8.

Clearly written and inviting exposition of the worldly and mathematical manifestations of self-similarity. For the literate but not highly scientific audience. **(p)**

From Cardinals to Chaos: Reflections on the Life and Legacy of Stanislaw Ulam. Ed: Necia Grant Cooper. New York: Cambridge University Press, 1989. 319pp. $75.00; $24.95. ISBN 0-521-36494-9; 0-521-36734-4.

A beautifully produced tribute to Ulam and his wide-ranging contributions to mathematics, physics, and biology. **(p)**

Fundamentals of Mathematics. Ed: H. Behnke et al. Transl: S. H. Gould. Cambridge, MA: MIT Press, 1974. $75.00 set. ISBN 0-262-02048-V. *Vol.I: Foundations of Mathematics. The Real Number System and Algebra*. x + 549pp. $32.50. ISBN 0-262-02048-3. *Vol. II: Geometry*. xi + 685pp. $32.50. ISBN 0-262-02069-6. *Vol. III: Analysis*. xiii + 541pp. $32.50. ISBN 0-262-02049-1.

Translation of Second (1962) edition of *Grundzüge der Mathematik*. This classic provides extensive coverage and is highly recommended. **(a,p,s)**

Gödel, Escher, Bach: An Eternal Golden Braid. Douglas R. Hofstadter. New York: Basic Books, 1979. xxi + 777pp. $34.95. ISBN 0-465-02685-0.

A brilliant and creative synthesis of the diverse mirrors with which art and science reflect reality. **(a,p)**

Infinity and the Mind: The Science and Philosophy of the Infinite. Rudy Rucker. Boston: Birkhäuser, 1982. x + 342pp. $23.95. ISBN 0-817-63034-1.

A wide-ranging discourse on the infinite, mixing traditional mathematical aspects with philosophical roots and pseudo-science. Exposition is both popular and technical. **(a,p)**

Innumeracy: Mathematical Illiteracy and Its Consequences. John Allen Paulos. New York: Hill & Wang, 1989. 135pp. $16.95. ISBN 0-8090-7447-8.

An outspoken expose of public obliviousness to quantitative meanings. The author links innumeracy with the spread of pseudoscience. **(a,p)**

Islands of Truth: A Mathematical Mystery Cruise. Ivars Peterson. New York: W. H. Freeman, 1990. xvii + 325pp. $19.95. ISBN 0-7167-2113-9.

Innovative and pleasure-filled guide reveals to the mathematical tourist many mysteries of modern mathematics. **(p)**

It Seems I Am a Jew: A Samizdat Essay. Grigori Freiman. Transl: Melvyn B. Nathanson. Carbondale, IL: Southern Illinois University Press, 1980. xvi + 96pp. $9.95. ISBN 0-8093-0962-9.

A moving, detailed, personal account of the purge of Jews from contemporary Russian mathematics. **(p)**

Journey Through Genius: The Great Theorems of Mathematics. William Dunham. New York: Wiley, 1990. xiii + 300pp. $10.95. ISBN 0-471-50030-5.

Twelve mathematical masterpieces from Hippocrates' quadrature of the lune to Cantor's theorem on non-denumerability, each set in historical context and explained in modern language. Excellent source material for popular audiences. **(a,p)**

Judgment Under Uncertainty: Heuristics and Biases. Ed: Daniel Kahneman, Paul Slovic, Amos Tversky. New York: Cambridge University Press, 1982. xiii + 555pp. $44.50; $14.95. ISBN 0-521-24064-6; 0-521-28414-7.

Interesting collection of articles focusing on psychological studies of how people estimate probabilities and the types of biases produced by these heuristic methods. Good set of references. **(a,p)**

Kahn on Codes: Secrets of the New Cryptology. David Kahn. New York: Macmillan, 1983. viii + 343pp. $19.95. ISBN 0-02-560640-9.

A collection of essays on codes on topics ranging from the historical to the contemporary. **(a,p)**

Das Kontinuum und andere Monographien. Hermann Weyl. New York: Chelsea, 1973. 1 v. (various pagings). $18.95. ISBN 0-8284-0134-9.

Reprint of 4 works: H. Weyl's *Das Kontinuum*, H. Weyl's *Mathematische Analyse des Raumproblems*, E. Landau's *Darstellung und Begruendung einiger neuerer Ergebnisse der Funktionentheorie*, and B. Riemann's *Ueber die Hypothesen, welche der Geometrie zu Grunde Liegen.* **(a)**

Learning and Doing Mathematics. John Healey Mason. New York: Macmillan, 1988. vii + 82pp. $10.00. ISBN 0-333-44942-8.

Text is based on a unit of the Open University Foundations Course in mathematics. **(p)**

Littlewood's Miscellany. Ed: Bela Bollobas. New York: Cambridge University Press, 1986. 200pp. $32.50;$11.95. ISBN 0-521-33058-0; 0-521-33702-X.

An edited reprinting of Littlewood's *A Mathematician's Miscellany* and *The Mathematician's Art of Work*. **(p)**

M. C. Escher: Art and Science. Ed: H. S. M. Coxeter et al. New York: Elsevier Science, 1986. xiii + 402pp. $50.00. ISBN 0-444-70011-0.

Thirty-seven papers form the proceedings of a 1985 international interdisciplinary conference. Papers deal with the links between Escher's work and symmetry, visual perception, geometry, etc. **(a,p)**

The Mathematical Experience. Philip J. Davis, Reuben Hersh. Boston: Birkhäuser, 1980. xix + 440pp. $29.95. ISBN 3-8176-3018-X.

Must reading: this is a marvelous montage of history, reflection, commentary and philosophy that reveals the roots and importance of contemporary mathematics to the general reader. **(p)**

Mathematical Gems III. Ross Honsberger. Dolciani Mathematical Expositions;9. Washington, DC: Mathematical Association of America, 1985. 250pp. $28.00. ISBN 0-88385-313-2.

Eighteen vignettes of number theory, combinatorics, and geometry which will delight anyone who has studied college mathematics. **(p)**

Mathematical Impressions. Anatoliĭ T. Fomenko. Providence, RI: American Mathematical Society, 1990. 184pp. $45.00. ISBN 0-8218-0162-7.

A collection of nearly 80 starkly beautiful drawings and paintings conveying Fomenko's romantic visions. Mathematical theorems and human emotions serve as inspiration. **(p)**

Mathematical Models from the Collections of Universities and Museums. Commentary. Ed: Gerd Fischer. Braunschweig, Germany: Friedr. Vieweg, 1986. viii + 83pp. $92.00. ISBN 3-528-08991-1.

A boxed set of 2 anniversary volumes, celebrating 200 years of Vieweg-Verlag, containing 132 photos and commentary on the collections of models of mathematical surfaces created in the latter part of the 19th century in the leading European research institutes. Models are grouped by mathematical type. **(a,p)**

Mathematical People: Profiles and Interviews. Ed: Donald J. Albers, G. L. Alexanderson. Boston: Birkhäuser, 1986. xvi + 372pp. $12.95. ISBN 0-8092-4976-6.

Twenty-five refreshingly candid portraits of contemporary mathematicians, from which one can gain insight into the creative process of mathematics. **(p)**

Mathematical Research Today and Tomorrow: Viewpoints of Seven Fields Medalists. Ed: C. Casacuberta, M. Castellet. Lecture Notes in Mathematics;1525. New York: Springer-Verlag, 1991. 112pp. $23.00. ISBN 0-387-56011-4.

The text of seven invited lectures by Fields Medalists given in June 1991 at the Institut d'Estudis Catalans' Symposium on the Current State and Prospects of Mathematics in Barcelona. Includes bibliographic references. **(a)**

Mathematical Sciences: Some Research Trends. Washington, DC: National Academy Press, 1988. viii + 126pp. Complimentary. No ISBN.

A synopsis of selected highlights on research in core mathematics, applied mathematics, and in statistical sciences. Three overviews are supplemented by eight in-depth essays ranging from soliton theory to resampling methods. **(a)**

Mathematical Snapshots. Third American Edition, Revised and Enlarged. H. Steinhaus. New York: Oxford University Press, 1983. 311pp. $8.95. ISBN 0-19-503267-5.

Reprint of 1969 Revised third edition of 1938 classic uses visual reasoning to "stretch the imagination without unduly straining the mind." **(p)**

Mathematical Time Exposures. I. J. Schoenberg. Washington, DC: Mathematical Association of America, 1982. ix + 270pp. $18.00; $30.00. ISBN 0-88385-438-4.

Eighteen chapters on convex sets, skew polygons, billiard ball paths inside a cube, finite Fourier series, etc. Not all chapters use calculus. **(p)**

The Mathematical Tourist: Snapshots of Modern Mathematics. Ivars Peterson. New York: W. H. Freeman, 1988. xv + 240pp. $17.95. ISBN 0-7167-1953-3.

A superb example of popular mathematical writing. The author has adapted his reports for *Science News* from the last 6 years to give insight into modern mathematics. **(p)**

Mathematical Writing. Donald E. Knuth, Tracy Larrabee, Paul M. Roberts. MAA Notes;14. Washington, DC: Mathematical Association of America, 1989. 115pp. $12.50. ISBN 0-88385-063-X.

Candid lecture notes full of good advice on mathematical writing, literate programming, publication, refereeing, etc. **(a)**

Mathematicians in Academia: 1975-2000. Charlotte V. Kuh, Roy Radner. Washington, DC: Conference Board of the Mathematical Sciences, 1980. v + 109pp. $3.00.

A detailed demographic and econometric projection of mathematics faculty in four-year colleges and universities. Caveats and appendices allow skeptics to analyze the model and its assumptions as well as its conclusions. Interesting reading. **(a)**

Mathematics, a Human Endeavor: A Book for Those who Think They Don't Like the Subject. Second Edition. Harold R. Jacobs. New York: W. H. Freeman, 1982. xiii + 649pp. $23.95. ISBN 0-7167-1326-8.

Careful, thorough revision of 1971 classic. **(p)**

Mathematics and Humor. John Allen Paulos. Chicago: University of Chicago Press, 1982. 116pp. $10.95. ISBN 0-226-65025-1.

A brief but serious inquiry into common elements in mathematics and humor. For example, both are forms of intellectual play that depend on distinct rules. The author uses self reference paradoxes and catastrophe theory as mathematical paradigms of various types of humor. **(p)**

Mathematics and Science. Ed: Ronald E. Mickens. Teaneck, NJ: World Scientific, 1990. viii + 342pp. $80.00. ISBN 981-02-0233-4.

"The purpose of the book is to present the opinions of a number of distinguished individuals who have given careful thought as to why mathematics is so "unreasonably effective" when applied to the analysis of the natural sciences."—Preface. **(a,p)**

Mathematics and the Physical World. Morris Kline. New York: Dover, 1981. 482pp. $8.95. ISBN 0-486-24104-1.

Corrected reprint of 1959 original, a series of surveys of various topics of applied mathematics. **(p)**

Mathematics and the Unexpected. Ivar Ekeland. Chicago: University of Chicago Press, 1988. xiii + 146pp. $19.95. ISBN 0-226-19989-4.

Translation of a popular French monograph explaining to the layman how deterministic mechanics can lead to random motion and relating these two to ancient world views reflected by Homer. **(p)**

Mathematics as a Cultural System. Raymond L. Wilder. Elmsford, NY: Pergamon Press, 1981. xii + 182pp. $34.00. ISBN 0-08-025796-8.

The author has been writing on the cultural aspects of mathematics for over 30 years and uses his wide knowledge of the historical development of mathematics to illustrate his ideas. This book is more overtly anthropological and less historical in nature than some of his earlier work. **(p)**

Mathematics in Western Culture. Morris Kline. New York: Oxford University Press, 1976. x + 484pp. $14.95. ISBN 0-19-500714-X.

Reprint of 1953 original which discusses the cultural significance of mathematics in the development of Western civilization. **(p)**

Mathematics: Its Content, Methods, and Meaning. Ed: A. D. Aleksandrov, A. N. Kolmogorov, M. A. Lavrent'ev. Transl: S. H. Gould, T. Bartha, K. Hirsch. Cambridge, MA: MIT Press, 1969. $41.95 paperbound set. ISBN 0-262-51014-6. *Volume One.* xi + 376pp. *Volume Two.* xi + 394pp. *Volume Three.* xi + 372pp.

Paperback edition of 1963 MIT reprint of 1962-1963 AMS translation of 1956 Russian original. This is one of the best surveys of mathematics ever written even though it may be a little old fashioned. **(a,p)**

Mathematics, Its Power and Utility. Karl J. Smith. Pacific Grove, CA: Brooks-Cole, 1983. xviii + 558pp. $41.95. ISBN 0-534-01190-X.

Emphasis on problem formulation and solving through development in identifiable situations. Extensive bibliography on math anxiety. **(p)**

Mathematics: People, Problems, Results. Ed: Douglas M. Campbell, John C. Higgins. Belmont, CA: Wadsworth, 1984. $27.95; $39.95set. ISBN 0-534-22884-5; 0-534-02879-9. *Volume I.* xvi + 304pp. *Volume II.* iv + 275pp. *Volume III.* iv + 292pp.

Splendid collection of 90 expository articles, mostly by distinguished contemporary mathematicians, on all aspects of mathematics. All selections are reprints. Lacks index. **(p)**

Mathematics: Queen and Servant of Science. E. T. Bell. Washington, DC: Mathematical Association of America, 1987. xxiv + 437pp. $15.95. ISBN 0-88385-447-3.

Reprint of Bell's 1951 classic, a meandering tour of traditional mathematics. **(a,p)**

Mathematics: The Loss of Certainty. Morris Kline. New York: Oxford University Press, 1980. 366pp. $9.95. ISBN 0-19-503-0850.

A popular historical account of the "rise and decline of the majesty of mathematics." Kline's literate and lucid tale moves beyond the "loss of truth" to suggest two issues for a new foundation of mathematics. **(p)**

Mathematics: The Man-Made Universe; An Introduction to the Spirit of Mathematics. Third Edition. Sherman K. Stein. New York: W. H. Freeman, 1976. xvi + 415pp. $28.95. ISBN 0-7167-0465-X.

Latest revision of a widely used and well reviewed book. **(p)**

Mathematics Tomorrow. Ed: Lynn Arthur Steen. New York: Springer-Verlag, 1981. vi + 282pp. $29.80. ISBN 0-387-90564-2.

This collection of essays by all kinds of mathematicians provides a basis for discussion of the near crisis in mathematical and scientific education. **(p)**

Matters Mathematical. Second Edition. I. N. Herstein, I. Kaplansky. New York: Chelsea, 1978. viii + 246pp. $15.95. ISBN 0-8284-0300-7.

Reprint, with minor additions and corrections, of 1974 original: a series of lectures for nonmajors in mathematics. **(p)**

The Merging of Disciplines: New Directions in Pure, Applied, and Computational Mathematics. Ed: Richard E. Ewing, Kenneth I. Gross, Clyde F. Martin. New York: Springer-Verlag, 1986. xvi + 214pp. $32.50. ISBN 0-387-96414-2.

Proceedings of the 1985 symposium honoring Gail Young on the occasion of his 70th birthday. The result is a spectacular collection of thoughtful papers on all aspects of mathematics. **(a)**

Moving Beyond Myths: Revitalizing Undergraduate Mathematics. Washington, DC: National Academy Press, 1991. ix + 65pp. $7.95. ISBN 0-309-04489-8. (Comes with *Actions for Renewing U. S. Mathematical Sciences Departments* [see p. 144]).

The long-awaited summary analysis and recommendations from the National Research Council's Committee on Mathematical Sciences in the Year 2000. **(a)**

Number, the Language of Science. Fourth Edition, Revised and Augmented. Tobias Dantzig. New York: Free Press, 1967. ix + 340pp. $14.95. ISBN 0-02-906990-4.
Reprint of 1954 original in which the author aimed to restore to mathematics its cultural content and to present the evolution of number as the profoundly human story which it is. **(a,p)**

A Panorama of Pure Mathematics, as Seen by N. Bourbaki. Jean Dieudonné. Pure and Applied Mathematics;97. San Diego, CA: Academic Press, 1982. x + 289pp. $53.00. ISBN 0-12-215560-2.
A remarkable survey of those areas of mathematics which have been exposed in the 30+ years of the Séminaire Bourbaki. **(a)**

Papers in Mathematics. Ed: Paul R. Meyer. Annals of the New York Academy of Sciences;321. New York: New York Academy of Sciences, 1979. 101pp. $22.00. ISBN 0-89766-025-0.
Diverse collection of expository papers, addressed to mathematicians who are not specialists in the fields of the papers. **(a)**

Paradoxes of Gambling Behaviour. Willem Albert Wagenaar. Hillsdale, NJ: Lawrence Erlbaum Associates, 1988. vii + 126pp. $34.95. ISBN 0-86377-080-0.
Psychological and statistical investigation based on extensive research with roulette, blackjack, and lottery players. **(p)**

Perspectives in Mathematics. Ed: W. Jager, J. Moser, R. Remmert. Boston: Birkhäuser, 1984. 587pp. $114.80. ISBN 3-764316-24-1.
Twenty-two papers outlining recent developments in mathematics, prepared in honor of the 40th anniversary of the Mathematics Research Institute in Oberwolfach. Expensive. **(a)**

Powers of Ten. Philip and Phyllis Morrison. New York: W. H. Freeman, 1982. 159pp. $32.95. ISBN 0-7167-1409-4.
Stunning recreation on paper of the famous short film "Powers of Ten:" each right-hand page contains a view of a picnic in Chicago at different distances, while the left-hand page offers commentary and other illustrations at each scale. Opening essay on matters of size, concluding commentary, and references. **(p)**

The Problems of Mathematics. Ian Stewart. New York: Oxford University Press, 1987. 257pp. $24.95;$14.95. ISBN 0-19-219201-9; 0-19-289182-0.
A lively, non-technical portrait of important themes in mathematics, from primes to computability, each traced through history. **(p)**

Proceedings of the International Congress of Mathematicians, August 16-24, 1983, Warszawa. Ed: Czeslaw Olech, Zbigniew Ciesielinski. New York: Elsevier, 1984. $215.50 set. ISBN 0-444-86661-2. *Volume 1.* lxii + 848pp. *Volume 2.* x + 881pp.

Reports on the work of 1982 Fields medalists, the Nevanlinna Prize winner, 12 of the plenary addresses, and 123 section addresses of the Congress. Expensive. **(a)**

Proceedings of the International Congress of Mathematicians, August 3-11, 1986, Berkeley, CA. Ed: Andrew M. Gleason. Providence, RI: American Mathematical Society, 1987. $195.00 set. ISBN 0-8218-0110-4. *Volume 1.* ci + 872pp. *Volume 2.* iii + 837pp.

These volumes contain well-written expository articles by prominent researchers on central questions in mathematics, coverage of the ceremonies, addresses, lists, etc. Expensive. **(a)**

Renewing U. S. Mathematics: A Plan for the 1990s. Washington, DC: National Research Council, National Academy Press, 1990. xiv + 133pp. $15.00. ISBN 0-309-04228-3.

"David II," this sequel to title below recommends continued effort to achieve goals of 1984 document, but modified by a special focus on human resources issues. **(a)**

Renewing U. S. Mathematics: Critical Resource for the Future. Washington, DC: National Research Council, 1984. xiv + 207pp. Complimentary. No ISBN.

Report of the "David" Committee, documenting a significant deterioration in federal support for mathematics research. **(a)**

Role of Mathematics in Science. M. M. Schiffer, Leon Bowden. New Mathematical Library;30. Washington, DC: Mathematical Association of America, 1984. xi + 207pp. $16.00. ISBN 0-88385-630-1.

A delightful mathematical excursion through classical science (mainly mechanics) to illustrate and teach the role of mathematics as a sixth sense for scientific intuition. **(a,p)**

Symmetrie. Second Edition. Hermann Weyl. Wissenschaft und Kultur;11. Boston: Birkhäuser, 1981. 157pp. $15.95. ISBN 0-8176-1280-7.

Reprint of 1955 original. **(a)**

Theorems and Counterexamples in Mathematics. Bernard R. Gelbaum, John M. H. Olmsted. New York: Springer-Verlag, 1990. xxxiv + 305pp. $42.00. ISBN 0-387-97342-7.

Larger in scope than the authors' classic *Counterexamples in Analysis* with guide and glossary/index for ease of access. **(a)**

Winning Ways for Your Mathematical Plays. Elwyn R. Berlekamp, John H. Conway, Richard K. Guy. San Diego, CA: Academic Press, 1982. *V.1: Games in General.* xxxii + 437pp. $113.00 ISBN 0-12-091150-7. *V.2: Games in Particular.* i-xxxi + 427-850 + i-xixpp. $108.00. ISBN 0-12-091152-3.

Serious mathematics in whimsical dress: strategic analyses of games. Casual reading is difficult due to specialized terminology and notations. **(a,p)**

Winning Women into Mathematics. Ed: Patricia Clark Kenschaft, Sandra Keith. Washington, DC: Mathematical Association of America, 1991. ix + 78pp. $11.00. ISBN 0-88385-453-8.

A varied collection of materials documenting the increasing role of women in mathematics as well as the continuing impediments they face. **(p)**

The World of Mathematics. James R. Newman. Redmond, WA: Microsoft Press, 1988. $99.95;$50.00 set. ISBN 1-55615-149-7;1-55615-148-9. *Volume One*. xxx + 711pp. *Volume Two*. ix + 677pp. *Volume Three*. viii + 602pp. *Volume Four*. viii + 449pp.

Computer reproduction of Newman's 1956 anthology of mathematics. Comprehensive index in each volume. **(a,p)**

ALGEBRA

Abstract Algebra. Second Edition. I. N. Herstein. New York: Macmillan, 1990. xix + 293pp. $39.75. ISBN 0-02-353822-8.

Minor revisions have been made to this excellent text. It provides a treatment of groups, rings, and fields which is insightful yet cognizant of today's less well (and differently) prepared students. **(a)**

Abstract Algebra: A Computational Approach. Charles C. Sims. New York: Wiley, 1984. xv + 491pp. $32.95. ISBN 0-471-09846-9.

A worthwhile text due to the addition of APL. **(a)**

Algebra. Michael Artin. New York: Prentice-Hall, 1991. xviii + 618pp. $56.25. ISBN 0-13-004763-5.

An innovative text that builds on concrete material (e.g., symmetry, linear groups) and combines linear algebra with groups, rings, and fields. **(a)**

Algebra. P. M. Cohn. New York: Wiley, 1977. *Volume 1*. xv + 410pp. $23.95 paperbound. ISBN 0-471-16431-3. *Volume 2*. xiii + 483pp. $49.95. ISBN 0-471-01823-6.

This classic graduate-level text provides an introduction to a broad range of topics in modern algebra. **(a,p,s)**

Algebra. Jacob K. Goldhaber, Gertrude Ehrlich. Melbourne, FL: Krieger, 1980. xiv + 418pp. $29.50. ISBN 0-88275-765-2.

Corrected reprint of the 1970 original, a text for a first year graduate course in modern algebra. **(p)**

Algebra. Thomas W. Hungerford. Graduate Texts in Mathematics;73. New York: Springer-Verlag, 1989. xxiii + 502pp. $42.00. ISBN 0-387-90518-9.

Fourth reprinting of 1974 original, a basic text for an algebra course at the beginning graduate level. **(a)**

Algebra. Second Edition. Serge Lang. Reading, MA: Addison-Wesley, 1984. xv + 714pp. $40.76. ISBN 0-201-05487-6.

Differs from the first edition by the addition of several topics concerning commutative and homological algebra and tensor products. Additional new exercises. **(a,p)**

Algebra. B. L. van der Waerden. Transl: John R. Schulenberger. New York: Springer-Verlag, 1991. $29.95 each. *Volume I.* xiv + 265pp. ISBN 0-387-97424-5. *Volume II.* xii + 284pp. ISBN 0-387-97425-3.

Reprint of seventh edition of this classic work. It still sets the standard for the discipline. **(a,p)**

Algebra and Its Applications. Ed: H. L. Manocha, J. B. Srivastava. Lecture Notes in Pure and Applied Mathematics;91. New York: Dekker, 1984. xii + 395pp. $110.00. ISBN 0-8247-7165-6.

Collection of 48 papers on topics of current research, presented at a 1981 New Delhi symposium. Extensive references. Expensive. **(a)**

Algebra I: Basic Notions of Algebra. Ed: A. I. Kostrikin, I. R. Shafarevich. Encyclopedia of Mathematical Sciences;11. New York: Springer-Verlag, 1990. 258pp. $59.80. 0-387-17006-5.

Excellent survey allows the reader to obtain a sense of the scope and depth of algebra and a feeling for the algebraic point of view in mathematics. **(a,p)**

Algebra I, Chapters 1-3. Nicolas Bourbaki. (Elements of Mathematics) New York: Springer-Verlag, 1989. xxiii + 708pp. $69.00. ISBN 0-387-19373-1.

Reprint of Addison-Wesley's 1974 translation of the first three chapters of Bourbaki's *Algèbre*, (see below) covering algebraic structures, linear algebra, and tensor algebras. **(a,p)**

Algebra II, Chapters 4-7. Nicolas Bourbaki. (Elements of Mathematics) Transl: P. M. Cohn, J. Howie. New York: Springer-Verlag, 1990. vii + 461pp. $98.00. ISBN 0-387-19375-8.

Translation from the French of the chapters on polynomials and rational functions, commutative fields, ordered groups and fields, and modules over principal ideal domains. Exercises, historical notes, bibliography, and index. **(a,p)**

Algèbre, Chapter 1 à 3. Nouvelle Édition. Nicolas Bourbaki. (Éléments de Mathématique) Paris: Diffusion C.C.L.S., 1982. (Available from Scientific & Medical Publications of France, New York, NY) xiii + 166 + 210 + 258pp. $120.00. ISBN 2-903684-01-4.

Reprint of 1970 edition. (See above for translation of chapters 1-3, Algebra I, Chapters 1-3.) **(a)**

Algèbre, Chapitre 4 à 7. Nicolas Bourbaki. (Éléments de Mathématique) Paris: Masson, 1981. (Available from Scientific & Medical Publications of France, New York, NY) vii + 422pp. $68.00. ISBN 2-225-68574-6.

(See above for translation of chapters 4-7, Algebra II, Chapters 4-7.) **(a)**

Algèbre, Chapitre 10: Algèbre Homologique. Nouvelle Édition. Nicolas Bourbaki. (*Éléments* de Mathematique) Paris: Masson, 1980. (Available from Scientific & Medical Publications of France, New York, NY) vii + 216pp. $79.00. ISBN 2-225-65516-2.

This, the final chapter of Bourbaki's *Algèbre*, describes the homological properties of modules needed especially for the applications of commutative algebra and complex analysis in the further volumes of Éléments de Mathématique. (**a**)

Algèbre Commutative, Chapitres 1 a 4. Nicolas Bourbaki. (Éléments de Mathématique) Paris: Masson, 1985. (Available from Scientific & Medical Publications of France, New York, NY) 362pp. $79.00. ISBN 2-225-80268-8.

Includes bibliographical references and index. (See below for English translation of chapters 1-7, *Commutative Algebra.*) (**a**)

Algèbre Commutative, Chapitres 5 a 7. Nicolas Bourbaki. (Éléments de Mathématique) Paris: Masson, 1985. (Available from Scientific & Medical Publications of France, New York, NY) 351pp. $54.00. ISBN 2-225-80269-6.

Includes bibliographical references and index. (See below for English translation of chapters 1-7, *Commutative Algebra.*) (**a**)

Algèbre Commutative, Chapitres 8 et 9. Nicolas Bourbaki. (Éléments de Mathématique) Paris: Masson, 1983. (Available from Scientific & Medical Publications of France, New York, NY) 200pp. $42.00. ISBN 2-225-78716-6.

The chapters on dimension and complete Noetherian local rings. (**a**)

Applied Abstract Algebra. Rudolf Lidl, Gunter Pilz. New York: Springer-Verlag, 1984. xviii + 545pp. $49.20. ISBN 0-387-96035-X.

This text covers applications of Boolean algebras, finite fields and semigroups including switching circuits, coding theory and automata. (**a**)

Basic Algebra I. Second Edition. Nathan Jacobson. New York: W. H. Freeman, 1985. xviii + 499pp. $41.95. ISBN 0-7167-1480-9.

Important chamges have been made in the chapter on Galois theory as well as adding an appendix which lists topics for independent study. (**a**)

A Book of Abstract Algebra. Charles C. Pinter. New York: McGraw-Hill, 1982. xv + 351pp. $37.80. ISBN 0-07-050130-0.

An excellent algebra text characterized by clear writing and an outstanding collection of exercises. (**p**)

Classical Galois Theory: With Examples. Fourth Edition. Lisl Gaal. New York: Chelsea, 1988. viii + 248pp. $17.95. ISBN 0-8284-1268-5.

Reprint of 1971 original. (**a**)

Commutative Algebra. Second Edition. Hideyuki Matsumura. Mathematics Lecture Note Series;56. Reading, MA: Addison-Wesley, 1980. xv + 313pp. $34.50 paperbound. ISBN 0-8053-7026-9.

Minor revisions and an appendix have been added. (**a**)

Commutative Algebra. Oscar Zariski, Pierre Samuel, I. S. Cohen. Graduate Texts in Mathematics;28-29. New York: Springer-Verlag, 1975-1976. *Vol. I.* xi + 329pp. $55.00. ISBN 0-387-90089-6. *Vol. II.* x + 414pp. $49.80. ISBN 0-387-90171-X.

1986 reprinting of volume I, corrected from 1958 original with formulation and strengthening of Theorem 29. Volume II has been reprinted in 1991 from 1960 original. (**a**)

Commutative Algebra, Chapters 1-7. Nicolas Bourbaki. (Elements of Mathematics) New York: Springer-Verlag, 1989. xxiv + 625pp. $69.00. ISBN 0-387-19371-5.

Reprint of Addison-Wesley's 1972 translation of Éléments de Mathématique, Algèbre Commutative. Includes seven chapters beginning with flat modules and concluding with divisors. (See above for chapters 8-9, Algèbre Commutative.) (**a,p**)

Commutative Normed Rings. I. Gelfand, D. Raikov, G. Shilov. New York: Chelsea, 1964. 306pp. $18.95. ISBN 0-8284-0170-5.

Translation of 1960 Russian original with a supplementary chapter. (**a**)

Commutative Rings. Revised Edition. Irving Kaplansky. Chicago: University of Chicago Press, 1974. viii + 182pp. $12.00. ISBN 0-226-42454-5.

Minor corrections from the first edition, although a brief new "Notes" section has been added. (**a**)

A Concrete Introduction to Higher Algebra. Lindsay Childs. New York: Springer-Verlag, 1988. 340pp. $38.00. ISBN 0-387-90333-X.

Number theory and theory of equations are discussed in first two-thirds of book while the remainder focuses on fields. (**a,p**)

Contemporary Abstract Algebra. Joseph A. Gallian. Lexington, MA: D. C. Heath, 1986. xviii + 468pp. $35.50. ISBN 0-669-09325-4.

All-inclusive, light-hearted text for a beginning course in abstract algebra. (**p**)

Cyclotomic Fields I and II. Combined Second Edition. Serge Lang. Graduate Texts in Mathematics;121. New York: Springer-Verlag, 1990. xvii + 433pp. $49.50.

Corrected edition now includes a purely cyclotomic proof of "the main conjecture." (**a**)

Fields and Rings. Second Edition. Irving Kaplansky. Chicago: University of Chicago Press, 1972. x + 206pp. $12.50. ISBN 0-226-42450-2.

This book contains 3 lecture notes previously only available from Chicago's Department of Mathematics: Theory of Rings, Notes on Ring Theory, and Homological Dimension of Rings and Modules. (**a**)

A First Undergraduate Course in Abstract Algebra. Fourth Edition. Abraham P. Hillman, Gerald B. Alexanderson. Belmont, CA: Wadsworth, 1988. xiv + 541pp. $45.95. ISBN 0-534-08844-9.

The authors have thoroughly revised the sections on Boolean algebras and their applications as well as rewritten the treatment of Euclid's algorithm. **(p)**

Galois Theory. Second Edition, With Additions and Revisions. Emil Artin. Ed: Arthur N. Milgram. Notre Dame Mathematical Lectures;2. Notre Dame, IN: University of Notre Dame Press, 1966. iii + 82pp. $3.95. ISBN 0-268-00108-1.

"New composition with corrections." **(a)**

Galois Theory. Harold M. Edwards. Graduate Texts in Mathematics;101. New York: Springer-Verlag, 1984. xiii + 152pp. $32.00. ISBN 0-387-90980-X.

An exciting, historical approach to the work of Galois on the solvability of equations by radicals. **(a)**

Galois Theory. Joseph Rotman. New York: Springer-Verlag, 1990. xii + 108pp. $29.00. 0-387-97305-2.

A considered and cohesive introduction to the main results of Galois theory. Especially suitable for self-study. **(a,p)**

Galois Theory. Ian Stewart. New York: Routledge, Chapman and Hall, 1973. xxv + 226pp. $22.50 paperbound. ISBN 0-412-10800-3.

Historical commentary is interwoven with an ordinary introduction of Galois theory in this unusual, superb monograph. **(a)**

Homology. Third Corrected Printing. Saunders Mac Lane. Die Grundlehren der mathematischen Wissenschaften;114. New York: Springer-Verlag, 1975. x + 422pp. $44.50. ISBN 0-387-03823-X.

This classic text develops the subject from the special to the general. Extensive, though dated, bibliography. **(a)**

Homotopy Theory. S. T. Hu. Pure and Applied Mathematics;8. San Diego, CA: Academic Press, 1959. xiii + 347pp. $81.00. ISBN 0-12-358450-7.

Comprehensive standard reference work. **(a)**

Introduction to Algebra. A. I. Kostrikin. Transl: Neal Koblitz. New York: Springer-Verlag, 1982. xiii + 575pp. $58.00. ISBN 0-387-90711-4.

This Soviet abstract algebra text contains interesting applications, historical notes, and many perceptive comments. **(a)**

Introduction to Commutative Algebra. M. F. Atiyah, I. G. Macdonald. Reading, MA: Addison-Wesley, 1969. viii + 128pp. $34.36. ISBN 0-201-00361-9.

This concise book focuses on commutative rings and prime ideals. **(a)**

Introduction to Finite Fields and Their Applications. Rudolf Lidl, Harald Niederreiter. New York: Cambridge University Press, 1986. viii + 407pp. $44.50. ISBN 0-521-30706-6.

A textbook edition of *Finite Fields*, with greater emphasis on applications and less on theoretical topics. **(a)**

An Introduction to Homological Algebra. Joseph J. Rotman. Pure and Applied Mathematics;85. San Diego, CA: Academic Press, 1979. xi + 376pp. $59.95. ISBN 0-12-599250-5.

Very readable introduction to this subject. **(a)**

Introduction to Modern Algebra. Fifth edition. Neal H. McCoy. Rev: Gerald J. Janusz. Dubuque, IA: Wm. C. Brown, 1992. xii + 273pp. $59.45. ISBN 0-697-06865-X.

New edition of a standard, well-regarded text for a first abstract algebra course. Good range of difficulty on exercises. More worked examples. Short bibliography has been updated. **(a)**

An Introduction to Noncommutative Noetherian Rings. K. R. Goodearl, R. B. Warfield. London Mathematical Society Student Texts;15. New York: Cambridge University Press, 1989. xvii + 303pp. $54.50;19.95. ISBN 0-521-36086-2;0-521-36925-8.

Intended as a largely self-contained introduction to the subject at the second-year graduate level. Exercises, notes, bibliography. **(a)**

Lectures in Abstract Algebra. Nathan Jacobson. Graduate Texts in Mathematics;30-32. New York: Springer-Verlag, 1975-1976. *Vol. I: Basic Concepts.* xii + 217pp. $39.80. ISBN 0-387-90181-7. *Vol. II: Linear Algebra.* xii + 280pp. $42.00. ISBN 0-387-90123-X. *Vol. III: Theory of Fields and Galois Theory. Second Corrected Printing.* xi + 323pp. $46.00. ISBN 0-387-90168-X.

Reprint of 1951-1964 Van Nostrand original. **(a)**

Lectures on Rings and Modules. Second Edition. Joachim Lambek. New York: Chelsea, 1976. viii + 183pp. $16.95. ISBN 0-8284-1283-9.

Reprint of 1966 original with some errors corrected. Full treatment of the complete ring of quotients. **(a)**

Lehrbuch der Algebra. Third Edition. Heinrich Weber. New York: Chelsea, 1979. $150.00 set. ISBN 0-8284-0144-6. *Erster Band.* xv + 703pp. *Zweiter Band.* xvi + 855pp.

Corrected reprint of 1962 edition. Expensive. **(a)**

Local Fields. Jean-Pierre Serre. Transl: Marvin Jay Greenberg. Graduate Texts in Mathematics;67. New York: Springer-Verlag, 1979. 241pp. $49.00. ISBN 0-387-90424-7.

This translation of *Corps Locaux*, with an updated bibliography (now dated) and improved problem sets, presents local class field theory from a cohomological viewpoint. **(a)**

A Modern Course on the Theory of Equations. David E. Dobbs, Robert Hanks. Washington, NJ: Polygonal, 1980. vii + 216pp. $18.00. ISBN 0-936-42803-1.

An introduction to the theory of polynomial equations. **(a)**

Polynomials. E. J. Barbeau. New York: Springer-Verlag, 1989. xxii + 441pp. $59.00. ISBN 0-387-96919-5.

Beautifully composed exercise sets designed to entice the student into the discovery of the basic ideas in the theory of equations. **(a,p)**

Quadratic and Hermitian Forms. Winfried Scharlau. Die Grundlehren der mathematischen Wissenschaften;270. New York: Springer-Verlag, 1985. x + 421pp. $89.00. ISBN 0-387-13724-6.

Sophisticated introduction to some of the main parts of the theory. Point of view is algebraic which leads to a number of new results. References. **(a)**

Ring Theory. Student Edition. Louis H. Rowen. San Diego, CA: Academic Press, 1991. xxvii + 623pp. $55.00. ISBN 0-12-599841-4.

A sourcebook of the results and proofs which make up the general structural theory of rings. Comprehensive bibliography. **(a)**

Rings and Ideals. Neal H. McCoy. Carus Mathematical Monographs;8. Washington, DC: Mathematical Association of America, 1948. xii + 216pp. $22.50. ISBN 0-88385-008-7.

A clear exposition of the basic definitions and theorems concerning rings plus the more special aspects of the theory, i.e., commutative rings, direct sums and matrix rings. References. **(a)**

Rings of Continuous Functions. Leonard Gillman, Meyer Jerison. Graduate Texts in Mathematics;43. New York: Springer-Verlag, 1976. xiii + 300pp. $39.00. ISBN 0-387-90198-1.

Reprint of 1960 Van Nostrand original. **(a)**

Topics in Algebra. Second Edition. I. N. Herstein. New York: Wiley, 1975. xi + 388pp. $67.95. ISBN 0-471-01090-1.

Excellent text revised to include direct products and more on Sylow theorems and Galois theory. **(a)**

Universal Algebra. Revised Edition. P. M. Cohn. Mathematics and Its Applications;6. Dordrecht: D. Reidel, 1981. (Available from Kluwer Academic, Norwell, MA) xv + 412pp. $33.00 paperbound. ISBN 90-277-1254-9.

Essentially unaltered reissue of the 1966 book with 4 new chapters (category theory and universal algebra, model theory and universal algebra, miscellaneous further results, and algebra and model theory) in a 70-page supplement. **(a)**

ALGEBRAIC GEOMETRY

Algebraic Curves: An Introduction to Algebraic Geometry. William Fulton, Richard Weiss. Reading, MA: Addison-Wesley, 1989. xix + 226pp. $27.75. ISBN 0-201-51010-3.

Reprint of 1969 original. **(a)**

Algebraic Geometry. Revised Edition. Robin Hartshorne. Graduate Texts in Mathematics;52. New York: Springer-Verlag, 1987. xvi + 496pp. $49.80. ISBN 0-387-90244-9.

Corrected third printing. An introductory text using methods of schemes and cohomology. Bibliography. **(a,s)**

Geometric Algebra. E. Artin. New York: Wiley, 1988. x + 214pp. $34.95. ISBN 0-471-60839-4.

Reprint of 1957 original. Most of the book is devoted to the study of algebraic structures arising from the various geometries. The approach is algebraic rather than geometric. **(a)**

Geometry of Algebraic Curves. Volume I. E. Arbarello et al. Die Grundlehren der mathematischen Wissenschaften;267. New York: Springer-Verlag, 1985. xvi + 386pp. $55.00. ISBN 0-387-90997-4.

Advanced work on algebraic curves on compact Riemann surfaces. **(a)**

History of Algebraic Geometry. Jean Dieudonné. Transl: Judith D. Sally. Pacific Grove, CA: Brooks-Cole, 1985. xii + 186pp. $46.95. ISBN 0-534-03723-2.

A concise survey of the evolution of algebraic geometry with a final chapter on open problems and current work. **(a)**

Plane Algebraic Curves. Egbert Brieskorn, Horst Knorrer. Transl: John Stillwell. Boston: Birkhauser, 1986. vi + 721pp. $69.70. ISBN 0-8176-3326-X.

This wonderful introduction to the topic ranges from elementary methods to resolution of singularities. **(a,p)**

Principles of Algebraic Geometry. Phillip Griffiths, Joseph Harris. New York: Wiley, 1978. xii + 813pp. $79.95. ISBN 0-471-32792-1.

A major expository presentation of much of the core of modern algebraic geometry. Good references and index. **(a)**

Relative Invariants of Sheaves. A. Verschoren. Pure and Applied Mathematics;104. New York: Dekker, 1987. xi + 249pp. $89.75. ISBN 0-8247-7734-4.

This introduction to sheaf theory combines methods from relative invariants of rings with classical techniques of sheaf theory. **(a)**

A Scrapbook of Complex Curve Theory. C. Herbert Clemens. New York: Plenum Press, 1980. ix + 186pp. $37.50. ISBN 0-306-40536-9.

Excellent orientation to the modern theory of Jacobian varieties. **(a)**

Traite des Courbes Speciales Remarquables Planes et Gauches. Francisco Gomes Teixeira. New York: Chelsea, 1971. $95.00 set. ISBN 0-8284-0255-8. *Tome I.* viii + 403pp. *Tome II.* viii + 488pp. *Tome III.* x + 419pp.

Corrected reprint of the 1908-1915 French edition. This was first published in 1905 in Spanish as *Tratado de las Curvas Especiales Notables.* **(a)**

ALGEBRAIC TOPOLOGY

Algebraic Topology. First Corrected Edition. Edwin H. Spanier. New York: Springer-Verlag, 1981. xvi + 528pp. $49.00. ISBN 0-387-90646-0.

Reprint of 1966 classic text. Very useful reference which squeezes a huge amount of algebraic topology into a single volume. **(a)**

Algebraic Topology, A First Course. Marvin J. Greenberg, John R. Harper. Mathematics Lecture Note Series;58. Reading, MA: Addison-Wesley, 1981. xi + 311pp. $35.50. ISBN 0-8053-3357-9.

Revision of Greenberg's *Lectures on Algebraic Topology*. Bibliography. **(a)**

Differential Forms in Algebraic Topology. Raoul Bott, Loring W. Tu. Graduate Texts in Mathematics;82. New York: Springer-Verlag, 1982. xiv + 331pp. $42.00. ISBN 0-387-90613-4.

Reprinted in 1986, this book is an excellent, although sometimes not easy to read, presentation of algebraic topology via differential forms. **(a)**

Graphs, Surfaces, and Homology: An Introduction to Algebraic Topology. Second Edition. P. J. Giblin. New York: Routledge, Chapman and Hall, 1981. xvii + 329pp. $22.50. ISBN 0-412-23900-0.

A number of references have been added and some discussions have been changed in light of work done or published since the 1977 first edition. **(a)**

A History of Algebraic and Differential Topology, 1900-1960. Jean Dieudonné. Boston: Birkhäuser, 1989. xxi + 648pp. $79.00. ISBN 0-8176-3388-X.

This encyclopedic work describes simplicial techniques and homology, the first applications of simplicial methods and of homology, and homotopy and its relation to homology. **(a)**

An Introduction to Algebraic Topology. Joseph L. Rotman. Graduate Texts in Mathematics;119. New York: Springer-Verlag, 1988. xiii + 433pp. $49.80. ISBN 0-387-96678-1.

Well-written text which catches the interplay of geometry and algebra and balances the study of homotopy and (co)homology. **(a,s)**

ALGORITHMS

Algorithms. Robert Sedgewick. Reading, MA: Addison-Wesley, 1983. viii + 551pp. $43.25. ISBN 0-201-06672-6. *Survey of some of the most important algorithms in computer science.* **(a,p)**

Algorithms and Data Structures: Design, Correctness, Analysis. Jefferson H. Kingston. Reading, MA: Addison-Wesley, 1990. ix + 315pp. $43.25 ISBN 0-201-41705-7.

Text emphasizes use of abstract data types, proofs of correctness, and appropriate algorithm design techniques. Unusual in its treatment of amortized complexity analysis, splay trees, and Fibonacci heaps. **(a)**

Algorithms from P to NP. Volume 1: Design & Efficiency. B. M. E. Moret, H. D. Shapiro. Redwood City, CA: Benjamin/Cummings, 1991. xv + 576pp. $35.95. ISBN 0-8053-8008-6.

This work illustrates algorithm design techniques through application of the design of efficient algorithms for some well-known problems in combinatorial optimization. **(a)**

Algorithmics: Theory and Practice. Gilles Brassard, Paul Bratley. New York: Prentice-Hall, 1988. xvi + 361pp. $48.00. ISBN 0-13-023243-2.

A detailed and readable examination of the algorithms of classical computer science, including searching, matrix multiplication, number theory, complexity, etc. **(a)**

Computational Geometry, An Introduction. Franco P. Preparata, Michael Ian Shamos. New York: Springer-Verlag, 1985. xii + 390pp. $49.50. ISBN 0-387-96131-3.

Comprehensive introduction to the study of geometrical algorithms, emphasizing algorithm design and analysis of worst-case complexity. **(a)**

Computer Algebra: Symbolic and Algebraic Computation. Ed: B. Buchberger et al. Computing Supplementum;4. New York: Springer-Verlag, 1982. vii + 283pp. $45.00. ISBN 0-387-81684-4.

This collection of 16 survey articles brings together the disparate results on efficiency and implementation of algebraic algorithms. Extensive references. **(a)**

Introduction to Algorithms. Thomas H. Cormen, Charles E. Leiserson, Ronald L. Rivest. Cambridge, MA: MIT Press, 1990. xvii + 1028pp. $49.95. ISBN 0-262-03141-8.

A comprehensive introduction to the study of computer algorithms. This is a useful reference work as well as textbook. **(a,p)**

Mathematics for the Analysis of Algorithms. Third Edition. Daniel H. Greene, Donald E. Knuth. Progress in Computer Science and Applied Logic;1. Boston: Birkhäuser, 1990. viii + 132pp. $24.95. ISBN 0-8176-3515-7.

Based on materials for an Stanford University advanced course in computer science, this revision adds some new appendices, improves the exposition, and makes minor corrections. **(a)**

ANALYSIS

Abstract Harmonic Analysis. Edwin Hewitt, Kenneth A. Ross. Die Grundlehren der mathematischen Wissenschaften;115,152. New York: Springer-Verlag, 1970-1979. *Volume I: Structure of Topological Groups, Integration Theory, Group Representations. Second Edition.* vii + 519pp. out-of-print. ISBN 0-387-09434-2. *Volume II: Structure and Analysis for Compact Groups, Analysis on Locally Compact Abelian Groups.* viii + 771pp. $89.00. ISBN 0-387-04832-4.

This work is intended for both the beginning reader and specialist. Some sections contain historical notes. Appendices, indices, and extensive bibliography. **(a)**

Advanced Mathematical Analysis: Periodic Functions and Distributions, Complex Analysis, Laplace Transform and Applications. Third Corrected Printing. Richard Beals. Graduate Texts in Mathematics;12. New York: Springer-Verlag, 1987. x + 230pp. $39.95. ISBN 0-387-90065-9.

This book lends itself to independent study due to its clear organization and detailed proofs. **(a)**

Almost Periodic Functions. Harald August Bohr. Transl: Harvey Cohn, F. Steinhardt. New York: Chelsea, 1980. 113pp. $10.95. ISBN 0-8284-0027-X.

Translation of *Fastperiodische Funktionen.* **(a)**

The Analysis of Linear Partial Differential Operators. Lars Hormander. Die Grundlehren der mathematischen Wissenschaften;256-257,274-275. New York: Springer-Verlag, 1983-1990. *I: Distribution Theory and Fourier Analysis. Second Edition.* ix + 391pp. $69.00. ISBN 0-387-52343-X. *II: Differential Operators with Constant Coefficients.* viii + 391pp. $87.10. ISBN 0-387-12139-0. *III: Pseudo-differential Operators.* viii + 524pp. $89.00. ISBN 0-387-13828-5. *Volume IV: Fourier Integral Operators.* vii + 351pp. $89.00. ISBN 0-387-13829-3.

A monumental work by a master. **(a,s)**

Analytic Functions of a Complex Variable. David Raymond Curtiss. Carus Mathematical Monograph;2. Washington, DC: Mathematical Association of America, 1926. $21.00. ISBN 0-88385-002-8.

Offers the fundamental principles of the theory of functions of a complex variable. **(a)**

Analytic Inequalities. D. S. Mitrinović, P. M. Vasić. Die Grundlehren der mathematischen Wissenschaften;165. New York: Springer-Verlag, 1970. xi + 400pp. $79.00. ISBN 0-387-04837-5.

A fascinating book on inequalities in analysis. Over 450 special inequalities are included; all results are referenced. **(a)**

Asymptotic Analysis. Second Edition. J. D. Murray. Applied Mathematical Sciences;48. New York: Springer-Verlag, 1984. vii + 164pp. $32.00. ISBN 0-387-90937-0.

An earlier version of this book was published in 1964 by Clarendon Press. This resource contains the basic and elementary subjects of asymptotic methods of functions defined by integrals or as solutions of ordinary differential equations. **(a)**

Basic Hypergeometric Series. George Gasper, Mizan Rahman. Encyclopedia of Mathematics and Its Applications;35. New York: Cambridge University Press, 1990. xx + 287pp. $59.50. ISBN 0-521-35049-2.

Excellent and very informative book on the subject. After basic series and some special cases are introduced, the authors discuss the latest results on the general theory and its extensions. Notes at the end of each chapter point the reader to related topics. **(a)**

Calculus of Variations. Gilbert Ames Bliss. Washington, DC: Mathematical Association of America, 1925. xiii + 189pp. $21.00. ISBN 0-88385-001-X.

Presents special problems that prepare the reader for the general theory of the calculus of variations. List of references. **(a)**

Calculus of Variations. Revised English Edition. I. M. Gelfand, S. V. Fomin. Transl: Richard A. Silverman. New York: Prentice-Hall, 1963. vii + 232pp. $56.00. ISBN 0-13-112292-4.

Improvements have been made in the text of the original 1961 Russian version. **(a)**

Calculus of Variations and Partial Differential Equations of the First Order. Third Edition. C. Caratheodory. Transl: Robert B. Dean. New York: Chelsea, 1982. xix + 402pp. $25.00. ISBN 0-8284-0318-X.

New edition of the great classic with known misprints corrected. **(a)**

The Calculus of Variations in the Large. Marston Morse. Colloquium Publications;18. Providence, RI: American Mathematical Society, 1934. ix + 368pp. $43.00. ISBN 0-8218-1018-9.

Seventh printing of this classic published in 1986. **(a)**

Calculus on Manifolds: A Modern Approach to Classical Theorems of Advanced Calculus. Revised and Corrected Edition. Michael Spivak. Reading, MA: Addison-Wesley, 1968. xii + 146pp. $19.96. ISBN 0-8053-9021-9.

This book is recommended for the mature mathematician as the exposition is lacking in sufficient illustrative examples to be useful in the normal undergraduate class. **(a)**

Chebyshev Polynomials: From Approximation Theory to Algebra and Number Theory. Second Edition. Theodore J. Rivlin. New York: Wiley, 1990. xiii + 249pp. $49.95. ISBN 0-471-62896-4.

This edition includes new applications and exercises as well as a new chapter on algebraic and number theoretic properties of the Chebyshev polynomials. **(a)**

Conformal Mapping. L. Bieberbach. Transl: F. Steinhardt. New York: Chelsea, 1986. 234pp. $8.95. ISBN 0-8284-0090-3. Translation of the 1949 original, *Einführung in die Konforme Abbildung.* **(a)**

Constructive Analysis. Errett Bishop, Douglas Bridges. Die Grundlehren der mathematischen Wissenschaften; 279. New York: Springer-Verlag, 1985. xii + 477pp. $48.00. ISBN 0-387-15066-8.

An important contribution to the fields of real and functional analysis. **(a)**

Convex Analysis. R. Tyrrell Rockafellar. Princeton Mathematical Series;28. Princeton, NJ: Princeton University Press, 1970. xviii + 451pp. $55.00. ISBN 0-691-08069-0.

Exposition of a branch of non-linear analysis suited to optimization problems and differing from classical analysis in that differentiability assumptions are replaced by convexity assumptions. **(a)**

A Course of Modern Analysis: An Introduction to the General Theory of Infinite Processes and of Analytic Functions, with an Account of the Principal Transcendental Functions. Fourth Edition. E. T. Whittaker, G. N. Watson. New York: Cambridge University Press, 1927. 608pp. $47.50 paperbound. ISBN 0-521-09189-6.

1988 printing of reprint of 1902 original. **(a)**

Elementary Classical Analysis. Jerrold E. Marsden et al. New York: W. H. Freeman, 1974. xiv + 549pp. $34.95. ISBN 0-7167-0452-8.

A text on multivariable calculus with chapters on uniform convergence and Fourier analysis. Each chapter contains a substantial section in which the proofs of the theorems are given in full. **(a,p)**

Éléments de la Théorie des Fonctions Elliptiques. Second Edition. Jules Tannery, Jules Molk. New York: Chelsea, 1972. 4v. in 2. $59.50 set. ISBN 0-8284-0257-4. *Tome I: Introduction, Calcul Différentiel. Tome II: Calcul Différentiel.* x + 246 + vi + 299pp. *Tome III: Calcul Intégral. Tome IV: Calcul Intégral.* vii + 267 + vii + 303pp.

Reprint. Volume 4 has letter from Tannery to Hermite. **(a)**

Foundations of Analysis: The Arithmetic of Whole, Rational, Irrational, and Complex Numbers. Second Edition. Edmund Landau. Transl: F. Steinhardt. New York: Chelsea, 1960. 136pp. $12.95. ISBN 0-8284-0079-2.

Reprint of the translation of *Grundlagen der Analysis.* **(a)**

Fourier Analysis. T. W. Korner. New York: Cambridge University Press, 1988. xii +591pp. $95.00 ISBN 0-521-25120-6.

A very well-written and extensive account of the theory and applications of Fourier analysis. Well worth the price. **(a,s)**

Fourier Series, a Modern Introduction. Second Edition. R. E. Edwards. Graduate Texts in Mathematics;64,85. New York: Springer-Verlag, 1979-1982. *Volume 1.* xii + 224pp. out-of-print. ISBN 0-387-90412-3. *Volume 2.* xi + 369pp. $55.00. ISBN 0-387-90651-7.

Mostly minor corrections in this edition, except for a substantial revision of chapters 14-15 of volume 2. **(a)**

Fourier Series and Boundary Value Problems. Third Edition. Ruel V. Churchill, James Ward Brown. New York: McGraw-Hill, 1978. viii + 271pp. $46.95. ISBN 0-07-010843-9.

A classic. Fourth edition now out of print. **(a,s)**

From Gauss to Painlevé: A Modern Theory of Special Functions. Katsunori Iwasaki et al. Aspects of Mathematics;E16. Braunschweig: Vieweg, 1991. x + 347pp. $56.00. ISBN 3-528-06355-6.

The authors give an excellent presentation of Painlevé functions, both historically and mathematically, in a form accessible to any mathematician or physicist. **(a)**

Functions of Several Variables. Second Edition. Wendell Fleming. New York: Springer-Verlag, 1977. xi + 411pp. $42.00. ISBN 3-540-90206-6.

Changes since 1965 First edition include a new chapter on elementary topology, additional applications to thermodynamics and mechanics, and new proofs of the inverse function and divergence theorems. **(a)**

Harmonic Analysis. Henry Helson. Reading, MA: Addison-Wesley, 1983. xii + 190pp. $43.25. ISBN 0-201-12752-0.

Excellent book aimed at a broad scientific audience. Assumes only an elementary knowledge of functional analysis. The author carefully introduces and proves classical facts in Fourier analysis and then shows how striking results in various distinct areas can by obtained through application. **(a,p)**

Inequalities. Fourth Printing. Edwin F. Beckenbach, Richard Bellman. Ergebnisse der Mathematik und ihrer Grenzgebiete;30. New York: Springer-Verlag, 1983. xi + 198pp. $49.00. ISBN 0-387-03283-5.

Reprint of 1965 Second revised printing of 1961 original. "The authors limit their attention 'to those topics that have particularly delighted and intrigued them' without making any 'attempt to be encyclopedic' either in their text or bibliographies."—Preface. **(a)**

Inequalities. Second Edition. G. H. Hardy, J. E. Littlewood, G. Pólya. New York: Cambridge University Press, 1988. xii + 324pp. $19.95. ISBN 0-521-35880-9.

A paperback reprint of this famous classic, containing the statement and proof of all the standard inequalities of analysis. **(a,p)**

Inequalities: Theory of Majorization and Its Applications. Albert W. Marshall, Ingram Olkin. Mathematics in Science and Engineering;143. San Diego, CA: Academic Press, 1979. xx +569pp. $49.50. ISBN 0-124-73750-1.

In this well-written text the theory of majorization is fundamental to a number of applications and provides a natural context for proving some of the inequalities. **(a)**

Infinite Sequences and Series. Konrad Knopp. Transl: Frederick Bagemihl. New York: Dover, 1956. 186pp. $4.95. ISBN 0-486-60153-6.

For the reader of the foundations of real and complex analysis, the importance of this little book is its coverage of the most important topics in the classical theory of series. **(p)**

Integral Equations. Harry Hochstadt. New York: Wiley, 1989. viii + 282pp. $28.95. ISBN 0-471-50404-1.

Reprint of the 1973 classic. **(p)**

Integral Equations. F. G. Tricomi. New York: Dover, 1985. viii+ 238pp. $6.00. ISBN 0-486-64828-1.

Unaltered reprint of 1957 classic. **(p)**

Integral Transforms and Their Applications. Second Edition. B. Davies. Applied Mathematical Sciences;25. New York: Springer-Verlag, 1985. xv + 411pp. $28.00. ISBN 0-387-96080-5.

Presents the classical integral tranforms of Fourier, Hankel, Laplace, and Mellin as well as several convolution and integro-differential operators. **(s)**

Intégration, Chapitre IX: Intégration sur les Espaces Topologiques Séparés. Nicolas Bourbaki. (Éléments de Mathématique) Paris: Diffusion C.C.L.S., 1982. (Available from Scientific & Technical Publications of France, New York, NY) 133pp. $35.00. ISBN 2-903684-06-5.

Reprint of 1969 edition. Bourbaki extends the theory of measure and integration to functions defined on topological spaces that are not necessarily locally compact. **(a)**

Interpolation and Approximation. Philip J. Davis. New York: Dover, 1975. xiv + 393pp. $8.95. ISBN 0-486-62495-1.

Reprint, with minor corrections, of 1963 original. Lists 26 books on approximation which have appeared since 1963. **(p)**

Introduction to Fourier Analysis and Generalised Functions. M. J. Lighthill. New York: Cambridge University Press, 1958. viii + 79pp. $15.95 paperbound. ISBN 0-521-09128-4.

Although intended for undergraduates, this short monograph provides a really deep and searching account of Fourier series and integrals, including a systematic treatment of their asymptotic behavior for large values of the argument. **(a,p,s)**

An Introduction to Harmonic Analysis. Second Corrected Edition. Yitzhak Katznelson. New York: Dover, 1976. xiv + 264pp. $7.95. ISBN 0-486-63331-4.

Reprint of 1968 edition, proceeding from Fourier series and Fourier analysis to commutative Banach algebras. **(a,p)**

Introduction to Tauberian Theory: From Tauber to Wiener. J. van de Lune. CWI Syllabus;12. Amsterdam: Mathematisch Centrum, 1986. iii + 102pp. Dfl. 16.50. ISBN 90-6196-309-5.

Introduction to the history of Tauberian theory, beginning with connections to Abel's theorem, and proceeding through the general theorems of Wiener and Pitt. **(a)**

An Introduction to Variational Inequalities and Their Applications. David Kinderlehrer, Guido Stampacchia. Pure and Applied Mathematics;88. San Diego, CA: Academic Press, 1980. xiv + 313pp. $75.00. ISBN 0-12-407350-6.

This work can either be used as a graduate level text or a reference for engineers, economists, etc. Bibliography. **(p)**

Isoperimetric Inequalities in Mathematical Physics. G. Pólya, G. Szegö. Annals of Mathematics Studies;27. Millwood, NY: Kraus Reprint, 1965. xvi + 279pp. $26.00. ISBN 0-527-02743-X.

Reprint of 1951 Princeton University Press original. **(a)**

Leçons sur l'Intégration. Third Edition. Henri Lebesgue. New York: Chelsea, 1973. xii + 340pp. $16.95. ISBN 0-8284-0267-1.

Unabridged reprint with minor corrections of 1928 second edition. **(a)**

Lectures on Bochner-Riesz Means. Katherine Michelle Davis, Yang-Chun Chang. London Mathematical Society Lecture Notes Series;114. New York: Cambridge University Press, 1987. ix + 150pp. $27.95. ISBN 0-521-31277-9.

"Self-contained exposition of the geometric theory of Bochner-Riesz means."—Preface. **(a)**

Lectures on the Calculus of Variations. Third Edition. Oskar Bolza. New York: Chelsea, 1973. ix + 269pp. $14.95. ISBN 0-8284-1145-X.

Reprint of 1904 and 1961 editions in which addenda are incorporated into text and references updated. **(a)**

Measure and Category: A Survey of the Analogies between Topological and Measure Spaces. Second Edition. John C. Oxtoby. Graduate Texts in Mathematics;2. New York: Springer-Verlag, 1980. ix + 106pp. $39.00. ISBN 0-387-90508-1.

"In this edition, a set of Supplementary Notes and Remarks has been added at the end, grouped according to chapter."—Preface. **(a)**

Measure Theory. Paul R. Halmos. Graduate Texts in Mathematics;18. New York: Springer-Verlag, 1988. xi + 304pp. $39.80. ISBN 0-387-90088-8.

Reprint of 1950 original. **(a)**

Multiple Integrals in the Calculus of Variations and Nonlinear Elliptic Systems. Mariano Giaquinta. Annals of Mathematics Studies;105. Princeton, NJ: Princeton University Press, 1983. vii + 296pp. $50.00;$21.95. ISBN 0-691-08330-4;0-691-08331-2.

Discusses the existence and differentiability of stationary points of regular functionals in the calculus of variations. Extensive bibliography. **(a)**

n-Widths in Approximation Theory. Allan Pinkus. Ergebnisse der Mathematik und ihrer Grenzgebiete;7. New York: Springer-Verlag, 1985. x + 291pp. $85.00. ISBN 0-387-13638-X.

Develops basic theory of n-widths and applies it to Hilbert spaces, spaces of analytic functions, etc. Extensive bibliography. **(a)**

Noncommutative Harmonic Analysis. Michael E. Taylor. Mathematical Surveys and Monographs;22. Providence, RI: American Mathematical Society, 1990. xvi + 328pp. $78.00. ISBN 0-8218-1523-7.

Reprint of 1986 original, a clearly-written survey of applications of Lie groups to analysis with the relationship between harmonic analysis and linear partial differential equations as the unifying theme. Extensive bibliography. **(a)**

Potential Theory in Modern Function Theory. Second Edition. Masatsugu Tsuji. New York: Chelsea, 1975. x + 590pp. $29.50. ISBN 0-8284-0281-7.

Reprint of 1959 original. **(a)**

Principles of Applied Mathematics: Transformation and Approximation. James P. Keener. Reading, MA: Addison-Wesley, 1988. xv + 560pp. $38.36. ISBN 0-201-15674-1.

Well-written text develops and applies spectral theory of operators and asymptotic analysis to a wide range of problems. References. **(a)**

Principles of Mathematical Analysis. Third Edition. Walter Rudin. New York: McGraw-Hill, 1976. x + 342pp. $43.96. ISBN 0-07-054235-X.

Latest edition of the classic text greatly expands the material on functions of several variables. References. **(a,p)**

Problems in Mathematical Analysis. Piotr Biler, Alfred Witkowski. Pure and Applied Mathematics;132. New York: Dekker, 1990. v + 227pp. $49.75. ISBN 0-8247-8312-3.

More than 1300 problems in analysis, most of them neither standard nor easy, covering the usual topics. **(a)**

Problems in Real and Complex Analysis. Bernard R. Gelbaum. New York: Springer-Verlag, 1992. ix + 488pp. $59.00. ISBN 0-387-97766-X.

Doubling an earlier work, *Problems in Analysis*, this text adds a new section of problems on complex analysis, covering most of the principle topics in theory of functions of a complex variable. **(a)**

Real and Complex Analysis. Third Edition. Walter Rudin. New York: McGraw-Hill, 1987. xiv + 416pp. $44.84. ISBN 0-07-054234-1.

While this edition retains the familiar features of the earlier ones, e.g., stressing the connections between real and complex analysis, the treatment of differentiation differs. **(a)**

Tata Lectures on Theta. David Mumford, with Madhav Nori, Peter Norman. Progress in Mathematics;28,43,97. Cambridge, MA: Birkhauser, 1983-1991. *I: Introduction and Motivation: Theta Functions in One Variable; Basic Results on Theta Functions in Several Variables*. xiii + 235pp. o.p. ISBN 0-8176-3109-7. *II: Jacobian Theta Functions and Differential Equations*. xiv + 272pp. $34.50;$31.35. ISBN 0-8176-3109-7;0-8176-3110-0. *III: Tata Lectures on Theta*. vii + 202pp. $49.50. ISBN 0-8176-3440-1.

Three volume set surveying theory of theta functions. **(a)**

Theory and Application of Infinite Series. Konrad Knopp. Transl: R. C. H. Young. New York: Dover, 1990. xii + 563pp. $12.95. ISBN 0-486-66165-2.

Reprint of the 1951 original, a classic. **(a)**

Theory of Functions. Second Edition. E. C. Titchmarsh. New York: Oxford University Press, 1939. x + 454pp. $35.00. ISBN 0-19-853349-7.

Systematic treatise on the theory of functions. A classic. **(a)**

Theory of Functions of a Complex Variable. Second Edition. A. I. Markushevich. Transl/Ed: Richard A. Silverman. New York: Chelsea, 1977. xxiii + 446 + 325 + 367pp (3 volumes in 1). $35.00. ISBN 0-8284-0296-5.

Revised translation of *Teoriia analiticheskikh Funktsii.* **(a)**

Theta Functions, Bowdoin 1987. Ed: Leon Ehrenpreis, Robert C. Gunning. Proceedings of the Symposia in Pure Mathematics;49:1-2. Providence, RI: American Mathematical Society, 1989. $158.00 set. ISBN 0-8218-1485-0. *Part 1.* x + 718pp. $105.00. ISBN 0-8218-1483-4. *Part 2.* x + 366pp. $61.00. ISBN 0-8218-1484-2.

Collection of papers illustrating the range of applications of theta functions. Expensive. **(a)**

Treatise on Analysis. J. Dieudonné. Pure and Applied Mathematics;10:I-VII. San Diego, CA: Academic Press, 1969-1988. *Volume I: Foundations of Analysis. Enlarged and Second Printing.* xviii + 387pp. $33.75. ISBN 0-12-215550-5. *Volume II. Enlarged and Corrected Printing.* Transl: I. G. Macdonald. xv + 453pp. $67.00. ISBN 0-12-215502-5. *Volume III.* Transl: I. G. Macdonald. xvii + 388pp. $68.50. ISBN 0-12-211503-3. *Volume IV.* Transl: I. G. Macdonald. xv + 444pp. $67.00. ISBN 0-12-211504-1. *Volume V.* Transl: I. G. Macdonald. xii + 243pp. $49.00. ISBN 0-12-215505-X. *Volume VI.* Transl: I. G. Macdonald. x + 239pp. $47.50. ISBN 0-12-215506-8. *Volume VII.* Transl: Laura Fainsilber. xiv + 366pp. $69.95. ISBN 0-12-215507-6.

English translation of the French original, *Éléments d'Analyse.* **(a,s)**

Trigonometric Series. Second Edition. A. Zygmund. New York: Cambridge University Press, 1988. xiii + 364pp. $49.50 for volumes I-II combined. ISBN 0-521-35885-X.

Written in 1935 and expanded to two volumes in 1959, this book still stands as a classic in the field of Fourier analysis. **(a)**

Undergraduate Analysis. Second Corrected Printing. Serge Lang. New York: Springer-Verlag, 1989. xiii + 545pp. $45.00. ISBN 0-387-90800-5.

This text is a revised version of Lang's 1968 *Analysis I.* It is designed for a first course in analysis, centering around those properties concerned with uniform convergence, uniform limits, and uniformity in general. **(p)**

Walsh Series and Transforms: Theory and Applications. B. Golubov, A. Efimov, V. Skvortsov. Mathematics and Its Applications. Soviet Series;64. Norwell, MA: Kluwer Academic, 1991. xiii + 368pp. $169.00. ISBN 0-7923-1100-0.

Translation of 1987 Russian work on the theoretical foundation of Walsh series and transforms. Appendices with background information, brief commentaries, and historical references. Extensive list of references. Expensive. **(a)**

APPLICATIONS

Actuarial Mathematics. Newton L. Bowers, Jr. et al. Itasca, IL: Society of Actuaries, 1986. xxiii + 624pp. $65.00. ISBN 0-938959-10-7.

This new textbook on actuarial mathematics presents the modern foundations and develops some powerful tools for the actuary of the future. The stochastic approach is used in the text. But the work also integrates contingency theory into a structure combining risk theory, economics, and population theory. **(a)**

Advanced Engineering Mathematics. Sixth Edition. Erwin Kreyszig. New York: Wiley, 1988. xviii + 1294pp. $62.50. ISBN 0-471-85824-2.

This edition of the well-known textbook on mathematical topics for engineers has been simplified by rewriting various sections and by placing more emphasis on applications, algorithms, and examples. **(a)**

Adventures in Celestial Mechanics: A First Course in the Theory of Orbits. Victor G. Szebehely. Austin, TX: University of Texas Press, 1989. xiv + 175pp. $27.50. ISBN 0-292-75105-2.

Enthusiastic telling of orbit mechanics and celestial mechanics, emphasizing the basics. Annotated list of major references. **(a)**

The Anthropology of Numbers. Thomas Crump. New York: Cambridge University Press, 1990. x + 200pp. $49.50. ISBN:0-521-38045-6.

First scholarly work to examine the role of mathematics in culture. Thorough bibliography; comprehensive index. An excellent resource. **(a,p)**

Brains, Machines, and Mathematics. Second Edition. Michael A. Arbib. New York: Springer-Verlag, 1987. xvi + 202pp. $34.00. ISBN 0-387-96539-4.

Revision of innovative 1962 work outlining a theory of automata to explain the processing of information in brain-like machines. **(a)**

Catastrophe Theory. Third Revised and Expanded Edition. V. I. Arnol'd. Transl: R. K. Thomas. New York: Springer-Verlag, 1992. ix + 108pp. Price not available. ISBN 0-387-54811-4.

A masterful, elementary exposition of catastrophe theory, from Whitney's singularity theory to Dynkin diagrams. **(p,s)**

Celestial Mechanics. Harry Pollard. Carus Mathematical Monographs;18. Washington, DC: Mathematical Association of America, 1976. 134pp. $21.00. ISBN 0-88385-019-2.

A brief and very solid introduction to mechanics in general and the central force problem, perturbation theory theory, the n-body problem, and Hamilton-Jacobi theory in particular. **(a)**

Coding and Information Theory. Second Edition. R. W. Hamming. New York: Prentice-Hall, 1986. xii + 259pp. $52.00. ISBN 0-13-139072-4.

Clear and concise elementary introduction to error-detecting and error-correcting codes, to Shannon's theory of information, and to algebraic coding theory. **(a)**

A Diary of Information Theory. Alfred Renyi. New York: Wiley, 1987. ix + 125pp. $48.95. ISBN 0-471-90971-8.

Reprint of 1984 Hungarian edition. As an imaginary rendering of a student's first lectures on information theory, it gives Renyi's thoughts on information theory, mathematics, pedagogy, science and styles of learning. **(p)**

The Dynamics of Physiologically Structured Populations. Ed: J. A. J. Metz, O. Diekmann. Lecture Notes in Biomathematics;68. New York: Springer-Verlag, 1986. xii + 511pp. $61.00. ISBN 0-387-16786-2.

Based on notes from a 1983 colloquium. Notes contain a systematic exposition of the present state of the mathematical theory. **(p)**

Electromagnetic Theory. Third Edition. Oliver Heaviside. New York: Chelsea, 1971. $99.50 set. ISBN 0-8284-0237-X. *Volume I.* xxxiv + 504pp. *Volume II.* xvi + 547pp. *Volume III.* ix + 666pp.

Reprint. Includes Heaviside's writings on vector algebra and analysis, generalized differentiation, and divergent series. **(a)**

The Financial Theory of Pricing Property-Liability Insurance Contracts. Stephen P. D'Arcy, Neil A. Doherty. S. S. Huebner Foundation Monograph Series;15. Homewood, IL: R. D. Irwin, 1988. xii + 99pp. $19.95. ISBN 0-256-06124-6.

"To provide actuaries, regulators, and insurance academics with a single source that explains the basic financial theories and techniques applicable to property-liability insurance pricing."—Preface. **(a)**

Finite Mathematics with Business Applications. Second, Reference Edition. John G. Kemeny, et al. New York: Prentice-Hall, 1972. $44.00. ISBN 0-13-317321-6.

Revision to acclaimed First edition includes new chapters on decision theory and Markov decision processes. **(a,p)**

The Geometry of Genetics. A. M. Findley, S. P. McGlynn, G. L. Findley. New York: Wiley, 1989. xi + 156pp. $44.95. ISBN 0-471-05617-0.

This book is concerned with the application of mathematics to molecular genetics and evolutionary biology. All necessary mathematics and biology is introduced as needed. **(a)**

Graduation: The Revision of Estimates. Dick London. Winsted, CT: ACTEX, 1985. viii + 183pp. $25.00 paperbound. ISBN 0-936031-00-X.

Text is primarily for the benefit of students preparing for actuarial exams. Chapters deal with the theory, assumptions, and mathematics of particular graduation approaches, with exercises at end of each chapter. Bibliography. **(a)**

Introduction to Applied Mathematics. Gilbert Strang. Wellesley, MA: Wellesley-Cambridge Press, 1986. xii + 758pp. $47.50. ISBN 0-9614088-0-4.

A dramatic, innovative text that integrates continuous and discrete methods with an algorithmic spirit link to readily available scientific software packages. **(a)**

An Introduction to Catastrophe Theory. P. T. Saunders. New York: Cambridge University Press, 1980. xii + 144pp. $37.50;$18.95. ISBN 0-521-23042-X; 0-521-29782-6.

A brief elementary exposition designed for those who are competent in calculus. It includes complete descriptions of elementary catastrophes with applications to physics, social science, and biology. **(p)**

Introduction to Modern Statistical Mechanics. David Chandler. New York: Oxford University Press, 1987. xiv + 274pp. $55.00. ISBN 0-19-504276-X.

Major modern topics such as Monte Carlo sampling, renormalization groups, and the fluctuation-dissipation theorem are emphasized in this text. **(a)**

Introduction to the Theory of Error-Correcting Codes. Vera Pless. New York: Wiley, 1982. xi + 169pp. $35.95. ISBN 0-471-08684-3.

An introduction to linear block codes for mathematics, computer science, and electrical engineering students which assumes knowledge of linear algebra. **(a)**

The Invisible Hand: Economic Equilibrium in the History of Science. Bruna Ingrao, Giorgio Israel. Transl: I. McGilvray. Cambridge, MA: MIT Press, 1990. xiii + 491pp. $47.50. ISBN 0-262-09028-7.

Explores the history of efforts to mathematicize economics. **(a)**

Life Contingencies: A Guide for the Actuarial Student. Robert W. Batten. New York: Prentice-Hall, 1990. x + 353pp. $46.67. ISBN 0-13-539859-2.

Illustrative examples and exercises drive home the myriad of detailed concepts of this subject to allow students to pass the Society of Actuaries' difficult exam in life contingencies. **(a)**

Mathematical Bioeconomics: The Optimal Management of Renewable Resources. Second Edition. Colin W. Clark. New York: Wiley,1990. xiii + 386pp. $49.95. ISBN 0-4715-0883-7.

Revision of 1977 work, featuring economic models of resource management, optimal control, and models. **(p)**

Mathematical Methods in Engineering. Ed: Glyn A. O. Davies. Handbook of Applicable Mathematics;5. New York: Wiley, 1984. xiv + 458pp. $57.95. ISBN 0-471-10331-4.

Guidebook of fourteen chapters, each with a reading list, on branches of engineering which rely on mathematical techniques. **(p)**

Mathematical Methods in the Physical Sciences. Second Edition. Mary L. Boas. New York: Wiley, 1983. xx + 793pp. $64.95. ISBN 0-471-04409-1.

Intended for students who want to develop in a short time, basic competence in the areas of mathematics needed in junior- to graduate-level courses in physics, chemistry, and engineering. More problems in this edition. **(p)**

Mathematical Methods of Classical Mechanics. Second Edition. V. I. Arnol'd. Transl: K. Vogtmann, A. Weinstein. Graduate Texts in Mathematics;60. New York: Springer-Verlag, 1989. xvi + 508pp. $49.80. ISBN 0-387-96890-3.

Based on a course given by the author in 1966-68, but brought up-to-date with three new appendices on his and co-authors' work on Poisson structures, etc. **(a)**

Mathematics and Medicine: How Serious is the Injury? William Sacco et al. Providence, RI: Janson, 1987. viii + 53pp. $9.95. ISBN 0-939765-06-3.

Self-teaching pamphlet showing how mathematical indices which measure the severity of patient illness/injury can be defined, constructed, and applied. **(a)**

Mathematics for Biomedical Applications. Stanton A. Glantz. Berkeley, CA: University of California Press, 1979. xii + 423pp. $45.00. ISBN 0-520-03599-2.

Highly recommended book for those interested in biomedical applications of differential equations. **(a)**

Mathematics in Action. O. G. Sutton. New York: Dover, 1984. viii + 236pp. $6.95. ISBN 0-486-24759-7.

Reprint of 1966 edition, this light treatment of mathematics applied to engineering is dated but quite readable. **(p)**

The Mathematics of Combustion. Ed: John D. Buckmaster. Philadelphia, PA: Society for Industrial and Applied Mathematics, 1985. xii + 254pp. $36.50. ISBN 0-89871-053-7.

Five papers on the subject, each with references. **(a)**

Mathematics of Compound Interest. Marjorie V. Butcher, Cecil J. Nesbitt. Ann Arbor, MI: Ulrich, 1971. xii + 324pp. $22.50. ISBN 0-9603000-1-5.

The purpose of the book is to give a comprehensive, precise treatment of the mathematical theory of the subject. Intended either for the classroom or individual study. **(a,p)**

Mathematics of Projectiles in Sport. Neville de Mestre. Australian Mathematical Society Lecture Series;6. New York: Cambridge University Press, 1990. xi + 175pp. $24.95. ISBN 0-521-39857-6.

All aspects of projectiles, derived from military sources, but applied to sports projectiles. Extensive references. **(a)**

Modern Cellular Automata: Theory and Applications. Kendall Preston, Jr., Michael J. B. Duff. New York: Plenum, 1984. xviii + 340pp. $75.00. ISBN 0-306-41737-5.

An introduction to cellular logic and cellular automata devoted particularly to parts of those subjects which deal with the manipulation of pictorial data. **(a)**

Multidimensional Systems Theory: Progress, Directions and Open Problems in Multidimensional Systems. Ed: N. K. Bose. Dordrecht: D. Reidel, 1985. xv + 264pp. $101.65. ISBN 90-277-1764-8.

Recent developments in multidimensional systems theory, intended for engineers and mathematicians. **(a)**

Negotiation Games: Applying Game Theory to Bargaining and Arbitration. Steven J. Brams. New York: Routledge Chapman and Hall, 1990. xviii + 297pp. $15.95;$49.50. ISBN 0-415-90338-6;0-415-90337-8.

A model-based survey of negotiations, rooted in game theory, but extended to common systems of bargaining and arbitration. **(p)**

The New Ambidextrous Universe: Symmetry and Asymmetry from Mirror Reflections to Superstrings. Third Revised Edition. Martin Gardner. New York: W. H. Freeman, 1990. xiv + 392pp. $19.95. ISBN 0-7167-2092-2.

Latest revision to Gardner's classic 1964 non-technical exposition of the fall of parity adds 5 more chapters on twistors, superstrings, etc. **(p)**

New Directions in Applied Mathematics. Ed: Peter J. Hilton, Gail S. Young. New York: Springer-Verlag, 1982. ix + 163pp. $49.00. ISBN 0-387-90604-5.

Seven lectures on modern topics in applied mathematics presented at Case Western Reserve University in April 1980 plus an essay on the nature and pedagogy of applied mathematics by Hilton. **(a)**

Optimal Shape Design for Elliptic Systems. Olivier Pironneau. New York: Springer-Verlag, 1984. xii + 168pp. $62.00. ISBN 0-387-12069-6.

The author presents the French school of applied mathematics in this applications-oriented study of physical systems. **(a)**

A Primer of Population Biology. Edward O. Wilson, William H. Bossert. Sunderland, MA: Sinauer Associates, 1971. 192pp. $12.95. ISBN 0-87893-926-1.

An introduction to the methods of population biology, e.g., mathematical model building, measurement techniques, and problem solving. **(p)**

Principles of Thermodynamics and Statistical Mechanics. D. F. Lawden. New York: Wiley, 1987. xiii + 154pp. $48.95. ISBN 0-471-91172-0.

Based on a course to mathematics students at the University of Aston, Birmingham, this volume emphasizes the mathematical development of thermodynamics. **(a)**

Problems in Theoretical Physics. Willi-Hans Steeb. Mannheim, Germany: Bibliographisches Institut, 1990. *Volume I: Introductory Problems.* 197pp. $18.55. ISBN 3-411-14451-3. *Volume II: Advanced Problems.* 290pp. $21.65. ISBN 3-411-14611-7.

The first volume should be familiar to those who have completed courses in linear algebra and complex analysis while the second volume contains 136 advanced problems in ordinary and partial differential equations, vector and tensor analysis, differential forms, etc. Both contain useful asides and clear solutions. **(p)**

Public Program Analysis: A New Categorical Data Approach. Ron N. Forthofer, Robert G. Lehnen. New York: Lifetime Learning, 1981. xx + 294pp. $31.95. ISBN 0-534-97974-2.

Describes and illustrates the least squares approach for analyzing categorical data in the areas of public health and public affairs. **(p)**

Quantum Mechanics in Simple Matrix Form. Thomas F. Jordan. New York: Wiley, 1986. xi + 259pp. $39.95. ISBN 0-471-81751-1.

A completely self-contained exposition of QP−PQ = ih/2 pi and all that it implies using only precalculus mathematics of complex numbers, matrix algebra, and expected value. **(p)**

Quantum Mechanics of Atoms and Molecules. Walter Thirring. A Course in Mathematical Physics;3. Transl: Evans M. Harrell. New York: Springer-Verlag, 1981. viii + 300pp. $49.00. ISBN 0-387-81620-8.

An axiomatic introduction to quantum mechanics which illustrates the use of modern functional analysis in mathematical physics. **(a)**

Rational Thermodynamics. Second Edition. C. Truesdell. New York: Springer-Verlag, 1984. xvii + 578pp. $120.00. ISBN 0-387-90874-9.

First edition has been significantly enlarged with the addition of an extensive new historical introduction plus 22 appendices by individuals whose work has advanced the field in the last fifteen years. **(a)**

Sound Propagation in Stratified Fluids. Calvin H. Wilcox. Applied Mathematical Sciences;50. New York: Springer-Verlag, 1984. ix + 198pp. $36.00. ISBN 0-387-90986-9.

A highly mathematical and usually clear work presenting the theory of the propagation of transient sound waves in fluids whose densities and sound speeds are functions of depth. **(a)**

Space Kinematics and Lie Groups. Adolf Karger, Josef Novak. New York: Gordon & Breach, 1985. xv + 422pp. $98.00. ISBN 2-88124-023-2.

A translation, divided into 2 somewhat independent parts, one for those who deal with concrete problems and the other for the theoretically oriented, with a beginning chapter introducing differentiable manifolds and Lie groups. **(a)**

Spacetime and Singularities, An Introduction. Gregory L. Nabor. London Mathematical Society Student Texts;11. New York: Cambridge University Press, 1988. ix + 178pp. $17.95. 0-521-33612-0.

A single-minded attempt to explain Hawking's Theorem. **(p)**

Studies in Mathematical Economics. Ed: Stanley Reiter. Studies in Mathematics;25. Washington, DC: Mathematical Association of America, 1986. xiii + 441pp. $46.50. ISBN 0-88385-127-X.

Eight well-known mathematical economists present essays on mathematical methods in modern economic theory. **(a)**

Superpower Games: Applying Game Theory to Superpower Conflict. Steven J. Brams. New Haven, CT: Yale University Press, 1985. xvi + 176pp. $9.95;$30.00. ISBN 0-300-03364-8; ISBN 0-300-03323-0.

An important and innovative analysis of deterrence, arms race, etc. modelled on games such as Chicken and Prisoners' Dilemma. **(a,p)**

Symplectic Techniques in Physics. Victor Guillemin, Shlomo Sternberg. New York: Cambridge University Press, 1984. xi + 468pp. $27.95. ISBN 0-521-24866-3.

Applications in physics are an integral part of this introduction to symplectic geometry from a modern point of view. **(a)**

The Theory of Interest. Second Edition. Stephen G. Kellison. Homewood, IL: R. D. Irwin, 1991. xiv + 446pp. $43.95. ISBN 0-256-09150-1.

This edition is a substantially revised and expanded treatment of the theory of interest. "With a few minor exceptions, all the material in the first edition has been retained and updated. In addition, a significant amount of new material [e.g., an introduction to the economic and financial theory of interest] and has been added."—Preface. **(a)**

The Theory of Error-Correcting Codes. F. J. MacWilliams, N. J. A. Sloane. North-Holland Mathematical Library;16. New York: Elsevier, 1978. xviii + 762pp. $65.00. ISBN 0-444-85193-3.

Encyclopedic work with over 31 classes of codes considered. The authors are concerned with the construction and properties of codes, bounds on the size and minimum distance of codes and the algebraic and combinatorial construction of codes. Bibliography. **(a)**

Topics in the Theory of Voting. Philip D. Straffin, Jr. Boston: Birkhäuser, 1980. ix + 69pp. $5.00. ISBN 3-7643-3017-1.

UMAP monograph surveying three aspects of the theory of voting: measures of voting powers, voting methods when there are more than two alternatives, and voting schemes based on intensities of preference. **(p)**

Toward a Formal Science of Economics: The Axiomatic Method in Economics and Econometrics. Brent P. Stigum. Cambridge, MA: MIT Press, 1990. xiv + 1033pp. $45.00. ISBN 0-262-19284-5.

Encyclopedic treatment of foundations for a formal methodological basis of econometric theory. **(a)**

A Treatise on the Analytical Dynamics of Particles and Rigid Bodies. Fourth Edition. E. T. Whittaker. New York: Cambridge University Press, 1988. xvii + 456pp. $27.95. ISBN 0-521-35883-3.

Paperback edition of the standard work on the subject. **(a)**

The Ubiquity of Chaos. Ed: Saul Krasner. Washington, DC: American Association for the Advancement of Science, 1990. viii + 247pp. $34.95 ISBN 0-87168-350-4.

Nineteen papers on diverse applications of chaos presented at AAAS meetings over the past several years. **(a)**

Variational Methods in Engineering. Ed: C. A. Brebbia. New York: Springer-Verlag, 1985. 750pp. $159.00. ISBN 0-387-15496-5.

Over 40 papers from the Second International Conference on Variational Methods in Engineering held in 1985. Expensive. **(a)**

Vibration and Coupling of Continuous Systems: Asymptotic Methods. J. Sanchez Hubert, E. Sanchez Palencia. New York: Springer-Verlag, 1989. xv + 421pp. $93.00. ISBN 0-387-19384-7.

Of interest to engineers and theorists concerned with mechanics as well as specialists in computational methods. **(a)**

Wave Propagation in Solids and Fluids. Julian L. Davis. New York: Springer-Verlag, 1988. x + 386pp. $82.50. ISBN 0-387-96739-7.

A systematic account of the mathematical methods of wave propagation. Bibliography limited to mostly well-known books and a few papers. **(a)**

Worked Problems in Applied Mathematics. N. N. Lebedev, I. P. Skalskaiа̂, Y.S. Ufliând. Transl: Richard A. Silverman. New York: Dover, 1979. xi + 429pp. $8.50. ISBN 0-486-63730-1.

Unabridged reprint of the 1965 English edition with over 600 problems. This is a good sourcebook, even though the bibliography is now dated. **(p)**

CALCULUS

Advanced Calculus. Third Edition. Wilfred Kaplan. Reading, MA: Addison-Wesley, 1984. xiv + 721pp. $48.76. ISBN 0-201-11680-4.

Rigorous text, presupposing calculus and providing introductory linear algebra. Much emphasis on physical motivation and applications. References. **(a,s)**

Applied Calculus: An Intuitive Approach for Management, Life, and Social Sciences. Richard L. Faber, Marvin I. Freedman, James L. Kaplan. Saint Paul, MN: West, 1986. xiii + 607pp. $44.50. ISBN 0-314-85235-2.

A text for students preparing for careers in business or social science with many examples drawn from these fields. Authors illustrate almost every concept with several intuitive examples designed to explain the "why" of calculus. **(a,p)**

Brief Calculus with Applications. James W. Burgmeier, Monte B. Boisen, Jr., Max D. Larsen. New York: McGraw-Hill, 1990. xvi + 428pp. $25.48. ISBN 0-07-557411-X.

Emphasizes examples and applications. Many exercises, figures, and photographs are provided to motivate new concepts. Annotated bibliography. **(a)**

Calculus. Second Edition. Michael Spivak. Houston: Publish or Perish, 1980. xii + 647pp. $35.00. ISBN 0-914098-77-2.

A great book is made better with the addition of appendices on polar coordinates, uniform continuity, parameterized curves, Riemann sums, and applications of integration. **(a,p)**

Calculus. Gilbert Strang. Wellesley, MA: Wellesley-Cambridge Press, 1991. xv + 655pp. $50.00. ISBN 0-9614088-2-0.

Fresh and innovative look at calculus. **(a)**

The Calculus: A Genetic Approach. Otto Toeplitz. Chicago: University of Chicago Press, 1981. xiv + 192pp. $12.95. ISBN 0-226-80667-7.

Reprint of 1963 edition, a translation of Toeplitz's posthumous *Entwicklung der Infinitesimalrechnung*. Appended exercises are essential to book. **(a)**

Calculus and Analytic Geometry. Third Edition. C. H. Edwards, Jr., David E. Penney. New York: Prentice-Hall, 1990. xv + 926 + 79pp. $51.75. ISBN 0-13-111204-X.

Shorter and crisper than previous edition, simpler problems, and historical comments. **(a)**

Calculus and Analytic Geometry. Seventh Edition. George B. Thomas, Jr., Ross L. Finney. Reading, MA: Addison-Wesley, 1988. xxv + 1136pp. $50.36. ISBN 0-201-16320-9.

Still the grandfather of them all, containing chapter reviews, historical notes, and quick-reference charts that group formulas and describe problem-solving strategies. **(a,p,s)**

Calculus and Its Applications. Fifth Edition. Larry J. Goldstein, David C. Lay, David I. Schneider. New York: Prentice-Hall, 1989. xix + 667pp. $56.00. ISBN 0-13-110537-X.

While lacking theory, this text is nevertheless more substantial than most. **(a)**

Calculus & Mathematica. Part 1: Preliminary Edition. Donald P. Brown, Horacio Porta, J. Jerry Uhl. Reading, MA: Addison-Wesley, 1991. *Basics, Tutorials, and Literacy Sheets*. xxi + 426pp. ISBN 0-201-57270-2. *Give It a Try*. 286pp. ISBN 0-201-57271-0.

Contains 30 lesson units covering the first two terms of calculus, both on disk and in printed form. The 3 components (basics, tutorial, and "give it a try" homework section) are designed using the **Mathematica** notebooks. **(a)**

Calculus for a New Century: A Pump, Not a Filter. Ed: Lynn Arthur Steen. MAA notes;8. Washington, DC: Mathematical Association of America, 1988. xiv + 258pp. $15.00. ISBN 0-88385-058-3.

Expanded proceedings of a 1987 colloquium intended to stimulate national discussion about revitalizing calculus. **(a)**

Calculus for the Managerial, Life, and Social Sciences. Second Edition. S. T. Tan. Boston: PWS-Kent, 1990. xiv + 568pp. $35.00. ISBN 0-534-92145-0.

Intuitive and informal text with many examples and problems. **(p)**

Calculus, One and Several Variables. Sixth Edition. S. L. Salas, Einar Hille. New York: Wiley, 1990. xv + 1094 + 113 + 7pp. $59.95. ISBN 0-471-61195-6.

This edition features numerous changes: an early introduction of trigonometry, better balanced exercise sets, more motivational pieces, fuller treatment of numerical techniques, and a middle-of-the-book chapter on modelling and differential equations. **(a)**

Differential and Integral Calculus. Edmund Landau. Transl: Melvin Hausner, Martin Davis. New York: Chelsea, 1980. 372pp. $23.95. ISBN 0-8284-0078-4.

Reprinting of Chelsea's 1965 edition of the translation of *Einführung in die Differentialrechnung und Integralrechnung*. **(a)**

The Fractional Calculus: Theory and Applications of Differentiation and Integration to Arbitrary Order. Keith B. Oldham, Jerome Spanier. Mathematics in Science and Engineering;111. San Diego, CA: Academic Press, 1974. xiii + 234pp. $75.00. ISBN 0-12-525550-0.

While fractional calculus is almost as old as calculus itself, this is the first book devoted entirely to the subject. It is restricted to calculus in a single variable and is oriented towards applications. Annotated bibliography. **(a)**

Integration in Finite Terms: Liouville's Theory of Elementary Methods. Joseph Fels Ritt. Ann Arbor, MI: Books on Demand, 1990. vi + 100pp. $27.50. ISBN 0-317-08490-9.

Photocopy of 1948 original. **(a)**

Introduction to Calculus and Analysis. Richard Courant, Fritz John. New York: Springer, 1989. *Vol. I*. xxiii + 661pp. $45.00. ISBN 0-387-97151-3. *Vol. II*. xxiii + 954pp. $59.00. ISBN 0-387-97151-3.

Reprint of 1965-74 edition, which was written especially for American students. It was a welcome addition to the literature as Courant's 1927 *Vorlesungen über Differential- und Integralrechnung* and its English editions have exerted a major influence on the teaching of calculus in the U.S. Volume 1 is concerned with functions of a single variable and volume 2 is *the* book for obtaining geometrical and physical understanding of Gauss-Green-Stokes theorems. **(a,s)**

Prof. E. McSquared's Expanded Intergalactic Version: A Calculus Primer. Howard Swann, John Johnson. Providence, RI: Janson, 1989. 261pp. $15.50. ISBN 0-939765-12-8.

New edition of aclassic. This primer on calculus makes it understandable through wildly funny cartoon characters. **(p)**

Schaum's Outline of Theory and Problems of Advanced Calculus. Murray R. Spiegel. New York: McGraw-Hill, 1963. 384pp. $13.95. ISBN 0-07-060229-8.

"This book has been designed for use either as a supplement to all current standard textbooks or as a textbook for a formal course in advanced calculus."—Preface. **(a,p)**

Schaum's Outline of Theory and Problems of Beginning Calculus. Elliott Mendelson. New York: McGraw-Hill, 1985. 344pp. $11.50. ISBN 0-07-041465-3.

This outline is limited to the essentials of calculus. It carefully develops, giving all the steps, the principles of differentiation and integration. Useful as a review or self-contained text. **(a,p)**

Schaum's Outline of Theory and Problems of Differential and Integral Calculus. Third Edition. Frank Ayres, Jr., Elliott Mendelson. New York: McGraw-Hill, 1990. 484pp. $12.95. ISBN 0-07-02662-9.

Revisions of this well-known review include analytic geometry treated from the beginning, exponential and loagarithmic functions treated in two places, and terminology, notation and standards of rigor updated. **(a,p)**

Schaum's 3000 Solved Problems in Calculus. Elliott Mendelson. New York: McGraw-Hill, 1988. v + 442pp. $19.95. ISBN 0-07-040-7.

Excellent source for extra examples or test problems in calculus sequence. **(a,p)**

Second Year Calculus: From Celestial Mechanics to Special Relativity. David M. Bressoud. New York: Springer-Verlag, 1991. xi + 386pp. $29.95. ISBN 0-387-97606-X.

An innovative work that relates the mathematics of three and four-dimensional vector calculus to the history of physics that motivated so much of its development. Use as text or "bedtime reading." **(a,p)**

Toward a Lean and Lively Calculus. Ed: Ronald G. Douglas. MAA Notes;6. Washington, DC: Mathematical Association of America, 1986. xxvii + 249pp. $12.50. ISBN 0-88385-056-7.

Report of the 1986 Calculus Conference held at Tulane University. Papers identify problems in the way in which calculus is taught and outline alternative curricula, teaching strategies, and implementation options. **(a)**

Two-Dimensional Calculus. Robert Osserman. Melbourne, FL: Krieger, 1977. xvii + 456pp. $37.50. ISBN 0-88275-473-4.

Reprint of 1968 original. **(a)**

Vector Calculus. Third Edition. Jerrold E. Marsden, Anthony J. Tromba. New York: W. H. Freeman, 1988. xiv + 655pp. $43.95. ISBN 0-7167-1856-1.

A well-written text which gets better with each edition. **(a)**

COMBINATORICS

Applied Combinatorics. Fred S. Roberts. New York: Prentice-Hall, 1984. xviii + 606pp. $56.00. ISBN 0-13-039313-4.

Versatile text emphasizing algorithmic thinking and providing illustrations and applications to a wide variety of fields. **(a,p,s)**

Applied Combinatorics. Second Edition. Alan Tucker. New York: Wiley, 1984. xi + 447pp. $59.95. ISBN 0-471-86371-8.

Strategies of counting and analysis of graphs presented mainly by example rather than by formal proof. This edition puts the graph theory chapters first and gives more attention to computer science's use of combinatorics. **(a,p)**

Challenging Mathematical Programs with Elementary Solutions. A. M. I͡Aglom, I. M. I͡Aglom. Transl: James McCawley, Jr. New York: Dover, 1987. $6.95 each. *Volume I: Combinatorial Analysis and Probability Theory*. ix + 231pp. ISBN 0-486-65536-9.

Volume II: Problems in Various Branches of Mathematics. ix + 214pp. ISBN 0-486-65537-7.

Reprint of the translation of *Non-Elementary Problems in an Elementary Exposition* from the Russian. **(a)**

Combinatorial Algorithms: An Update. Herbert S. Wilf. CBMS-NSF Regional Conference Series in Applied Mathematics;55. Philadelphia: Society for Industrial and Applied Mathematics, 1989. v + 47pp. $12.00. ISBN 0-89871-231-9.

1987 lecture notes providerecent research results on various coding and tree algorithms. **(a)**

Combinatorial Algorithms: Theory and Practice. Edward M. Reingold, Jurg Nievergelt, Narsingh Deo. New York: Prentice-Hall, 1977. xii + 433pp. $59.00. ISBN 0-13-152447-X.

This book is a comprehensive exposition of both theoretical and practical aspects of combinatorial algorithms. The authors present algorithms in an extremely clear form. Extensive index makes it useful as a reference book. **(s)**

Combinatorial Enumeration. I. P. Goulden, D. M. Jackson. New York: Wiley, 1983. xxiv + 569pp. $83.95. ISBN 0-471-86654-7.

Encyclopedic account of the various combinatorial interpretations of the Lagrange inversion formula, the basic techniques of ordinary and exponential generating functions, with applications to a wide variety of enumeration problems. **(a)**

Combinatorial Mathematics. Herbert John Ryser. Washington, DC: Mathematical Association of America, 1963. xiv + 154pp. $21.00. ISBN 0-88385-014-1.

This clear presentation is devoted mainly to existence problems including several basic, original contributions from the author. Appropriate bibliography for each chapter as well as a good index. **(a)**

Combinatorial Search. Martin Aigner. New York: Wiley, 1988. 368pp. $99.00. ISBN 0-471-92142-4.

This book gives an introduction to the basic ideas and presents a collection of the most interesting search problems. **(a)**

Combinatorial Theory. Second Edition. Marshall Hall, Jr. New York: Wiley, 1986. xv + 440pp. $64.95. ISBN 0-471-09138-3.

While preserving the structure of the First edition, the master of combinatorics updates his 1967 work with additional sections and proofs plus a new chapter on theory of error-correcting codes. **(a)**

Combinatorics of Finite Sets. Ian Anderson. New York: Oxford University Press, 1989. xvi + 250pp. $21.95 paperbound. ISBN 0-19-853379-9.

Readable exposition of most of the fundamental ideas makes this a suitable text at the graduate level. **(a)**

Combinatorics: Set Systems, Hypergraphs, Families of Vectors and Combinatorial Probability. Bela Bollobas. New York: Cambridge University Press, 1986. xii + 177pp. $14.95 paperbound. ISBN 0-521-33703-8.

Text gives a thorough grounding in set systems and hypergraphs. Many exercises. References. **(a)**

Constructive Combinatorics. Dennis Stanton, Dennis White. New York: Springer-Verlag, 1986. x + 183pp. $25.00. ISBN 0-387-96347-2.

A unique and innovative text featuring algorithmic explanations for combinatorical phenomena. **(a)**

Convexity and Related Combinatorial Geometry. Ed: David C. Kay, Marilyn Breen. Lecture Notes in Pure and Applied Mathematics;76. New York: Dekker, 1982. viii + 243pp. $89.75. ISBN 0-8247-1278-1.

Twenty papers from a 1980 conference on convexity and combinatorics. **(a)**

The Equivalence of Some Combinatorial Matching Theorems. Philip J. Reichmeider. Washington, NJ: Polygonal, 1984. 123pp. $15.50. ISBN 0-936428-09-0.

This work emphasizes the classical theorems of Hall, König, Dilworth, Menger, Ford, and Fulkerson. **(a)**

The Four Color Problem: Assaults and Conquest. Thomas L. Saaty, Paul C. Kainen. New York: McGraw, 1977. ix + 217pp. $6.00. ISBN 0-0705-4382-8.

Longer treatment of Saaty's 1972 *American Mathematical Monthly* paper as well as an extensive discussion of Appel and Haken's 1976 computer-aided solution. Bibliography. **(a)**

An Introduction to Combinatorial Analysis. John Riordan. Princeton, NJ: Princeton University Press, 1978. xii + 244pp. $37.50. ISBN 0-691-02365-4.

New printing of the valuable 1968 classic. **(a)**

Introduction to Combinatorial Theory. R. C. Boase, B. Manvel. New York: Wiley, 1984. ix + 237pp. $45.95. ISBN 0-471-89614-4.

Excellent introductory text. **(p)**

Packing and Covering in Combinatorics. Ed: A. Schrivjer. Mathematical Centre Tracts;106. Amsterdam: Mathematisch Centrum, 1979. 313pp. $41.50. ISBN 90-6196-180-7.

This book is based on lectures given during a study week in 1978 at the Mathematical Center. Each chapter provides an excellent introduction to results, both classical and recent, of a wide variety of combinatorial problems which can be viewed as problems of packing and covering. There is an emphasis not only on results, but on methods and techniques used to obtain them. **(a)**

Ramsey Theory. Second Edition. Ronald L. Graham, Bruce L. Rothschild, Joel H. Spencer. New York: Wiley, 1990. xi + 196pp. $49.95. ISBN 0-471-50046-1.

Well-written book on this cohesive subdiscipline of combinatorial analysis (i.e., that within any sufficiently large system some regularity must always exist). This edition gives a complete treatment of Shelah's proof as well as the original proof of van der Waerden. (**a**)

Rudiments of Ramsey Theory. Ronald L. Graham. CBMS Regional Conference Series in Mathematics;45. Providence, RI: American Mathematical Society, 1981. v+65pp. $5.60 ISBN O-8218-1696-9.

Based on lectures given at a 1979 CBMS regional research conference at St. Olaf College.

COMPLEX ANALYSIS

Analytic Function Theory. Second Edition. Einar Hille. New York: Chelsea, 1973. *Volume I*. xi + 308pp. $22.00. ISBN 0-8284-0269-8. *Volume II*. xii + 496pp. $24.00. ISBN 0-8284-0270-1.

Reprint of the 1959 classic. (**a**)

Applied and Computational Complex Analysis. Peter Henrici. New York: Wiley, 1974-1986. $189.00 set. ISBN 0-471-84241-9. *Volume 1: Power Series–Integration–Conformal Mapping–Location of Zeros*. xv + 682pp. $68.95;$36.95. ISBN 0-471-37244-7;0-471-60841-6. *Volume 2: Special Functions–Integral Transforms–Asymptotics–Continued Fractions*. ix + 662pp. $39.95. ISBN 0-471-01525-3. *Volume 3: Discrete Fourier Analysis–Cauchy Integrals–Construction of Conformal Maps–Univalent Functions*. xvi + 637pp. $36.95. ISBN 0-471-08703-3.

Highly-regarded 3 volume set devoted to the theory of functions of one or more complex variables and its applications. A principal objective of Henrici is to develop solutions in algorithmic form and adaptable, if possible, to machine computation. Expensive. (**s**)

Basic Complex Analysis. Second Edition. Jerrold E. Marsden, Michael J. Hoffman. New York: W. H. Freeman, 1987. xii + 604pp. $43.95. ISBN 0-7167-1814-6.

Substantial revision with new material added, parts rewritten, enlarged collection of many exercises. A recommended text for complex analysis courses and independent study. (**p**)

Complex Analysis: A Functional Analysis Approach. D. H. Luecking, L. A. Rubel. New York: Springer-Verlag, 1984. vi + 176pp. $29.00. ISBN 0-387-90993-1.

Highly recommended for both the specialist and those students with a semester of complex analysis. Much material found in no other text, often related to Rubel's research. (**a**)

Complex Analysis: An Introduction to the Theory of Analytic Functions of One Complex Variable. Third Edition. Lars V. Ahlfors. New York: McGraw-Hill, 1979. xiv + 331pp. $45.95. ISBN 0-07-00657-1.

Major changes have been made from the Second edition: a new, simpler proof of the general form of Cauchy's theorem, an introduction to the Riemann zeta function, and a rewritten, modern treatment of global analytic functions in the terminology of germs and sheaves. (**a**)

Complex Analysis and Applications. Second Edition. William R. Derrick. Pacific Grove, CA: Brooks-Cole, 1984. xiii + 347pp. $53.95. ISBN 0-534-02853-5.

A detailed and relatively rigorous treatment of standard topics, with many worked examples. (**a**)

Complex Analysis: The Geometric Viewpoint. Steven G. Krantz. Carus Mathematical Monographs;23. Washington, DC: Mathematical Association of America, 1990. xv + 212pp. $25.95. ISBN 0-88385-026-5.

Intended for a broad audience and assuming only one semester of complex analysis, this book explains how complex analysis can be studied using methods of geometry. (**a**)

Complex Variables. Second Edition. Stephen D. Fisher. Belmont, CA: Wadsworth, 1990. xiv + 427pp. $58.95. ISBN 0-534-13260-X.

Minor revision to his introduction to theory, methods, and mathematical applications of analytic and harmonic functions, although with one substantial change: Cauchy's theorem is obtained more directly. (**a**)

Complex Variables. Norman Levinson, Raymond M. Redheffer. Oakland, CA: Holden-Day, 1970. 429pp. $39.95. ISBN 0-812-5104-5.

Well-written with an abundance of examples and problems , this book has become the standard text at the senior level. (**a**)

Complex Variables with Applications. Fourth Edition. Ruel V. Churchill, James Ward Brown. New York: McGraw-Hill, 1984. x + 339pp. $46.60. ISBN 0-07-010873-0.

This book covers the rudimentary theory in the first half, then applies theory to conformal mapping and its physical applications, Poisson-type integrals, and further function theory. (**a**)

Fonctions d'Une Variable Réelle: Théorie Élémentaire. Nouvelle Édition. Nicolas Bourbaki. (Éléments de Mathématique) Paris: Diffusion C.C.L.S., 1982. (Available from Scientific & Medical Publications of France, New York, NY) 329pp. $82.00. ISBN 2-903684-04-9.

Reprint of 1976 edition. Another volume exhibiting the same high standard as the rest of the Bourbaki series and distributing results between text and problems. (**a**)

Function Theory of Several Complex Variables. Steven G. Krantz. New York: Wiley, 1982. xiii + 437pp. $59.95. ISBN 0-471-09324-6.

Useful and well-written text introduces holomorphic functions on C^n. Contains recent results as well as computations never collected in book form before. (**a**)

Functions of One Complex Variable. Second Edition. John B. Conway. Graduate Texts in Mathematics;11. New York: Springer-Verlag, 1986. xiii + 317pp. $39.00. ISBN 0-387-90328-3.

Corrected fourth printing of 1978 Second edition. This revision includes new proofs and exercises and a bibliographical appendix. **(a)**

Holomorphic Functions and Integral Representations in Several Complex Variables. R. Michael Range. Graduate Texts in Mathematics;108. New York: Springer-Verlag, 1986. xix + 386pp. $49.80. ISBN 0-387-96259-X.

Readable introduction to complex analysis of several variables, emphasizing integral formulas and their applications to global behavior of complex analytic objects. Extensive historical commentary. **(a)**

An Introduction to Complex Analysis in Several Variables. Third Edition. Lars Hormander. North-Holland Mathematical Library;7. New York: Elsevier, 1989. xii + 254pp. $80.00. ISBN 0-444-88446-7.

Unique among monographs on various aspects of complex analysis in several variable variables in that it uses the analytic techniques based on the solution of the d bar Neumann problem as the main tool. **(a)**

An Introduction to Multicomplex Spaces and Functions. G. Baley Price. Pure and Applied Mathematics;140. New York: Dekker, 1991. xiii + 402pp. $99.75 ISBN 0-8247-8345-X.

Clear, readable exposition with useful asides to reader. **(a)**

Introduction to Riemann Surfaces. Second Edition. George Springer. New York: Chelsea, 1981. viii + 309pp. $19.95. ISBN 0-8284-0313-9.

Corrected reprint of 1957 original. **(a)**

Kleinian Groups and Uniformization in Examples and Problems. S. L. Krushkal, B. N. Apanasov, N. A. Gusevskiĭ. Translations of Mathematical Monographs;62. Providence, RI: American Mathematical Society, 1986. vii + 198pp. $76.00. ISBN 0-8218-4516-0.

Unified exposition of all Kleinian groups and uniformization of manifolds. References to primary sources instead of proofs. Extensive bibliography. **(a)**

The Logarithmic Integral. Paul Koosis. Cambridge Studies in Advanced Mathematics;12. New York: Cambridge University Press, 1988. I. xvi + 606pp. $89.50. ISBN 0-521-30906-9.

First volume of a two-volume enthusiastic and comprehensive introduction to the logarithmic integral and its many roles in real, complex, and functional analysis. Each volume is self-contained. **(a)**

New Constructions of Functions Holomorphic in the Unit Ball of Cn. Walter Rudin. CBMS Regional Conference Series in Mathematics;63. Providence, RI: American Mathematical Society, 1986. xvi + 78pp. $17.00. ISBN 0-8218-0713-7.

Notes from 1985 expository lectures of long-standing "existence of inner functions" problem. (**a**)

Papers on Fuchsian Functions. Henri Poincaré. Transl: J. Stillwell. New York: Springer-Verlag, 1985. 483pp. $49.00. ISBN 0-387-962158-8.

Five major papers on discontinuous groups and automorphic functions. (**a**)

Le Théorème de Picard-Borel et la Théorie des Fonctions Méromorphes. Rolf Nevanlinna. New York: Chelsea, 1974. x + 171pp. $13.95. ISBN 0-8284-0272-8.

Unabridged reprint of the 1939 classic. (**a**)

Théorie des Fonctions Algébriques. Third Edition. Paul Appell, Edouard Goursat, P. Fatou. New York: Chelsea, 1976-1978. *Tome I: Étude des Fonctions Analytiques sur une Surface de Riemann*. xxxv + 526pp. $35.00. ISBN 0-8284-0285-X. *Tome II: Fonctions Automorphes*. xiv + 521pp. $30.00. ISBN 0-8284-0299-X.

Unaltered reprint of 1929 original. (**a**)

Théorie des Fonctions Algébriques de Deux Variables Indépendantes. Second Edition. Émile Picard, Georges Simart. New York: Chelsea, 1971. xiv + 524pp. $39.50. ISBN 0-8284-0248-5.

One volume reprint of two volume 1897, 1906 work on integration of several real or complex variables over curves, singularities of algebraic surfaces. (**a**)

Theory of Complex Functions. Reinhold Remmert. Transl: Robert B. Burckel. Graduate Texts in Mathematics;122. New York: Springer-Verlag, 1991. xix + 453pp. $59.00. ISBN 0-387-97195-5.

A readable and enthusiastic introduction to the theory of complex functions of one variable. Exercises. (**a**)

Univalent Functions. Peter L. Duren. Die Grundlehren der mathematischen Wissenschften;259. New York: Springer-Verlag, 1983. xiv + 382pp. $76.00. ISBN 0-387-90795-5.

Lucid and well-motivated text with numerous historical references and interesting research-type problems with each chapter. (**a**)

COMPUTER LITERACY

A Computer Science Reader. Ed: Eric A. Weiss. New York: Springer-Verlag, 1988. xvi + 447pp. $45.00. ISBN 0-387-96544-0.

Articles and features selected from the first 14 issues of *Abacus*. A handy compendium of trends and controversies in computer science today. (**p**)

The Cult of Information: The Folklore of Computers and the True Art of Thinking. Theodore Roszak. New York: Pantheon Books, 1986. xii + 238pp. $17.95. ISBN 0-394-54622-9.

A passionate discourse on deliterious effects of the computer culture (but not of computers themselves). (**p**)

Electronic Life: How to Think about Computers. Michael Crichton. New York: Alfred Knopf, 1983. ix + 209pp. $12.95. ISBN 0-394-53406-9.

A delightfully opinionated series of brief comments, from A (Afraid of Computers) to Z (Zenith: The Final Days of Man) conveying an attitude towards computers that preserves the dignity of man. **(p)**

The Sachertorte Algorithm and Other Antidotes to Computer Anxiety. John Shore. New York: Penguin, 1986. xvi + 269pp. $7.95. ISBN 0-14-008037-6.

Reprint of 1985 original. A sound, informative, well-written and non-patronizing introduction to computers for laymen. **(a,p)**

The Turing Omnibus: 61 Excursions in Computer Science. A. K. Dewdney. New York: Computer Science Press, 1989. xiv + 415pp. $24.95. ISBN 0-7167-8154-9.

Fascinating short essays on computer science topics. **(a,p)**

Using Computers: The Human Factors of Information Systems. Raymond S. Nickerson. Cambridge, MA: MIT Press, 1986. xiv +434pp. $22.50. ISBN 0-262-14040-3.

Systematic survey of trends and research in the interface between people and computers. Extensive bibliography. Indices. **(a,p)**

Writing Space: The Computer, Hypertext, and the History of Writing. Jay David Bolter. Hillsdale, NJ: Lawrence Erlbaum Associates, 1991. xiii + 258pp. $22.50;$45.00. ISBN 0-8058-0428-5;0-8058-0427-7.

Excellent introductory work on computer literacy. **(a,p)**

COMPUTER SCIENCE AND THEORY OF COMPUTATION

ACM Turing Award Lectures: The First Twenty Years, 1966-1985. Reading, MA: Addison-Wesley, 1987. xviii + 483pp. $18.36. ISBN 0-201-54885-2.

A valuable set of essays by the first 23 winners of the Turing Award which give perspective on the past, present and future of computer science. **(a)**

Algorithms for Graphics and Image Processing. Theo Pavlidis. New York: W. H. Freeman, 1982. xvii + 416pp. $45.95. ISBN 0-716-78106-9.

This text emphasizes the mathematical tools used in pictorial representation processing by computers. **(a)**

Arithmetic Complexity of Computations. Shmuel Winograd. CBMS Regional Conference Series in Applied Mathematics;33. Philadelphia: Society for Industrial and Applied Mathematics, 1980. 93pp. $16.50. ISBN 0-89871-163-0.

A brief monograph based on lectures given at the University of Pittsburgh. The unifying theme of this work is computation of a bilinear form with its greatest application in signal processing. **(a)**

The Armchair Universe: An Exploration of Computer Worlds. A. K. Dewdney. New York: W. H. Freeman, 1988. xiii + 330pp. $21.95. ISBN 0-716-71938-X.

A pleasant, informal introduction to computer recreation, reprinted from the 1984-1987 Scientific American columns. **(p)**

The Art of Computer Programming. Donald E. Knuth. Reading, MA: Addison-Wesley, 1973-1981. *Volume 1: Fundamental Algorithms. Second Edition.* xxi + 640pp. $36.76. ISBN 0-201-03809-9. *Volume 2: Seminumerical Algorithms. Second Edition.* xiii + 688pp. $36.76. ISBN 0-201-03822-6. *Volume 3: Sorting and Searching.* xi + 722pp. $36.76. ISBN 0-201-03803-X.

Very complete standard reference work. First two volumes has been extensively revised and updated. Depth of mathematics used in the analyses makes them delightful reading. **(a,s,p)**

The Artificial Intelligence Debate: False Starts, Real Foundations. Ed: Stephen R. Graubard. Cambridge, MA: MIT Press, 1988. 311pp. $9.95. ISBN 0-262-57074-2.

Reprint of the winter 1988 issue of *Daedalus* featuring 14 papers defining and debating the prospects of artificial intelligence and speculating on promising directions for future research. A special focus is on "connectionism." **(p)**

Cipher Systems: The Protection of Communications. Henry Beker, Fred Piper. New York: Wiley, 1982. 427pp. $61.95.

Comprehensive introduction to cryptography, with references to computer science, electronics, and statistics. **(a,p)**

The Complexity of Robot Motion Planning. John Canny. Cambridge, MA: MIT Press, 1988. xvi + 197pp. $27.50. ISBN 0-262-03136-1.

In this outstanding dissertation, the author obtains exponentially faster solution algorithms by application of sophisticated mathematics, then proceeds to solve three long-open problems about robot motion complexity. **(a,p)**

The Computer Age: A Twenty Year View. Ed: Michael L. Dertouzos, Joel Moses. Cambridge, MA: MIT Press, 1979. xvi + 491pp. $14.95. ISBN 0-2620-4055-7.

With much hindsight it is interesting to gauge how well the 21 "veterans" of the computer revolution predicted its impact on science, industry, education,and society for the rest of this century. **(a,p)**

Elements of the Theory of Computation. Harry R. Lewis, Christos H. Papadimitriou. New York: Prentice-Hall, 1981. xiv + 466pp. $22.95. ISBN 0-132-73417-6.

Introductory text covering theory of automata and formal languages, computability, computational complexity, and mathematical logic. Bibliography. **(p,s)**

Graphics Gems. Ed: Andrew S. Glassner. San Diego, CA: Academic Press, 1990. xxix + 833pp. $49.95. ISBN 0-12-286165-5.

A collection of algorithms, programs, and mathematical techniques for the computer graphics programmer. **(p)**

An Introduction to Splines for Use in Computer Graphics and Geometric Modeling. Richard H. Bartels, John C. Beatty, Brian A. Barsky. San Mateo, CA: Morgan Kaufmann, 1987. xiv + 476pp. $49.95. ISBN 0-934613-27-3.

An intuitive introduction to parametric spline curves and parametric, tensor-product splines surfaces emphasizing those topics most applicable to computer graphics. **(a)**

Intelligence: The Eye, the Brain, and the Computer. Martin A. Fischler, Oscar Firschein. Reading, MA: Addison-Wesley, 1987. xiv + 331pp. $24.76. ISBN 0-201-12001-1.

A readable introduction to current research in artificial intelligence, covering cognition, vision, and representations in the very context of both the brain and the computer. **(p)**

An Introduction to Formal Language Theory. Robert N. Moll, Michael A. Arbib, A. J. Kfoury. New York: Springer-Verlag, 1988. x + 203pp. $49.00. ISBN 0-387-96698-6.

Covers families of languages and corresponding automata, parsing, computability, fixed point methods, but the text's unique feature is its introduction to formal theory of natural languages. **(a)**

An Introduction to Microcomputers. Volume 1: Basic Concepts. Revised Edition. Adam Osborne. Berkeley, CA: Osborne/McGraw-Hill, 1987. 320pp. $19.95. ISBN 0-0793-1064-8.

Very good technical introduction to machine organization in the framework of microcomputers. Good index. **(p)**

Knapsack Problems: Algorithms and Computer Implementations. Silvano Martello, Paolo Toth. New York: Wiley, 1990. xii + 296pp. $110.00. ISBN 0-471-92420-2.

Presents state-of-the-art exact and approximate algorithms for a number of important NP-hard integer programming problems as well as including theoretical and empirical results on computational complexity. Expensive. **(a)**

Logics of Programs. Ed: Edmund Clarke, Dexter Kozen. Lecture Notes in Computer Science;164. New York: Springer-Verlag, 1984. vi + 527pp. $39.00. ISBN 0-387-12896-4.

Thirty-four papers on topics relating logics and programs, including interval temporal logic, algebra of hierarchies, etc. **(a)**

Non-Uniform Random Variate Generation. Luc Devroye. New York: Springer-Verlag, 1986. xvi + 843pp. $82.00. ISBN 0-387-96305-7.

An immense work on computer generation of the random variate. While it does not cover Monte Carlo methods, it does nearly everything else. Annotated bibliography. **(a)**

Structural Complexity. Jose Luis Balcazar, Josep Diaz, Joaquim Gabarro. EATCS Monographs on Theoretical Computer Science;11,22. New York: Springer-Verlag, 1988-1990. I. x + 191pp. $37.00. ISBN 0-387-18622-0. II. ix + 283pp. $39.50. ISBN 0-387-52079-1.

This set of books focuses on a type of complexity theory which is less concerned with concrete algorithmic problems than with classes of problems and their unifying characteristics. **(a)**

COMPUTER SYSTEMS

Computation Structures. Stephen A. Ward, Robert H. Halstead, Jr. Cambridge, MA: MIT Press, 1990. xx + 789pp. $49.95. ISBN 0-2622-3139-5.

Interesting textbook used at MIT covers the design and implementation of computers from the level of logic gates to machine language. **(a)**

Computer Algebra: Systems and Algorithms for Algebraic Computation. J. H. Davenport, Y. Siret, E. Tournier. Transl: A. Davenport, J. H. Davenport. San Diego, CA: Academic Press, 1988. xix + 267pp. $39.95. 0-12-204230-1.

The first systematic exposition of algorithms and data representation used in computer algebra systems. Translated from the French original. **(a)**

Exploring Mathematics with Mathematica: Dialogs Concerning Computers and Mathematics. Theodore W. Gray, Jerry Glynn. Reading, MA: Addison-Wesley, 1991. 535pp. $46.25. ISBN 0-201-52809-6.

A glib, eclectic tour of the computer algebra system Mathematica showing off its power with both traditional and bizarre applications. **(a,p)**

Mathematica: A System for Doing Mathematics by Computer. Stephen Wolfram. Reading, MA: Addison-Wesley, 1991. xxii + 961pp. $35.96; $24.76. ISBN 0-201-51502-4;0-201-51507-5.

The "definitive" user guide to the latest symbolic computer package. It provides an extensive example-based survey of *Mathematica* behavior and organized so the reader only needs to learn the part needed for a particular calculation. Full reference guide and a thorough explanation of system design. **(a)**

Mathematica for the Sciences. Richard E. Crandall. Reading, MA: Addison-Wesley, 1991. xiii + 300pp. $26.36. ISBN 0-201-51001-4.

Collection of examples and exploratory exercises from mathematics and the sciences that illustrate the use of Mathematica as a problem-solving tool. **(s)**

Mathematica in Action. Stan Wagon. New York: W. H. Freeman, 1991. xiv + 419pp. $14.95; $24.95. ISBN 0-716-72229-1;0-716-72202-X.

Intended for the Mathematica novice, this book allows the reader to explore the virtually limitless capabilities of the program. Hundreds of diverse examples are worked in detail. **(a,p)**

CONTROL THEORY, DYNAMICAL SYSTEMS, AND GAME THEORY

The Beauty of Fractals: Images of Complex Dynamical Systems. H.-O. Peitgen, P. H. Richter. New York: Springer-Verlag, 1986. xii + 199pp. $39.00. ISBN 0-387-15851-0.

A stunning photographic and mathematical display of fractal geometry arising in dynamical systems. **(a,p,s)**

Chaos. Hao Bai-Lin. Teaneck, NJ: World Scientific, 1984. 592pp. $29.00;$62.00. ISBN 9971-966-51-4;9971-966-50-6.

An exposition of the occurrence of turbulence and chaos, especially in physical systems, followed by reprints of 41 key papers. **(p,s)**

Chaos and Fractals: The Mathematics Behind the Computer Graphics. Ed: Robert L. Devaney, Linda Keen. Proceedings of Symposia in Applied Mathematics;39. Providence, RI: American Mathematical Society, 1989. ix + pp. $33.00. ISBN 0-8218-0137-6.

Expanded lectures from the 1988 AMS Short Course revealing the mathematics behind the glitz. Many color plates. **(a,s)**

Chaos, Fractals, and Dynamics: Computer Experiments in Mathematics. Robert L. Devaney. Reading, MA: Addison-Wesley, 1990. x + 181pp. $22.00. ISBN 0-201-23288-X.

High-school-level introduction to quadratic iteration and the Julia and Mandelbrot sets using simple calculator and Basic programs. **(p)**

Chaotic Dynamics of Nonlinear Systems. S. Neil Rasband. New York: Wiley, 1990. x + 230pp. $42.95. ISBN 0-471-63418-2.

Nice text primarily intended for physics students on the mathematics of chaos. Extensive bibliography and very useful glossary. **(a,s)**

Chaotic Evolution and Strange Attractors. David Ruelle. New York: Cambridge University Press, 1989. xi + 96pp. $12.95; $39.50. ISBN 0-521-36830-8; 0-521-36272-5.

Series of short chapters investigating chaotic behavior of deterministic dynamical systems using time series analysis. Intended for scientists in a wide range of fields. **(a,s)**

Chaotic Vibrations: An Introduction for Applied Scientists and Engineers. Francis C. Moon. New York: Wiley, 1987. xv + 309pp. $49.95. ISBN 0-471-85685-1.

An applied view of chaotic phenomena, richly illustrated with physical systems, mathematical models, and standard examples. **(a,s)**

The Compleat Strategyst: Being a Primer on the Theory of Games of Strategy. J. D. Williams. New York: Dover, 1986. xvi + 268pp. $6.95. ISBN 0-486-25101-2.

An unabridged reprint of the 1966 revision of the 1954 original, an enduring classic. **(p)**

Differential Equations and Dynamical Systems. Lawrence Perko. Texts in Applied Mathematics;7. New York: Springer-Verlag, 1991. xii + 403pp. $39.00. ISBN 0-387-97443-1.

Emphasizing qualitative and geometric behavior, this work would make a good text for a second course in differential equations. Author covers linear systems, local aspects of nonlinear systems, and bifurcations of nonlinear systems. **(a)**

Dynamical Systems. Ed: D. V. Anosov, V. I. Arnol'd et al. Encyclopedia of Mathematical Sciences;1-6,39. New York: Sringer-Verlag, 1988-1993. *Volume I: Ordinary Differential Equations and Smooth Dynamical Systems.* ix + 233pp. $59.00. ISBN 0-387-17000-6. *Volume II: Ergodic theory with Applications to Dynamical Systems and Statistical Mechanics.* viii + 281pp. $59.95. ISBN 0-387-17001-4. *Volume III: Mathematical Aspects of Classical and Celestial Mechanics.* xiv + 291pp. $59.00. ISBN 0-387-17002-2. *Volume IV: Symplectic Geometry and its Applications.* vii + 283pp. $59.00. ISBN 0-387-17003-0. *Volume V: Bifuraction Theory and Catastrophe Theory.* 280pp. $59.00. ISBN 0-387-18173-3. *Volume VI: Singularity Theory I.* 245pp. $59.00. ISBN 0-387-50583-0. *Volume VII: Integrable Systems. Nonholonomic Dynamical Systems.* 350pp. $77.50. ISBN 0-387-18176-8. *Volume VIII: Singularity Theory II: Applications.* 235pp. $59.00. ISBN 0-387-53376-1.

From the Russian school, these are the first volumes of a set of books on dynamical systems, differential equations, and mathematical aspects of classical and celestial mechanics. Many examples motivate definitions and theorems. **(a)**

Dynamics. S. Neil Rasband. New York: Wiley, 1983. xi + 272pp. $43.95. ISBN 0-471-87398-5.

Classical dynamics for a non-traditional geometric viewpoint with perturbation and Lie algebra techniques. Selected references. **(a)**

Dynamics–The Geometry of Behavior. Ralph H. Abraham, Christopher D. Shaw. Santa Cruz, CA: Aerial Press, 1982-1988. *Part 1: Periodic Behavior.* x + 220pp. $34.00. ISBN 0-942344-01-4. *Part 2: Chaotic Behavior.* xi + 139pp. $15.00. ISBN 0-942344-02-2. *Part 3: Global Behavior.* xi + 123pp. $28.00. ISBN 0-942344-03-0. *Part 4: Bifurcation Behavior.* xi + 196pp. $38.00. ISBN 0-942344-04-9.

Innovative series, consisting of captioned four-color illustrations of the mathematical ideas of nonlinear dynamics. Companion Apple discs are available. **(a)**

Ergodic Theory. I. P. Kornfel'd, S. V. Fomin, I͡A. G. Sinai. Die Grundlehren der mathematischen Wissenschaften;245. New York: Springer-Verlag, 1982. x + 486pp. $94.00. ISBN 0-387-90580-4.

Translation of 1980 *Ergodischeskai͡a Teorii͡a.* Comprehensive survey, not for the beginner, ranging from the foundations of the subject to a discussion of progress over the last 20 years as seen from a Russian viewpoint. **(a)**

Ergodic Theory and Topological Dynamics. James R. Brown. Pure and Applied Mathematics;70. San Diego, CA: Academic Press, 1976. x + 190pp. $64.00. ISBN 0-12-137150-6.

This book is written for the person wanting to learn ergodic theory from scratch. Some mathematical errors. **(a)**

Exploring Complexity, An Introduction. Gregoire Nicolis, Ilya Prigogine. New York: W. H. Freeman, 1989. xi + 313pp. $16.95. ISBN 0-7167-1859-6.

Authors stress that concepts of complexity are universal and important to better understand the global environment. **(a,p)**

Foundations of Optimal Control. E. B. Lee, L. Markus. Melbourne, FL: Krieger, 1986. x + 576pp. $64.95. ISBN 0-89874-807-0.

Slightly revised reprint of 1967 original which gives the basic concepts of control theory via the qualitative theory of deterministic differential systems. **(a)**

Fractals Everywhere. Michael Barnsley. San Diego, CA: Academic Press, 1988. xii + 394pp. $44.50. ISBN 0-12-079062-9.

First text of fractal geometry which focuses on compression and reconstruction of images that can be approximated by algorithmic transformations that describe iterative relations among self-similar parts. **(a,s)**

Game Theory and Its Applications. Ed: William F. Lucas. Proceedings of Symposia in Applied Mathematics;24. Providence, RI: American Mathematical Society, 1981. viii + 125pp. $18.00. ISBN 0-8218-0025-6.

Lecture notes from a 1979 AMS Short Course in Biloxi. Topics cover cooperative games, non-cooperative games, valuation and measurement of power, and economic market games. **(a)**

Geometric Theory of Dynamical Systems: An Introduction. Jacob Palis, Jr., Welington de Melo. Transl: A. K. Manning. New York: Springer-Verlag, 1982. xii + 198pp. $45.00. ISBN 0-387-90668-1.

This graduate-level text is a beautiful treatment of the foundations of dynamical systems and is, for the most part, self-contained. **(a)**

Global Bifurcations and Chaos: Analytical Methods. Stephen Wiggins. Applied Mathematical Sciences;73. New York: Springer-Verlag, 1991. xiv + 494pp. $49.95. ISBN 0-387-96775-3.

First reprinting of a rigorous treatment of chaotic phenomena in nonlinear dynamical systems written for the applied scientist. **(s)**

An Introduction to Chaotic Dynamical Systems. Second Edition. Robert L. Devaney. Reading, MA: Addison-Wesley, 1989. xvi + 336pp. $34.36. ISBN 0-201-13046-7.

An example-oriented introduction to nonlinear dynamical systems via the study of iterates of maps in one and two dimensions. This edition adds new material on the orbit diagram and a new section on the Mandelbrot set. **(a)**

Introduction to Dynamic Systems: Theory, Models, and Applications. David G. Luenberger. New York: Wiley, 1979. xiv + 446pp. $67.95. ISBN 0-471-02594-1.

This text blends the classical differential equation approach with the state space approach, with much of the last half devoted to control theory. **(a)**

Introduction to Dynamics. Ian Percival, Derek Richards. New York: Cambridge University Press, 1982. 228pp. $24.95. ISBN 0-521-28149-0.

Good initial text on dynamical systems. Many practical problems presented either as solved examples or exercises. **(p)**

Iterated Maps on the Interval of Dynamical Systems. Pierre Collet, Jean-Pierre Eckmann. Progress in Physics;1. Boston: Birkhäuser, 1980. vii + 248pp. $39.50. ISBN 0-8176-3026-0.

Thorough and readable introduction to some aspects of the theory of one-dimensional dynamical systems. **(a)**

The Mathematics of Time: Essays on Dynamical Systems, Economic Processes, and Related Topics. Steve Smale. New York: Springer-Verlag, 1980. vi + 151pp. $29.80. ISBN 0-387-90519-7.

This collection of expository essays traces Smale's involvement in dynamical systems and mathematical economics over about 12 years, including an previously unpublished essay. **(a,p)**

Optimal Control of Systems Governed by Partial Differential Equations. J. L. Lions. Transl: S. K. Mitter. Die Grundlehren der mathematischen Wissenschaften;170. New York: Springer-Verlag, 1971. xi + 396pp. $69.00. ISBN 0-387-05115-5.

This translation of the 1968 French edition discusses the control of systems in which the state is a solution to a partial differential equation. **(a)**

Perspectives of Nonlinear Dynamics. E. Atlee Jackson. New York: Cambridge University Press, 1989-1990. *Volume 1.* xix + 495pp. $59.50. ISBN 0-521-34504-9. *Volume 2.* xviii + 632pp. $59.50. ISBN 0-521-35458-7.

Both volumes form an extensive introduction written by a physicist for graduate students in the sciences and engineering. The first volume is a collection of examples, results, and applications in nonlinear systems arranged in rough chronological order, while the second includes twist maps, solitons, Lyapunov exponents, and much more. Extensive bibliography. **(a,s)**

Report of the Panel on Future Directions in Control Theory: A Mathematical Perspective. Wendell H. Fleming. Philadelphia: Society for Industrial and Applied Mathematics, 1988. 98pp. Gratis. ISBN 0-89871-234-3.

A good overview of modern control theory with 11 vignettes providing specific illustrations. **(a)**

Smooth Dynamical Systems. M. C. Irwin. Pure and Applied Mathematics;94. San Diego, CA: Academic Press, 1980. x + 259pp. $112.00. ISBN 0-12-374450-4.

Introduction to the ideas and methods of the theory of dynamical systems, closely connected to the theory of ordinary differential equations. Expensive. **(a)**

DIFFERENTIAL EQUATIONS

Applications of Lie Groups to Differential Equations. Peter J. Olver. Graduate Texts in Mathematics;107. New York: Springer-Verlag, 1986. xxxvi + 497pp. $54.00. ISBN 0-387-96250-6.

This beautiful book gives an introduction to the classical treatment of applications of Lie groups to differential equations while serving at the same time as a textbook, giving the beginner access to mathematics developed over the last two decades. **(s)**

Difference Equations. Ronald E. Mickens. New York: Van Nostrand Reinhold, 1987. xii + 243pp. $41.95. ISBN 0-442-26076-8.

A presentation and explanation of mathematical methods for determining solutions to linear and nonlinear difference equations. **(p)**

Differential Equations: A Modelling Approach. Robert L. Borrelli, Courtney S. Coleman. New York: Prentice-Hall, 1987. xvii + 607 + 77pp. $46.00. ISBN 0-13-211533-6.

Intended as an introductory level textbook in differential equations, presented within the context of a modeling environment, for second- or third-year university students of science, mathematics, and engineering. **(a)**

Differential Equations and Their Applications. Third Edition, Short Version. Martin Braun. New York: Springer-Verlag, 1983. x + 341pp. $19.80. ISBN 0-387-90847-1.

This edition features a rewritten section on singular solutions of differential equations and a new section on Euler equations. **(p)**

Differential Equations, Dynamical Systems, and Linear Algebra. Morris W. Hirsch, Stephen Smale. Pure and Applied Mathematics;60. San Diego, CA: Academic Press, 1974. xi + 358pp. $39.50. ISBN 0-12-349550-4.

While this book is definitely designed to be a course textbook, this course does not exist at most universities. Thus, it appears to be most useful as supplementary reading in a differential equations course. Length and complexity of authors' presentation of linear algebra is offset by clarity of sections dealing with fundamental theory plus the sections on applications. **(a)**

Differential Equations: Geometric Theory. Second Edition. Solomon Lefschetz. New York: Dover, 1977. x + 390pp. $8.95. ISBN 0-486-63466-9.

Unabridged reprint of Second 1963 edition. **(a)**

Differential Inclusions: Set-Valued Maps and Viability Theory. Jean-Pierre Aubin, Arrigo Cellina. Die Grundlehren der mathematischen Wissenschaften;264. New York: Springer-Verlag, 1984. xiii + 342pp. $89.00. ISBN 0-387-13105-1.

Intended for the general reader, this book presents a new, systematic and unified approach to the essential ideas of differential inclusions and related topics, i.e., viability problems and optimal control. References. **(a)**

Differential-Gleichungen. Sophus Lie. New York: Chelsea, 1967. xiv + 568pp. $35.00. ISBN 0-8284-0206-X.

Corrected reprint of 1891 Leipzig edition, *Vorlesungen über Differentialgleichungen.* **(a)**

Elementary Differential Equations. Seventh Edition. Earl D. Rainville, Phillip E. Bedient. New York: Macmillan, 1989. xiii + 546pp. $44.25. ISBN 0-02-397860-0.

Another revision of an old friend. Its direct style covers all the basic methods of solving ordinary differential equations. **(p)**

Elementary Differential Equations and Boundary Value Problems. Fourth Edition. William E. Boyce, Richard C. DiPrima. New York: Wiley, 1986. xvi + 703pp. $20.95 paperbound. ISBN 0-471-87096-X.

A solid, dependable introductory text. Exposition has been improved by extensive rewriting. **(a,p)**

Elementary Differential Equations with Boundary Value Problems. Second Edition. C. H. Edwards, Jr., David E. Penney. New York: Prentice-Hall, 1989. xvii + 696pp. $58.50. ISBN 0-13-254061-4.

Expanded version of *Elementary Differential Equations with Applications*, adding chapters on Fourier series and boundary value problems. **(a)**

Geometrical Methods in the Theory of Ordinary Differential Equations. Second Edition. V. I. Arnol'd. Transl: Joseph Szucs. Die Grundlehren der mathematischen Wissenschaften;250. New York: Springer-Verlag, 1988. xiii + 351pp. $49.00. ISBN 0-387-96649-8.

The author explores simple examples in detail to reveal deep and complex ideas. A beautiful book which covers stability, perturbation theory, normal forms, and local bifurcation theory. **(a)**

Hyperbolic Boundary Value Problems. Reiko Sakamoto. Transl: Katsumi Miyahara. New York: Cambridge University Press, 1982. ix + 210pp. $42.50. ISBN 0-521-23568-5.

Treats the existence and uniqueness of solutions of hyperbolic initial value problems in partial differential equations of higher orders. **(a)**

Ill-Posed Problems for Integro-Differential Equations in Mechanics and Electromagnetic Theory. Frederick Bloom. Philadelphia, PA: Society for Industrial and Applied Mathematics, 1981. ix + 222pp. $44.00. ISBN 0-89871-171-1.

Treatment mainly based on variants of logarithmic convexity and concavity arguments. Well referenced and indexed. **(a)**

Introduction to Differential Equations. Second Edition. Richard K. Miller. New York: Prentice-Hall, 1991. xi + 618pp. $59.00. ISBN 0-13-478264-X.

Well-written text includes many carefully chosen problems of varying degrees of difficulty. **(a,p,s)**

An Introduction to Nonlinear Oscillations. Ronald E. Mickens. New York: Cambridge University Press, 1981. xiv + 224pp. $95.00. ISBN 0-521-22208-7.

Introduces some of the approximation techniques most commonly used in small-parameter problems in the theory of nonlinear oscillations, with a good collection of worked examples. (**a**)

Introduction to Perturbation Techniques. Ali Hasan Nayfeh. New York: Wiley, 1981. xiv + 519pp. $64.95. ISBN 0-471-08033-0.

Comprehensive work as well as a quick reference guide for practical asymptotic methods in differential equations. In this text, the author presents the material from his 1973 *Perturbation Methods* in a more elementary manner. (**a**)

Linear Differential Equations and Group Theory from Riemann to Poincaré. Jeremy Gray. Boston: Birkhäuser, 1986. xxv + 460pp. $53.50. ISBN 0-8176-3318-9.

This work traces the historical origins of the geometric view of complex function theory which culminated in Klein's and Poincaré's theory of automorphic functions. Extensive references. (**a**)

Linear Differential Equations with Periodic Coefficients. V. A. I͡Akubovich, V. M. Starzhinskiĭ. Transl: D. Louvish. New York: Wiley, 1975. (Available from Krieger, Melbourne, FL) $62.50 set. ISBN 0-470-96953-9. 1. xii + 1-386pp. 2. xii + 387-839pp.

Translation of 1972 original, a comprehensive and profound survey of the subject. Bibliography. (**a**)

Linear Turning Point Theory. Wolfgang Wasow. Applied Mathematical Sciences;54. New York: Springer-Verlag, 1985. ix + 246pp. $49.80. ISBN 0-387-96046-5.

Historical account as well as a demonstration of the classical amd modern methods for studying the behavior of solutions around singularities. (**a**)

The Method of Differential Approximation. I͡U. I. Shokin. Transl: K. G. Roesner. New York: Springer-Verlag, 1983. xiii + 296pp. $81.00. ISBN 0-387-12225-7.

Revised and expanded version of 1979 Russian work on differential approximation methods applied to difference schemes in the solution of equations of hyperbolic type. References. (**a**)

The Numerical Solution of Nonlinear Problems. Ed: Christopher T. H. Baker, Chris Phillips. Oxford: Clarendon Press, 1981. viii + 369pp. $67.50. ISBN 0-19-853354-3.

Proceedings of the annual summer school lectures at Liverpool University. Includes an extensive list of references. (**a**)

Ordinary Differential Equations. V. I. Arnol'd. Transl: Richard A. Silverman. Cambridge, MA: MIT Press, 1978. ix + 280pp. $14.95 paperbound. ISBN 0-262-51018-9.

Subject is treated in a new way with notions of vector field, phase space, phase flow and one parameter groups of transformations dominating the presentation. (**a**)

Ordinary Differential Equations. George F. Carrier, Carl E. Pearson. Classics in Applied Mathematics;6. Philadelphia: Society for Industrial and Applied Mathematics, 1991. x + 220pp. $25.50. ISBN 0-89871-265-3.

Reprint, with minor revisions, of 1968 original. This work covers all the usual topics, with beautiful, open-ended problems driving the discussion. **(a)**

Ordinary Differential Equations. Second Edition. Philip Hartman. Boston: Birkhäuser, 1982. xv + 612pp. $38.50. ISBN 0-8176-3068-6.

Reprint of corrected 1973 reprint of 1964 original. **(a)**

Quadratic Form Theory and Differential Equations. John Gregory. Mathematics in Science and Engineering:152. San Diego, CA: Academic Press, 1980. xii + 237pp. $57.00. ISBN 0-1230-1450-6.

Unified theory of quadratic forms including a new formal theory of approximation of quadratics forms/linear operators in Hilbert spaces. **(a)**

Solitons and the Inverse Scattering Transform. Mark J. Ablowitz, Harvey Segur. SIAM Studies in Applied Mathematics;4. Philadelphia, PA: Society for Industrial and Applied Mathematics, 1981. x + 425pp. $66.00. ISBN 0-89871-174-6.

Reprint of 1981 exposition of a method which can be used to obtain exact solutions to certain nonlinear evolution equations and some of its applications. **(a,s)**

Theory of Ordinary Differential Equations. Earl A. Coddington, Norman Levinson. Melbourne, FL: Krieger, 1984. xii + 429pp. $38.50. ISBN 0-89874-755-4.

Reprint of 1955 original. Elegant manner of presentation and choice of material (most of which is not found in standard textbooks on differential equations) distinguishes this work. **(a,s)**

Volterra and Functional Differential Equations. Ed: Kenneth B. Hannsgen et al. Lecture Notes in Pure and Applied Mathematics; 81. New York: Dekker, 1982. xii + 333pp. $99.75. ISBN 0-8247-1721-X.

Papers from 24 invited lectures at the 1981 Blacksburg conference on Volterra and functional differential equations. **(a)**

2500 Solved Problems in Differential Equations. Richard Bronson. New York: McGraw-Hill, 1989. v + 440pp. $19.95. ISBN 0-07-00797-X.

Minimal discussion of theory, but an excellent source of problems from any area normally encountered in a first course in differential equations. **(a,p)**

DIFFERENTIAL GEOMETRY

Analysis on Real and Complex Manifolds. Second Revised Edition. R. Narasimhan. North-Holland Mathematical Library;35. New York: Elsevier, 1986. xiv + 246pp. $58.25. ISBN 0-444-87776-2.

Based on 1964-65 Tata Institute lectures, chapters cover main theorems of differential topology, real and complex manifolds, and elliptic differential operators. (**a**)

Bieberbach Groups and Flat Manifolds. Leonard S. Charlap. New York: Springer-Verlag, 1986. xiii + 242pp. $49.00. ISBN 0-387-96395-2.

Designed as a textbook and basic reference on flat Riemannian manifolds. (**a**)

Complex Manifolds Without Potential Theory: With an Appendix on the Geometry of Characteristic Classes. Second Edition. Shiing-shen Chern. New York: Springer-Verlag, 1979. 152pp. $25.00. ISBN 0-387-90422-0.

This edition adds as an appendix Chern's lectures on the geometry of characteristic classes. (**a**)

A Comprehensive Introduction to Differential Geometry. Second Edition. Michael Spivak. Houston, TX :Publish or Perish, 1979. $135.00 set. ISBN 0-914098-83-7. *V. I.* xiii + 668pp. *V. II.* xiv + 423pp. *V. III.* xi + 466pp. *V. IV.* vii + 561pp. *V. V.* vii + 661pp.

This fine exposition of the core of differential geometry has been revised to make the whole set better unified. Expensive. (**a,s**)

Differential Analysis on Complex Manifolds. R. O. Wells, Jr. Graduate Texts in Mathematics;65. New York: Springer-Verlag, 1986. x + 260pp. $42.00. ISBN 0-387-90419-0.

Reprint of 1980 original. This corrected edition includes a completely revised chapter on compact complex manifolds. (**a**)

Differential Geometry, Gauge Theories, and Gravity. M. Gockler, T. Schucker. New York: Cambridge University Press, 1987. xii + 230pp. $59.50. ISBN 0-521-32960-4.

An introduction to those areas of modern differential geometry fundamental to particle physics and general relativity. (**a**)

Differential Geometry in the Large. Second Edition. Heinz Hopf. Lecture Notes in Mathematics;1000. New York: Springer-Verlag, 1989. vi + 184pp. $18.20. ISBN 0-387-51497-X.

Essentially unchanged text of the lecture notes from 1946 NYU and 1956 Stanford courses on global geometry of two dimensional surfaces in three dimensional Euclidean space. (**a**)

Elementary Differential Geometry. Barrett O'Neill. San Diego, CA: Academic Press, 1966. viii + 411pp. $49.95. ISBN 0-12-526750-9.

Traditional text includes as newer material "the geometry of a surface as seen by its inhabitants." (**a,p**)

Elementary Topics in Differential Geometry. John A. Thorpe. New York: Springer-Verlag, 1979. xiii + 253pp. $38.00. ISBN 0-387-90357-7.

The relatively simple approach of this book is made by using linear algebra via a vector field (not affine geometry) approach to co-dimensional one manifolds. **(a)**

Foundations of Differentiable Manifolds and Lie Groups. Frank W. Warner. Graduate Texts in Mathematics;94. New York: Springer-Verlag, 1987. ix + 272pp. $35.00. ISBN 0-387-90894-3.

Essentially unaltered reprint of 1971 original. **(a)**

Foundations of Differential Geometry. Shoshichi Kobayashi, Katsumi Nomizu. Interscience Tracts in Pure and Applied Mathematics;15:1-2. New York: Wiley, 1963-1969. *Vol. I.* xi + 329pp. $69.95. ISBN 0-470-49647-9. *Vol. II.* xv + 470pp. $76.95. ISBN 0-470-49648-7.

High quality of exposition make this set a convenient reference source and a very suitable text for a graduate course. **(a)**

Geometric Measure Theory: A Beginner's Guide. Frank Morgan. San Diego, CA: Academic Press, 1988. viii + 145pp. $29.95. ISBN 0-12-506855-7.

A profusely-illustrated text which focuses on the search of surfaces of minimal area with given boundary. **(p)**

An Introduction to Differentiable Manifolds and Riemannian Geometry. Second Edition. William M. Boothby. Pure and Applied Mathematics;120. San Diego, CA: Academic Press, 1986. xvi + 430pp. $79.95;$39.95. ISBN 0-12-116052-1;0-12-116053-X.

Proofs have been improved, historical notes enlarged, errors corrected, and references updated in this edition. **(a)**

Leçons sur la Théorie Générale des Surfaces et les Applications Géométriques du Calcul Infinitésimal. Third Edition. Gaston Darboux. New York: Chelsea, 1972. $125.00 set. ISBN 0-8284-0216-7. *Premièr Partie: Généralites, Coordonnées Curvilignes, Sufaces Minima.* xxiv + 601pp. *Deuxième Partie: Les Congruences et les Équations Linéaires aux Dérivées Partielles, Les Lignes Tracées sur les Surfaces.* xix + 567pp. *Troisième Partie: Lignes Géodésiques et Courbure Géodésique, Paramètres Différentiels, Déformation des Surfaces.* xvi + 501pp. *Quatrième Partie: Déformation Infiniment Petite et Représentation Sphérique.* xvi + 537pp.

Unabridged reprint of 1894-1915 originals, i.e., volumes 1-2 of the second edition and v.3-4 of the first edition. Darboux never got around to completing v.3-4 of the second edition. Expensive. **(a)**

Lecture Notes on Elementary Topology and Geometry. I. M. Singer, J. A. Thorpe. New York: Springer-Verlag, 1987. viii + 232pp. $37.50. ISBN 0-387-90202-3.

Third printing of the 1967 classic text, providing a unified view of topology. **(a)**

Lectures on Classical Differential Geometry. Second Edition. Dirk J. Struik. New York: Dover, 1988. viii + 232pp. $6.95. ISBN 0-486-65609-8.

Unabridged and unaltered reprint of 1961 edition. Emphasizes geometry, not vector algebra. Essential in any mathematics collection. **(p)**

Lectures on Differential Geometry. Second Edition. Shlomo Sternberg. New York: Chelsea, 1983. xi + 442pp. $29.50. ISBN 0-8284-0316-3.

This edition adds 2 appendices: Appendix III is complete reprint of a paper co-authored with Guillemin on transitive differential geoemtry, and IV is a brief survey by the author of relevant developments of the past 20 years, with references. **(a)**

The Method of Equivalence and Its Applications. Robert B. Gardner. CBMS-NSF Regional Conference Series in Applied Mathematics;58. Philadelphia, PA: Society for Industrial and Applied Mathematics, 1989. vii + 127pp. $21.75. ISBN 0-89871-240-8.

Author discusses cases where the diffeomorphisms may be described as solutions to first order systems of differential equations. **(a)**

Modern Geometry–Methods and Applications. B. A. Dubrovin, A. T. Fomenko, S. P. Novikov. Transl: Robert G. Burns. Graduate Texts in Mathematics;93,104,124. New York: Springer-Verlag, 1984-1990. *Part I: The Geometry of Surfaces, Transformation Groups, and Fields.* x + 464pp. $59.80. ISBN 0-387-90872-2. *Part II: The Geometry and Topology of Manifolds.* xv + 430pp. $59.00. ISBN 0-387-96162-3. *Part III: Introduction to Homology Theory.* ix + 416pp. $59.00. ISBN 0-387-97271-4.

Three-volume introduction to modern geometry, resulting from an extensive elaboration of lecture notes from a course at Moscow State University. Emphasis is on applications to other areas of mathematics and theoretical physics. **(a)**

Riemannian Geometry. S. Gallot, D. Hulin, J. Lafontaine. New York: Springer-Verlag, 1987. xi + 248pp. $29.00. ISBN 0-387-17923-2.

Clearly-written and quite self-contained text with a large number of exercises. For the reader somewhat familiar with differentiable manifolds. **(a)**

Schaum's Outline of Theory and Problems of Tensor Calculus. David C. Kay. New York: McGraw-Hill, 1988. 228pp. $11.95. ISBN 0-07-033484-6.

Intended for any student in need of mastering the basic concepts and methods of tensors with much elementary classical differential geometry met along the way. **(a,p)**

Semi-Riemannian Geometry: With Applications to Relativity. Barrett O'Neill. Pure and Applied Mathematics;103. New York: Academic Press, 1983. xiii + 468pp. $64.00. ISBN 0-12-526740-1.

An interesting first course in geometry for mathematics and physics students. **(a)**

Singularity Theory: Selected Papers. V. I. Arnol'd. London Mathematical Society Lecture Notes Series;53. New York:Cambridge University Press, 1981. 266pp. $27.50. ISBN 0-521-28511-9.

Reprint from *Russian Mathematical Surveys* of 7 seminal papers published between 1968 and 1978 on the classification of critical points in smooth functions. **(a)**

Solid Shape. Jan J. Koenderink. Cambridge, MA: MIT Press, 1990. xiv + 699pp. $49.00. ISBN 0-262-11139-X.

A largely expository text on the differential geometry of submanifolds of three-dimensional space. It contains some marvelous intuition which makes it well worth reading. (**a**)

Spinors and Space-Time. Roger Penrose, Wolfgang Rindler. New York: Cambridge University Press, 1984-1986. *Volume 1: Two-Spinor Calculus and Relativistic Fields*. x + 458pp. $29.95;$100.00. ISBN 0-521-33707-0;0-521-24527-3. *Volume 2: Spinor and Twistor Methods in Space-Time Geometry*. ix + 501pp. $37.50;$100.00. ISBN 0-521-34786-6;0-521-25267-9.

Volume 1 begins at a relatively elementary level and proceeds through Einstein-Maxwell equations and Yang-Mills equations. Volume 2 introduces the theory of twistors and studies applications of twistors and 2-spinors to the study of space time. (**a**)

Teichmuller Theory and Quadratic Differentials. Frederick P. Gardiner. New York: Wiley, 1987. xvii + 236pp. $46.95. ISBN 0-471-84539-6.

Treats the moduli space of inequivalent Riemann surface structures on a given topological surface. (**a**)

Tensor Analysis on Manifolds. Richard L. Bishop, Samuel I. Goldberg. New York: Dover, 1980. viii + 280pp. $6.95. ISBN 0-486-64039-6.

This classic contains the topology, linear algebra, and multivariate analysis necessary to introduce differential geometry with some physical applications. (**a**)

Tensor Analysis: Theory and Applications to Geometry and Mechanics of Continua. Second Edition. I. S. Sokolnikoff. Melbourne, FL: Krieger, 1990. xii + 361pp.

Reprint of 1964 original. (**a**)

Tensor Calculus. J. L. Synge, A. Schild. New York: Chelsea, 1978. ix + 324pp. $7.00. ISBN 0-486-63612-7.

Reprint of 1969 University of Toronto Press edition. Authors claim this book is intended as a general brief introduction to tensor calculus. (**a**)

Tensor Geometry, The Geometric Viewpoint and Its Uses. C. T. J. Dodson, T. Poston. New York: Wiley, 1979. xiii + 598pp. $34.95. ISBN 0-470-20468-0.

Outstanding multivariable calculus and differential geometry book. (**p**)

Theorie der konvexen Koerper. T. Bonnesen, W. Fenchel. Ergebnisse der Mathematik und ihrer Grenzgebiete;3:1. New York: Chelsea, 1948. vii+ 164pp. $8.95. ISBN 0-8284-0054-7.

Reprint of 1934 original. (**a**)

The Theory of Singularities and Its Applications. V. I. Arnol'd. New York: Cambridge University Press, 1991. 72pp. $19.75. ISBN 0-521-422809.

This exposition of singularity theory aims to illustrate by examples and key theorems. (**a**)

Vector and Tensor Analysis with Applications. A. I. Borisenko, I. E. Tarapov. Transl: Richard A. Silverman. New York: Dover, 1979. x + 257pp. $6.95. ISBN 0-486-63833-2.

Unabridged, corrected reprint of 1968 edition. An introduction to the algebra and calculus of vectors and tensors, with applications in fluid dynamics and electromagnetic theory. **(p)**

DISCRETE AND FINITE MATHEMATICS

Applications of Discrete Mathematics. Ed: John G. Michaels, Kenneth H. Rosen. New York: McGraw-Hill, 1991. x + 515pp. $17.95. ISBN 0-07-041823-3.

Contains 24 separate topics arranged into 3 sections: discrete structures and computing, combinatorics, and graph theory. Each topic has problems and suggested reading list. **(a)**

Discrete Mathematics. Kenneth P. Bogart. Lexington, MA: D. C. Heath, 1988. xvii + 659pp. $36.00. ISBN 0-669-08665-7.

More demanding and more comprehensive than most texts. Includes an appendix of suggested readings. **(a)**

Discrete Mathematics. Second Edition. Kenneth A. Ross, Charles R. B. Wright. New York: Prentice-Hall, 1988. xiv + 625pp. $55.00. ISBN 0-13-215427-7.

This substantial introduction to finite, algebraic, algorithmic, and discrete topics emphasizes abstract algebra more than similar books. In this edition the treatment of algorithms has been completely revised and the account of algebraic topics is not less theoretical. **(a)**

Discrete Mathematics and Its Applications. Second Edition. Kenneth H. Rosen. New York: McGraw-Hill, 1991. xxii + 661pp. $45.00. ISBN 0-07-053744-5.

An introductory text which stresses 5 important themes: mathematical reasoning, combinatorial analysis, discrete structures, applications and modeling, and algorithmic thinking. Sections on probability theory and number theory, new exercises, and review questions have been added to this edition. **(a)**

Discrete Mathematics in the First Two Years. Ed: Anthony Ralston. MAA Notes; 15. Washington, DC: Mathematical Association of America, 1989. 101pp. $10.00. ISBN 0-88385-064-8.

Reports from the six campuses, funded by the Sloan Foundation in 1984-86, which experimented with various ways of incorporating discrete mathematics into the standard first two-year curriculum. **(a)**

Finite Mathematics with Applications. Third Edition. A. W. Goodman, J. S. Ratti. New York: Macmillan, 1979. xiv + 584pp. $45.75. ISBN 0-02-344760-5.

This edition has added a chapter on statistics and more on linear programming and review problems. **(p)**

Graphs, Models, and Finite Mathematics. Joseph Malkevitch, Walter Meyer. New York: Prentice-Hall, 1974. x + 515pp. $65.00. ISBN 0-13-363465-5.

"This text is designed for courses in finite mathematics and mathematics for the liberal arts student. It offers two main novelties: an early and substantial treatment of graphs (newtworks) and an explicit treatment of the process of mathematical modeling."—Preface. **(a)**

Introduction to Discrete Mathematics. Robert J. McEliece, Robert B. Ash, Carol Ash. New York: Random House, 1989. 462pp. $27.50. ISBN 0-317-58264-X.

Authors use a unifying theme of algorithms to present chapters on set theory, combinatorics, graphs, etc. Each chapter has overview, summary, and historical footnotes. **(a)**

Introduction to Finite Mathematics. Third Edition. John G. Kemeny, J. Laurie Snell, Gerald L. Thompson. New York: Prentice-Hall, 1974. xi + 484pp. $56.00. ISBN 0-13-483834-3.

Attractive, contemporary text of finite mathematics. **(s)**

Methods of Finite Mathematics. John W. Brown, Donald R. Sherbert. New York: Wiley, 1989. xii + 668pp. $52.95. ISBN 0-471-63003-9.

This text covers a variety of topics, with numerous examples and exercises. Software package with manual is available. **(a)**

EDUCATION

Academic Preparation in Science: Teaching for Transition From High School to College. Second Edition. Jose P. Mestre, Jack Lochhead. Ed: Robert Orrill. New York: College Entrance Examination Board, 1990. ix + 130pp. $6.95. ISBN 0-87447-352-7.

Major revision of 1986 original, rewritten to reflect new foundations of science education provided by recent cognitive research. Mathematics is similar in analyses and strategies of learning. **(a,p)**

A Call for Change: Recommendations for the Mathematical Preparation of Teachers of Mathematics. Ed: James R.C. Leitzel. Washington, DC: Mathematical Association of America, 1991. xiv + 47pp. $7.00 ISBN 0-88385-072-9.

Emphasizes the need to improve how college mathematics is taught so prospective teachers will learn by example how they can meet the NCTM *Standards* for K-12 school teachers. **(a)**

Computer Environments for Children: A Reflection on Theories of Learning and Education. Cynthia Solomon. Cambridge, MA: MIT Press, 1988. viii + 183pp. $10.95 paperbound. ISBN 0-262-69125-6.

The author looks at the innovative uses of computers in elementary mathematics education stemming from research in the 1960s and 1970s. **(p)**

Computers and Mathematics: The Use of Computers in Undergraduate I Instruction. Ed: David A. Smith et al. MAA notes;9. Washington, DC: Mathematical Association of America, 1988. xi +147pp. $9.00. ISBN 0-88385-059-1.

Nineteen stimulating surveys exploring the use of computing across the undergraduate mathematics curriculum. **(a)**

Constructivist Views on the Teaching and Learning of Mathematics. Ed: Robert B. Davis, Carolyn A. Maher, Nel Noddings. Journal for Research in Mathematics Education;4. Reston, VA: National Teachers of Mathematics, 1990. ix + 210pp. $9.50. ISBN 0-87353-300-3.

Twelve papers looking into the evidence for the notion that children construct mathematical knowledge for themselves and its impact on teaching and learning.

Counting on You: Actions Supporting Mathematics Teaching Standards. Washington, DC: National Research Council, 1991. 17pp. Free. No ISBN.

A brief policy synopsis highlights recommended changes and intends to help mobilize public and corporate support for improvements in mathematics education. **(a,p)**

Curriculum and Evaluation Standards for School Mathematics. Reston, VA: National Council of Teachers of Mathematics, 1989. vii + 258pp. $25.00. ISBN 0-87353-273-2.

"Must" reading for anyone who teaches mathematics. Presents guidelines for the content of school mathematics curriculum and suggests means for evaluating student mastery of this content. **(p)**

Encouraging Girls in Mathematics: The Problem and the Solution. Lorelei R. Brush et al. Cambridge, MA: Abt Books, 1980. xv + 163pp. $30.00. ISBN 0-89011-542-7.

Both a record of the results of a detailed longitudinal study of why students do not wish to continue in mathematics as well as suggestions for teachers of mathematics. **(a,p)**

Escalante: The Best Teacher in America. Jay Mathews. New York:Henry Holt, 1988. ix + 322pp. $19.95. ISBN 0-8050-0450-5.

A fascinating, behind-the-scenes portrait of Jaime Escalante, his students, and fellow teachers at Garfield High School in East Los Angeles—a school which moved in ten years from intellectual bankruptcy to outstanding success in AP calculus and other AP courses. **(p)**

Estimation and Mental Computation. Ed: Harold L. Schoen, Marilyn J. Zweng. NCTM Yearbook;1986. Reston, VA: National Council of Teachers of Mathematics, 1986. viii + 248pp. $18.00. ISBN 0-87353-226-0.

Articles emphasize the interplay between estimation and conceptual understanding. **(a,p)**

Everybody Counts: A Report to the Nation on the Future of Mathematics Education. Washington, DC: National Research Council, 1989. xii + 114pp. $7.95. ISBN 0-309-03977-0.

A forceful consensus report from units of the National Academy of Sciences on mathematics education. **(a,p)**

Fantastiks of Mathematiks: Applications of Secondary Mathematics. Cliff Sloyer. Providence, RI: Janson, 1986. xi + 143pp. $13.95. ISBN 0-939765-00-4.

Forty problems and discussion guides designed to motivate students and teachers to work on a variety of mathematical applications. Useful for teachers of secondary mathematics and "non-technical" college mathematics. **(a,p)**

The Future of College Mathematics. Ed: Anthony Ralston, Gail S. Young. New York: Springer-Verlag, 1983. ix + 278pp. $42.00. ISBN 0-387-90813-7.

Proceedings of a 1982 summer conference analyzing the proposed need for a lower level undergraduate discrete mathematics alternative to calculus. **(a)**

Guidelines for the Continuing Mathematical Education of Teachers. MAA Notes;10. Washington, DC: Mathematical Association of America, 1988. ix + 83pp. $8.00. ISBN 0-88385-060-5.

Detailed, practical recommendations for in-service and master's degrees courses for practicing teachers of mathematics. **(a)**

Historical Topics for the Mathematics Classroom. John K. Baumgart. Reston, VA: National Council of Teachers of Mathematics, 1989. xv + 542pp. $28.00. ISBN 0-87353-281-3.

Updated from 1969 edition, including current references and chapter on contemporary mathematics. **(a)**

The Laboratory Approach to Teaching Calculus. Ed: L. Carl Leinbach et al. MAA Notes;20. Washington, DC: Mathematical Association of America, 1991. ix + 264pp. $20.00. ISBN 0-88385-074-5.

Thirty-three class-tested examples of diverse strategies to teach calculus with a laboratory approach. **(a)**

Making the Grade in Mathematics: Elementary School Mathematics in the United States, Taiwan, and Japan. Harold W. Stevenson et al. Reston, VA: National Council of Teachers of Mathematics, 1990. iv + 34pp. $9.50. ISBN 0-87353-292-9.

Concise summary, plainly written, of comparative research of childrens' mathematics achievement and attitudes about the role of work vs. ability, and their mothers' role and attitudes concerning mathematics education. **(p)**

Making Mathematics Work for Minorities: Framework for a National Action Plan 1990-2000. Washington, DC: National Research Council, 1990. 24pp. $3.95. No ISBN.

Highlights of issues and recommendations that emerged from a 1990 national symposium sponsored by the Mathematical Sciences Education Board. **(p)**

Math Panic. Laurie Buxton. Portsmouth, NH: Heinemann, 1991. xv + 260pp. $18.50. ISBN 0-435-08313-9.

Revised American edition of a 1981 British account of adults recalling the anxiety and despair they experienced with mathematics, both in school and at work. **(p)**

Mathematical Problem Solving. Alan H. Schoenfeld. San Diego, CA: Academic Press, 1985. xvi + 409pp. $79.95;37.50. ISBN 0-12-628870-4; 0-12-628871-2.

Observation-based analysis of how students learn to solve mathematics problems. Extensive references to the literature of educational research and cognitive psychology. (**a**)

Mathematicians and Education Reform. Ed: Naomi Fisher, Harvey Keynes, Philip Wagreich. CBMS Issues in Mathematics Education;1. Providence, RI: American Mathematical Society in cooperation with the Mathematical Association of America, 1990. viii + 217pp. $37.00. ISBN 08218-3500-9.

The proceedings of a July 1988 workshop focusing on contexts that enable university mathematicians to work effectively in mathematics education. (**a**)

Mathematics and Cognition. Ed: Pearla Nesher, Jeremy Kilpatrick. New York: Cambridge University Press, 1990. 180pp. $42.50. ISBN 0-521-36608-9.

Concise, well-referenced state-of-the-art survey of research in the psychology of mathematics education prepared by an affiliate of ICMI. (**a**)

The Mathematics Curriculum: Towards the Year 2000. Ed: John Malone, Hugh Burkhardt, Christine Keitel. Perth, Western Australia: Curtin University of Technology, 1989. 469pp. No price. ISBN 0-909848-95-5.

Seventy-five papers from around the world form the proceedings of a Theme Group from the 1988 International Congress on Mathematical Education in Budapest. (**a**)

The Mathematics Report Card: Are We Measuring Up? John A. Dossy et al. Princeton, NJ: Educational Testing Service, 1988. 143pp. $14.00. ISBN 0-88685-072-X.

Summary and analysis of the National Assessment of Educational Progress, a survey of mathematical knowledge of 9-, 13-, and 17-year-olds conducted every four years. (**p**)

Mathematics: Report of the Project 2061 Phase I Mathematics Panel. David Blackwell, Leon Henkin. Washington, DC: American Association for the Advancement of Science, 1989. xi + 47pp. $7.50. ISBN 0-87168-344-X.

This report offers an ambitious vision of which important ideas of mathematics should be understood by the next century's "typical adults." Innovative recommendations are included in the report. (**p**)

Mindstorms: Children, Computers, and Powerful Ideas. Seymour Papert. New York: Basic Books, 1980. viii + 230pp. $9.95. ISBN 0-465-04629-0.

Young children can explore subtle geometric, algorithmic, and computational ideas if they are placed in an environment with appropriate and enjoyable tools such as LOGO and Turtle Geometry. (**p**)

National Curricula in Mathematics. Geoffrey Howson. Leicester, England: Mathematical Association, 1991. vii + 238pp. $10.65;$30.00. ISBN 0-906-588-219;0-906-588-243.

A comparative study of school mathematics curricula in the European Community plus Hungary and Japan. (**a**)

On the Shoulders of Giants: New Approaches to Numeracy. Ed: Lynn Arthur Steen. Washington, DC: National Academy Press, 1990. vii + 232pp. $17.95. ISBN 0-309-04234-8.

Collection of 5 essays illustrate the potential of powerful mathematics to stimulate the school curricula. **(a,p)**

Overcoming Math Anxiety. Sheila Tobias. Boston: Houghton Mifflin, 1980. 284pp. $8.95. ISBN 0-395-29088-0.

Paperback edition of the classic brief on the sources of defeat and terror that afflict millions of people. **(p)**

A Problem Solving Approach to Mathematics for Elementary School Teachers. Third Edition. Rick Billstein, Shlomo Libeskind, Johnny Lott. Redwood City, CA: Benjamin-Cummings, 1987. x + 782pp. $37.95. ISBN 0-8053-0865-2.

Differs from previous editions through reorganization and the addition of a new chapter on computers. Book is written so material can be used by students with diverse mathematical backgrounds. **(a,p)**

Proceedings of the Sixth International Congress on Mathematical Education. Ed: Ann and Keith Hirst. Budapest: ICMI Secretariat; Janos Bolyai Mathematical Society, 1988. (Available from Mathematical Association of America, Washington, DC). 397pp. $45.00. ISBN 9-6380-2248-5.

Five plenary adresses followed by reports from the 30 action, theme, and topic groups of the 1988 ICME Congress in Budapest. **(a)**

Professional Standards for Teaching Mathematics. Reston, VA: National Council of Teachers of Mathematics, 1991. viii + 196pp. $25.00. ISBN 0-87353-307-0.

Companion volume to NCTM's 1989 standards for curriculum and evaluation, this volume describes 24 standards for teaching mathematics and its evaluation and for support and development of mathematics teachers. **(a)**

The Psychology of Mathematics for Instruction. Lauren B. Resnick, Wendy W. Ford. Hillsdale, NJ: L. Erlbaum Associates, 1981. vi + 226pp. $39.95. ISBN 0-89859-029-9.

Review of psychological research on mathematics learning and thinking processes with implications for teaching. **(a)**

Research Issues in the Learning and Teaching of Algebra. Ed: Sigrid Wagner, Carolyn Kieran. Research Agenda for Mathematics Education;4. Reston, VA: National Council of Teachers of Mathematics and Lawrence Erlbaum Associates, 1989. x + 287pp. $19.95. ISBN 0-87353-268-6.

Twenty-three papers from a 1987 University of Georgia conference on what algebra is and should be, how algebraic thinking relates to mathematical thinking, etc. **(a)**

Reshaping College Mathematics: A Project of the Committee on the Undergraduate Program in Mathematics. Ed: Lynn Arthur Steen. MAA Notes;13. Washington, DC: Mathematical Association of America, 1989. x + 124pp. $12.00. ISBN 0-88385-062-1.

Compendium of the last ten years' CUPM reports, covering the spectrum of undergraduate courses. **(a)**

Responses to the Challenge: Keys to Improved Instruction by Teaching Assistants and Part-Time Instructors. Ed: Bettye Anne Case. MAA Notes;11. Washington, DC: Mathematical Association of America, 1989. vi + 266pp. $15.00. ISBN 0-88385-061-3.

A compilation of information from a joint MAA-AMS project of policies, practices, successes, failures, and goals connected with the use of teaching assistants and part-time instructors. **(a)**

School Mathematics in the 1990s. Ed: Geoffrey Howson, Bryan Wilson. New York: Cambridge University Press, 1986. viii + 104pp. $39.50. ISBN 0-521-33614-7.

Based on a 1986 seminar in Kuwait, this volume addresses major issues facing all nations concerning school mathematics. **(p)**

Science Achievement in Seventeen Countries: A Preliminary Report. Elmsford, NY: Pergamon Press, 1988. xi + 125pp. $10.00. ISBN 0-08-036563-9.

First report of an international assessment of science education for students of ages 10, 14, 18. U.S. results in science are comparable for those in mathematics, i.e., very low. **(p)**

Science for All Americans. Washington, DC: American Association for the Advancement of Science, 1989. ix + 217pp. $24.95. ISBN 0-87168-341-5.

Summary report of Project 2061, the futuristic attempt by AAAS to articulate what everyone should know of mathematics, science, and technology. The vision of mathematics is substantially stronger than current average performance, but much weaker that that recommended by the project's own mathematics panel or the NCTM. **(p)**

Second Interantional Mathematics Study: Detailed Report for the United States. Ed: Kenneth J. Travers. Champaign, IL: Stipes, 1986. xix + 445pp. $21.00. ISBN 0-87563-299-8.

Thorough analysis of the U.S. data from the 1985 *Summary Report* and the public summary *The Underachieving Curriculum.* **(a,p)**

Second International Mathematics Study: Summary Report for the United States. Kenneth J. Travers. Champaign, IL: Stipes, 1985. xix + 97pp. $7.80. ISBN 0-87563-267-X.

Revision of a 1984 report, documenting a 1981-82 international study which placed U.S. 12th grade students in the lowest quartile internationally. **(p)**

A Sourcebook for College Mathematics Teaching. Ed: Alan H. Schoenfeld. Washington, DC: Mathematical Association of America, 1990. ii + 68pp. $10.00. ISBN 0-88385-068-0.

Collection of pointers to improve undergraduate mathematics instruction. Individual chapters are replete with references to research findings and other sources of supplementary information. **(a)**

Succeed With Math: Every Student's Guide to Conquering Math Anxiety. Sheila Tobias. New York: The College Board, 1987. xviii + 252pp. $12.95. ISBN 0-87447-259-8.

Designed to build self-confidence, this sequel to *Overcoming Math Anxiety* (see above) covers many of the mathematical skills needed to study freshman-level mathematics. **(a,p)**

Teaching and Learning Mathematics in the 1990s. Thomas J. Cooney, Christian R. Hirsch. NCTM Yearbook;1990. Reston, VA: National Council of Teachers of Mathematics, 1990. viii + 216pp. $18.00. ISBN 0-87353-285-6.

Twenty-eight articles, ranging from elementary through undergraduate teaching, are grouped into 7 parts. This book is devoted "to the people of mathematics education—teachers and students—and their changing roles in the face of calls for reform in the 1990s." **(a)**

Twice as Less: Black English and the Performance of Black Students in Mathematics and Science. Eleanor Wilson Orr. New York: W. W. Norton, 1987. 240pp. $15.95. ISBN 0-393-02392-3.

An original, controversial, and immensely important work analyzing how Black vernacular clashes with mathematical convention, e.g., "one-third of" becomes "three times less." **(p)**

The Underachieving Curriculum: Assessing U.S. Public School Mathematics From an International Perspective. Curtis C. McKnight et al. Champaign, IL: Stipes, 1987. xiv + 127pp. $8.00. ISBN 0-87563-298-X.

A forthright analysis of the results of the Second International Assessment of Mathematics. Covers achievement, attitudes, curriculum, opportunity to learn, and "deceptive explanations." **(p)**

Using Writing to Teach Mathematics. Ed: Andrew Sterrett. MAA Notes;16. Washington, DC: Mathematical Association of America, 1990. xvii + 139pp. $15.00. ISBN 0-88385-066-4.

Lucidly written compendium of 30 essays by those with considerable experience in requiring their students to write about mathematics. **(a)**

Visualization in Teaching and Learning Mathematics. Ed: Walter Zimmermann, Steve Cunningham. MAA Notes;19. Washington, DC: Mathematical Association of America, 1991. viii + 224pp. $20.00. ISBN 0-88385-071-0.

A delightful collection of well illustrated papers on various aspects of visualization in mathematics to stimulate innovation. **(a)**

FUNCTIONAL ANALYSIS

Asymptotic Theory of Finite Dimensional Normed Spaces. Vitali D. Milman, Gideon Schechtman. New York: Springer-Verlag, 1986. viii + 156pp. $21.30. ISBN 0-387-16769-2.

This brief monograph introduces results, problems, and methods in the theory of normed spaces. Foundational material is covered in 5 appendices. (**a**)

Banach Algebra Techniques in Operator Theory. Ronald G. Douglas. Pure and Applied Mathematics;49. San Diego, CA: Academic Press, 1972. xvi + 216pp. $59.95. ISBN 0-12-221350-5.

A discussion of certain advanced topics in operator theory as well as an excellent introduction to Banach algebras and general operator theory. (**a**)

Banach Algebras and Several Complex Variables. Second Edition. John Wermer. Graduate Texts in Mathematics;35. New York: Springer-Verlag, 1976. ix + 161pp. $42.00. ISBN 0-387-90160-4.

This edition adds sections on submanifolds of high dimension, generators, the fibers over a plan domain, and examples of hulls. The section on analytic structure has been revised and the bibliography updated. (**a**)

Banach-Mazur Distances and Finite-Dimensional Operator Ideals. Nicole Tomczak-Jaegermann. Pitman Monographs and Surveys in Pure and Applied Mathematics;38. New York: Wiley, 1989. xii + 395pp. $110.00. ISBN 0-470-20982-8.

Serious in-depth study into the modern theory of finite-dimensional Banach spaces, containing many notes and references as well as open problems. Expensive. (**a**)

Basic Operator Theory. Israel Gohberg, Seymour Goldberg. Boston: Birkhäuser, 1981. xiii + 285pp. $26.50. ISBN 3-7643-3028-7.

This book is intended to provide an introduction to functional analysis, with a special emphasis on operator theory and its applications. Exercises. (**a**)

Espaces Vectoriels Topologiques, Chapitres 1 à 5. Nicolas Bourbaki. (Éléments de Mathématique) Paris: Masson, 1981. (Available from Scientific & Medical Publications of France, New York, NY) vii + 368pp. $35.00 ISBN 2-225-68410-3.

Includes bibliography and indexes. (See p. 217 for English translation, *Topological Vector Spaces, Chapters 1-5.*).

Functional Analysis. George Bachman, Lawrence Narici. San Diego, CA: Academic Press, 1966. xiv + 530pp. $64.00. ISBN 0-12-070250-9.

This book provides an in-depth, axiomatic, and highly detailed treatment of functional analysis in its pure aspects. (**a**)

Functional Analysis. Second Edition. Walter Rudin. New York: McGraw-Hill, 1991. xv + 424pp. $41.32. ISBN 0-07-054236-8.

"I have added a few items which were easy to fit into the existing text: the mean ergodic theorem of von Neumann, the Hille-Yosida theorem on semigroups of operators, a couple of fixed point theorems, Bonsall's surprising application of the closed range theorem, and Lomonosov's spectacular invariant subspaces theorem."—Preface. (**s**)

Functional Equations in Several Variables. J. Aczel, J. Dhombres. Encyclopedia of Mathematics and Its Applications;31. New York: Cambridge University Press, 1989. xiv + 462pp. $100.00. ISBN 0-521-35276-2.

Authors emphasize applications of the theory to mathematics, information theory, and to the natural and social sciences. Historical notes. Extensive bibliography. Expensive. **(a)**

Fundamentals of the Theory of Operator Algebras. Richard V. Kadison, John R. Ringrose. Pure and Applied Mathematics;100. San Diego, CA: Academic Press, 1983-1986. *Volume I: Elementary Theory.* xv + 398pp. $34.50 paperbound. ISBN 0-12-393301-3. *Volume II: Advanced Theory.* xiv + 676pp. $39.50 paperbound. ISBN 0-12-393303-X.

These 2 volumes are devoted to teaching rather than encyclopedic treatment of C*-algebras and von Neumann algebras. Can be used for self-study or as a text. **(a)**

Green's Functions and Boundary Value Problems. Ivar Stakgold. New York: Wiley, 1979. xv + 638pp. $76.95. ISBN 0-471-81967-0.

One-volume version of the author's 1967-68 two-volume text *Boundary Value Problems of Mathematical Physics*, featuring more recent and varied applications. **(a)**

A Hilbert Space Problem Book. Second Edition, Revised and Enlarged. Paul R. Halmos. Graduate Texts in Mathematics;19. New York: Springer-Verlag, 1982. xvii + 369pp. $49.00. ISBN 0-387-90685-1.

Halmos chooses interesting problems (1/4 of the 260 problems are new), makes an outstanding presentation, and provides invaluable remarks on the problems. **(a)**

Introduction to Banach Spaces and their Geometry. Second, Revised Edition. Bernard Beauzamy. North-Holland Mathematics Studies;68. New York: Elsevier, 1985. xv + 338pp. $66.75;$22.25. ISBN 0-444-87878-5;0-444-87879-3.

Readable, self-contained graduate text, covering common examples and their geometry, metric properties, and geometry of super-reflexive spaces. This edition contains a fifth part, devoted to the local theory of Banach spaces. **(a)**

Introduction to Hilbert Space. Second Edition. Sterling K. Berberian. New York: Chelsea, 1976. xi + 206pp. $13.95. ISBN 0-8284-0287-6.

Reprint, with an addendum, of 1961 original. **(a)**

Introduction to Hp Spaces, with an Appendix on Wolff's Proof of the Corona Problem. Paul Koosis. London Mathematical Society Lecture Note Series;40. xv + 376pp. $42.50. ISBN 0-521-23159-0.

Written in a refreshing, conversational style, this book is an account of the theory of Hardy spaces in the unit disk and the upper half-plane. Very few exercises. **(a)**

Introduction to Various Aspects of Degree Theory in Banach Spaces. E. H. Rothe. Mathematical Surveys and Monographs;23. Providence, RI: American Mathematical Society, 1986. vi + 242pp. $70.00. ISBN 0-8218-1522-9.

Good survey of different topics in degree theory. **(a)**

Introductory Functional Analysis with Applications. Erwin Kreyszig. New York: Wiley, 1989. xiv + 688pp. $34.95 paperbound. ISBN 0-471-50459-9.

Reprint of 1978 original. **(p)**

Linear Operators. Nelson Dunford, Jacob T. Schwartz. Pure and Applied Mathematics;7. New York: Wiley, 1988. *Part I: General Theory.* xiv + 858pp. $39.95. ISBN 0-471-60848-3. *Part II: Spectral Theory, Selfadjoint Operators in Hilbert Space.* x + 859-1923pp. $36.95. ISBN 0-471-60847-5. *Part III: Spectral Operators.* xx + 1925-2592pp. $36.95. ISBN 0-471-60846-7.

Reprint of the 1958-1971 original, a comprehensive account of the modern theory of linear operators, mainly in Banach spaces, and of other applications of the theory of such operators to other parts of mathematics. **(a)**

Metric Linear Spaces. Stefan Rolewicz. Norwell, MA: Kluwer Academic, 1985. xii + 459pp. $160.00. ISBN 90-277-1480-0.

An encyclopedic, definitive treatment of non-locally convex spaces. Expensive. **(a)**

Multivariate Approximation Theory: Selected Topics. E. Ward Cheney. Philadelphia: Society for Industrial and Applied Mathematics, 1986. v + 68pp. $14.50. ISBN 0-89871-207-6.

Based on lectures given at a regional conference on approximation theory and numerical analysis at the University of Alaska. Extensive bilbiography. **(a)**

Nonlinear Functional Analysis and Its Applications. Eberhard Zeidler. Transl: Peter R. Wadsack, Leo F. Boron, Juergen Quandt. New York: Springer-Verlag, 1986- . *Volume I: Fixed-Point Theorems.* xxi + 897pp. $109.00 ISBN 0-387-90914-1. *Volume II/A: Linear Monotone Operators.* xi + 467pp. $119.00. ISBN 0-387-96802-4. *Volume II/B: Nonlinear Monotone Operators.* xv + 469-1202pp. $129.00. ISBN 0-387-97167-X. *Volume III: Variational Methods and Optimization.* xxii + 662pp. $98.00. ISBN 0-387-90915-X. *Volume IV: Applications to Mathematical Physics.* xxiii + 975pp. $129.00. ISBN 0-387-96499-1.

Four of a five-volume exposition of the main principles of nonlinear functional analysis and its applications to the natural sciences, economics, and numerical analysis. The presentation is self-contained and accessible to the nonspecialist. Comprehensive bibliography. Well worth reading. Expensive. **(a)**

Sequences and Series in Banach Spaces. Joseph Diestel. Graduate Texts in Mathematics;92. New York: Springer-Verlag, 1984. xiii + 261pp. $49.00. ISBN 0-387-90859-5.

Introduction, focusing on recent developments, to results and methods of structure theory of Banach spaces. **(a)**

Studies in Functional Analysis. Ed: R. G. Bartle. Studies in Mathematics;21. Washington, DC: Mathematical Association of America, 1980. xi + 227pp. $26.00. ISBN 0-88385-121-0.

Five fascinating expository articles on developments in functional analysis. **(a)**

Théorie des Opérations Linéaires. Second Edition. Stefan Banach. New York: Chelsea, 1963. xii + 259pp. $13.95. ISBN 0-8284-0110-1.

Reprint of 1932 edition, with corrections of errata and with an added note on Haar Measure, which originally appeared as an addendum to Saks' *Théorie de l'Intégrale.* (**a**)

Theory of Operator Algebras I. Masamichi Takesaki. New York: Springer-Verlag, 1979. vii + 415pp. $66.00. ISBN 0-387-90391-7.

An indispensible reference work for all researchers in operator algebras. (**a**)

Topological Vector Spaces, Chapters 1-5. Nicolas Bourbaki. (Elements of Mathematics) Transl: H. G. Eggleston, S. Madan. New York: Springer-Verlag, 1987. viii + 364pp. $79.00. ISBN 0-387-13627-4.

Translation of the French original (see p. 214), a brand new book which completely supersedes the version of 30 years ago: much of the material has been rearranged, rewritten, or replaced by more up-to-date exposition, reflecting the progress made in the field. (**a**)

Topological Vector Spaces. A. Grothendieck. Transl: Orlando Chaljub. New York: Gordon & Breach, 1973. x + 245pp. $89.00. ISBN 0-677-30020-4.

Translation of the 1954 classic, *Espaces Vectoriels Topologiques.* (**a**)

GEOMETRY

An Adventure in Multidimensional Space: The Art and Geometry of Polygons, Polyhedra, and Polytopes. Koji Miyazaki. New York: Wiley. ii + 112pp. $68.95. ISBN 0-471-81648-5.

Figures, photographs, and drawings illustrate a variety of views of the role of polygons, polyhedra, and 4-polytopes in 2-, 3-, and 4-dimensional worlds. (**a**)

The Adventures of Archibald Higgins: Here's Looking at Euclid (And Not Looking at Euclid). Jean-Pierre Petit. Transl: Ian Stewart. Providence, RI: Janson, 1987. 72pp. $7.95. ISBN 0-86576-092-6.

Translation of a French comic book interpretation of serious science, in this case, Archibald Higgins in the land of intrinsic geometry, geodesics, and hyperspheres. (**p**)

Basic Concepts of Geometry. Walter Prenowitz, Meyer Jordan. New York: Ardsley House, 1989. xix + 350pp. $34.95. ISBN 0-912675-48-9.

Paperbound reprint of the 1965 classic introduction to modern geometry. (**p**)

Beyond the Third Dimension: Geometry, Computer Graphics, and Higher Dimensions. Thomas F. Banchoff. New York: Scientific American Library, 1990. ix + 210pp. $22.95. ISBN 0-7167-5025-2.

A visually rich (full color plates and computer-generated drawings) and intellectually rewarding portrait of dimensions. (**p**)

A Budget of Trisections. Underwood Dudley. New York: Springer-Verlag, 1987. xvii + 169pp. $35.00. ISBN 0-387-96568-8.

A warm and witty account of the countless amateur mathematicians who pursue their dream of trisecting triangles. **(p)**

Build Your Own Polyhedra. Peter Hilton, Jean Pedersen. Reading, MA: Addison-Wesley, 1988. 175pp. $27.50. ISBN 0-201-22060-1.

A good book for self-study of polyhedra. **(p)**

A Catalog of Special Plane Curves. J. Dennis Lawrence. New York: Dover, 1972. xi + 218pp. $8.95. ISBN 0-486-60288-5.

A careful, systematic survey of the historical species of plane analytic geometry, illustrated with computer plots. **(a)**

Connections: The Geometric Bridge Between Art and Science. Jay Kappraff. New York: McGraw-Hill, 1991. xxi + 471pp. $19.50; $39.50. ISBN 0-07-034251-2;0-07-034250-4.

A spectacular presentation of design science with many exercises and problems and illustrated with over 200 figures. Extensive multidisciplinary bibliography that is well-referenced to the text. **(a,p)**

Convex Polytopes. Branko Grünbaum et al. Pure and Applied Mathematics;16. Ann Arbor, MI: Books on Demand, 1987. xiv + 456pp. $117.50. ISBN 0-317-08683-9.

Photocopy of 1967 original. Expensive. **(a)**

Convex Sets and Their Applications. Steven R. Lay. New York: Wiley, 1982. xvi + 244pp. $49.95. ISBN 0-471-09584-2.

Natural companion volume to Roberts' and Varberg's *Convex Functions* (neither listed), this beautiful book is less abstract than Valentine's *Convex Sets*. **(p)**

Elementary Geometry from an Advanced Standpoint. Second Edition. Edwin E. Moise. Reading, MA: Addison-Wesley, 1974. xv + 425pp. $37.56. ISBN 0-201-04793-4.

This rigorous presentation of Euclidean geometry now contains more problems. **(a)**

The Ethics of Geometry: A Genealogy of Modernity. David Rapport Lachterman. New York: Routledge, Chapman & Hall, 1989. xiv + 255pp. $17.95;$49.50. ISBN 0-415-90141-3;0-415-90053-0.

A scholarly treatment of the relationship between philosophy and mathematics. Extensive notes with references to both primary and secondary sources. **(p)**

Famous Problems of Elementary Geometry: The Duplication of the Cube, the Trisection of an Angle, the Quadrature of the Circle. Second Edition. Revised and Enlarged with Notes by Raymond Clare Archibald. Felix Klein. Transl: Wooster Woodruff Beman, David Eugene Smith. New York: Chelsea, 1956. xi + 92pp. $12.95. ISBN 0-8284-0108-X.

Unabridged, unaltered republication of 1930 Second edition. **(p)**

Foundations of Geometry. Second Edition. David Hilbert. Transl: Leo Unger. Peru, IL: Open Court, 1980. ix + 226pp. $9.00. ISBN 0-87548-164-7.
Fifth printing of the Tenth German edition which had been revised by Paul Bernays. **(p)**

The Foundations of Geometry and the Non-Euclidean Plane. George E. Martin. New York: Springer-Verlag, 1986. $32.00. ISBN 0-387-90694-0.
Reprint of an excellent 1975 geometry text. **(a,p)**

Fractals: Endlessly Repeated Geometrical Figures. Hans Lauwerier. Transl: Sophia Gill-Hoffstadt. Princeton, NJ: Princeton University Press, 1991. xiv + 209pp. $14.95; $49.50. ISBN 0-691-02445-6; 0-691-08551-X.
An engaging elementary introduction to the geometry of fractals in the plane along with simple algebraic analysis of their properties. **(a)**

Fundamental Concepts of Geometry. Bruce E. Meserve. New York: Dover, 1983. ix + 321pp. $7.95. ISBN 0-486-63415-9.
Slightly corrected reprint of 1955 classic. **(a,p)**

Geometric Inequalities. Nicholas D. Kazarinoff. New Mathematical Library;4. Washington, DC: Mathematical Association of America, 1961. 132pp. $10.00. ISBN 0-88385-604-2.
The author starts with simple arithmetic and algebraic inequalities and leads the reader through results including the arithmetic-geometric mean inequality to a far-ranging consideration of many geometric inequalities. **(a)**

Geometric Transformations. I. M. I͡Aglom. Transl: Allen Shields, A. Shenitzer. New Mathematical Library;8,21,24. Washington, DC: Mathematical Association of America, 1962-1973. *I.* 133pp. $10.00. ISBN 0-88385-608-5. *II.* 189pp. $ 11.75. ISBN 0-88385-621-2. *III.* 237pp. $12.00. ISBN 0-88385-624-7.
These 3 volumes range from Euclidean geometry, symmetry, and the other properties of isometric transformations to affine and projective transformations and their associated geometries. **(a)**

Geometry: A High School Course. Second Edition. Serge Lang, Gene Murrow. New York: Springer-Verlag, 1988. xii + 394pp. $39.00. ISBN 0-387-96654-4.
A text which may help revitalize high school geometry. **(p)**

Geometry and Convexity, A Study in Mathematical Methods. Paul J. Kelly, Max L. Weiss. New York: Wiley, 1979. x + 261pp. $39.95. ISBN 0-471-04637-X.
Integration of the methods of geometric metric topology, linear algebra, classical analysis and geometry in order to study convexity, convex bodies and surfaces, and applications. **(a)**

Geometry and the Imagination. D. Hilbert, S. Cohn-Vossen. Transl: P. Nemenyi. New York: Chelsea, 1990. ix + 357pp. $18.95. ISBN 0-8284-1087-9.
Translation of 1932 *Anschauliche Geometrie.* **(a,p)**

Geometry of Complex Numbers: Circle Geometry, Moebius Transformation, Non-Euclidean Geometry. Hans Schwerdtfeger. New York: Dover, 1979. xiii + 200pp. $6.95. ISBN 0-486-63830-8.

Corrected reprint of 1962 edition which shows the interrelatedness of geometry, analysis, and algebra. **(a,p)**

Geometry Revisited. H. S. M. Coxeter, S. L. Greitzer. New Mathematical Library;19. Washington, DC: Mathematical Association of America, 1967. xiv + 193pp. $13.95. ISBN 0-88385-619-0.

This brief book, intended for high school to university students, covers a variety of topics and includes exercises, references, and glossary. **(a)**

Introduction to Geometry. Second Edition. H. S. M. Coxeter. New York: Wiley, 1969. 485pp. $34.95. ISBN 0-471-18283-4.

A classic, the author enlivening and clarifying his text with references to literature, history, architecture, etc. Diagrams and illustrations numerous and beautiful. **(a,p)**

Journey into Geometries. Marta Sved. Washington, DC: Mathematical Association of America, 1991. xv + 182pp. $21.00. ISBN 0-88385-500-3.

A serious introduction to non-Euclidean geometry through a whimsical Socratic conversation between Lewis Carroll, Alice, and a contemporary inquisitor. Wonderful reading or for study of the exercises and problems. **(p)**

The Mathematical Description of Shape and Form. E. A. Lord, C. B. Wilson. New York: Halsted Press, 1986. 260pp. $63.95. ISBN 0-470-20043-X.

A unique survey of shapes and patterns backed up by mathematical representations from linear algebra and vector calculus. **(a)**

Mathematics and Optimal Form. Stefan Hildebrandt, Anthony Tromba. New York: W. H. Freeman, 1985. xvi + 215pp. $32.95. ISBN 0-7167-5009-0.

A stunning presentation of the geometry of optimal shapes (e.g., minimal paths and surfaces, soap films) together with the stories of the mathematicians who discovered them. Profusely illustrated with both drawings and photographs. **(a,p)**

Metric Affine Geometry. Ernst Snapper, Robert J. Troyer. New York: Dover, 1989. xx + 435pp. $10.95. ISBN 0-486-66108-3.

Reprint of 1971 edition, based on Snapper's 1964 Cornell course. Assumes a knowledge of linear algebra and elementary group theory. **(a)**

A Modern View of Geometry. Leonard M. Blumenthal. New York: Dover, 1980. xii + 191pp. $7.95. ISBN 0-486-63962-2.

Unabridged corrected reprint of 1961 original. **(p)**

Non-Euclidean Geometry. Fifth Edition. H. S. M. Coxeter. Mathematical Expositions;2. Toronto: University of Toronto Press, 1965. xv + 309pp. $35.00. ISBN 0-8020-1068-7.

The principal change for this edition is where Poincaré's conformal model for hyperbolic spaces is extended in the manner advocated by Liebmann."— Preface. **(p)**

The Non-Euclidean, Hyperbolic Plane: Its Structure and Consistency. Paul Kelly, Gordon Matthews. New York: Springer-Verlag, 1981. xiii + 333pp. $49.00. ISBN 0-387-90522-9.
Careful, thorough, and elementary treatment of hyperbolic geometry. **(a)**

Polyhedron Models. Magnus J. Wenninger. New York: Cambridge University Press, 1989. xii + 208pp. $24.95 paperbound. ISBN 0-521-09859-9.
Reprint of 1974 edition, which gives detailed instructions for the construction of 11 polyhedra. **(p)**

Projective Geometry. Second Edition. H. S. M. Coxeter. New York: Springer-Verlag, 1987. xii + 162pp. $35.00. ISBN 0-387-96532-7.
Slightly revised reprint of 1974 Second edition. **(p)**

Recent Advances in Geometric Inequalities. D. S. Mitrinović, J. E. Pečarić, V. Volenec. Norwell, MA: Kluwer Academic, 1989. xix + 710pp. $149.00. ISBN 90-277-2565-9.
This monumental work updates and extends Bottema's 1969 *Geometric Inequalities* to several thousand inequalities. Extensive references, various methods of proof, conjectures, and unsolved problems. Expensive. **(a)**

Results and Problems in Combinatorial Geometry. V. G. Boltianskiĭ, I. T͡S. Gohberg. New York: Cambridge University Press, 1985. 108pp. $37.50;$11.95. ISBN 0-521-26298-4; 0-521-26923-7.
English translation of an inviting discussion of a few problems from the theory of convex bodies, mainly connected with the partition of a set into smaller parts. **(a)**

Shapes, Space, and Symmetry. Alan Holden. New York: Dover, 1991. 200pp. $26.00 paperbound. ISBN 0-486-26851-9.
Reprint of 1971 original. Mostly photographs by Doug Kendall plus a description of ways to make these beautiful shapes. **(p)**

Spherical Models. Magnus J. Wenninger. New York: Cambridge University Press, 1979. xii + 147pp. $18.95. ISBN 0-521-29432-0.
Secrets of model construction are shown in clear diagrams and photographs. **(p)**

Surfaces. Second Edition. H. B. Griffiths. New York: Cambridge University Press, 1981. xii + 128pp. $44.50;$12.95. ISBN 0-521-23570-7;0-521-299772.
Revised primarily in its explanation of a twisted bridge. **(a)**

Symmetry, An Analytical Treatment. J. Lee Kavanau. Los Angeles, CA: Science Software Systems, 1980. xxx + 649pp. $24.95. ISBN 0-937292-00-1.
Striking new results on symmetry and classification of curves. **(a)**

Textual Studies in Ancient and Medieval Geometry. Wilbur Richard Knorr. Boston: Birkhäuser, 1989. xvii + 852pp. $119.00. ISBN 0-8176-3387-1.

Exposition of the problems of cube duplication and angle trisection by Pappus and Eutocius. Numerous notes and references. Expensive. **(a)**

Tilings and Patterns. Branko Grünbaum, G. C. Shephard. New York: W. H. Freeman, 1986. ix + 700pp. $59.95. ISBN 0-7167-1193-1.

Comprehensive and systematic treatment of the subject, focusing on classification and enumeration of tilings. **(a)**

Topological Geometry. Second Edition. Ian R. Porteous. New York: Cambridge University Press, 1981. 486pp. $42.50;$85.00. ISBN 0-521-29839-3;0-521-23160-4.

Corrected reprint of the First edition shows the relationship between affine geometry and multivariable calculus. **(a)**

Turtle Geometry: The Computer as a Medium for Exploring Mathematics. Harold Abelson, Andrea A. diSessa. Cambridge, MA: MIT Press, 1981. xx + 477pp. $37.50. ISBN 0-262-01063-1.

This book introduces both geometry as discovered by turtle (a mathematical creature that crawls across computer screens) and the principles of programming required to implement turtle geometry. **(p)**

GRAPH THEORY

Algorithmic Graph Theory and Perfect Graphs. Martin Charles Golumbic. San Diego, CA: Academic Press, 1980. xx + 284pp. $70.00. ISBN 0-12-289260-7.

An introduction to graph theory which emphasizes those problems which are related to the structure of various graphs. Algorithms related to each structure are described. References. **(a)**

Distance in Graphs. Fred Buckley, Frank Harary. Reading, MA: Addison-Wesley, 1990. xiii + 335pp. $32.76. ISBN 0-201-09591-2.

Introductory text based on Harary's classic Graph Theory (see below), updated and unified by the concept of distance in graphs. Extensive bibliography. **(a,p)**

Examples and Counterexamples in Graph Theory. Michael Capobianco, John C. Molluzzo. New York: Elsevier, 1978. xii + 259pp. $59.75. ISBN 0-444-00255-3.

Compilation of over 500 examples and counterexamples in graph theory, for the graph theorist or the social or physical scientist. **(a)**

Graph Algorithms. Shimon Even. New York: W. H. Freeman, 1979. ix + 249pp. $41.95. ISBN 0-7167-8044-5.

An algorithmic approach to graph theory intended for computer science students. **(a)**

Graph Theory. Frank Harary. Reading, MA: Addison-Wesley, 1969. ix + 274pp. $31.96. ISBN 0-201-02787-9.

Self-contained text strictly on graph theory, written with wit and clarity. **(a,p)**

Graph Theory with Applications. J. A. Bondy, U. S. R. Murty. New York: Elsevier, 1979. x + 264pp. $48.00. ISBN 0-444-19451-7.

Stressing constructive proofs, this work includes several carefully developed and meaningful applications of graph theory. **(a)**

Graphical Enumeration. Frank Harary, Edgar M. Palmer. San Diego, CA: Academic Press, 1973. xiv + 271pp. $81.00. ISBN 0-12-324245-2.

First time that all of the major methods of solving graph counting problems have appeared together. Careful explanations. **(a)**

Graphs and Algorithms. Michel Gondran, Michel Minoux. Transl: Steven Vajda. New York: Wiley, 1984. xix + 650pp. $200.00. ISBN 0-471-10374-8.

This translation of the 1979 French text, provides an excellent overview of graph theory and the combinatorial algorithms which link the abstract results and their practical implications. Excellent bibliography. Expensive. **(a,s)**

Graphs & Digraphs. Second Edition. Gary Chartrand, Linda Lesniak. Belmont, CA: Wadsworth, 1986. viii + 359pp. $53.95. ISBN 0-534-06324-1.

A major change in this edition is the integration of graph and digraph theory. **(a)**

Graphs and Their Uses. Oystein Ore. Rev: Robin J. Wilson. New Mathematical Library;34. Washington, DC: Mathematical Association of America, 1990. viii + 153pp. $16.95. ISBN 0-88385-600-X.

This classic has been revised with updated terminology, modern notation, and new material on interval graphs, travelling salesman problem, etc. A brief contemporary bibliography. **(a,p)**

Introduction to Graph Theory. Third Edition. Robin J. Wilson. New York: Wiley, 1985. viii + 166pp. $21.95. ISBN 0-582-44685-6.

A highly recommended elementary introduction to graph theory. This edition has expanded the section on applications, reorganized the material on connectivity and edge-colorings, and rewritten most of the exercises. **(a)**

Pearls in Graph Theory: A Comprehensive Introduction. Nora Hartsfield, Gerhard Ringel. San Diego, CA: Academic Press, 1990. ix + 246pp. $29.95. ISBN 0-12-328552-6.

Innovative text covering elementary concepts as well as a clear exposition of unusual and more advanced topics. Especially good chapters are those on planar maps and their colorings, graphs on surfaces, Map Color Theorem, and Heawood's Empire Problems. **(a)**

GROUP THEORY

Atlas of Finite Groups: Maximal Subgroups and Ordinary Characters for Simple Groups. J. H. Conway et al. New York: Oxford University Press, 1985. xxxiii + 252pp. $59.00. ISBN 0-19-853199-0.

Intended to convey "every interesting fact about every interesting finite group" by providing computer-generated tables for finite simple groups containing various constructions, information about subgroups, and a "compound character table." **(a)**

The Classification of Finite Simple Groups. Volume 1: Groups of Noncharacteristic 2 Type. Daniel Gorenstein. New York: Plenum, 1983. x + 487pp. $89.50. ISBN 0-306-41305-1.

An invaluable and comprehensive outline of the classification of finite simple groups. **(a)**

Finite Groups. Second Edition. Daniel Gorenstein. New York: Chelsea, 1980. xvii + 519pp. $29.50. ISBN 0-8284-0301-5.

Minor updating of 1968 edition. **(a)**

Finite Reflection Groups. Second Edition. L. C. Grove, C. T. Benson. Graduate Texts in Mathematics;99. New York: Springer-Verlag, 1985. x +133pp. $35.00. ISBN 0-387-96082-1.

This edition has added a chapter on classical invariant theory of finite reflection groups and an appendix listing the Schoenflies and international notational schemes for crystallographic point groups. **(a)**

Finite Simple Groups: An Introduction to their Classification. Daniel Gorenstein. New York: Plenum Press, 1982. x + 333pp. $42.50. ISBN 0-306-40779-5.

Beautifully written exposition of the classification of finite simple groups beginning with an overview, then a description of the simple grpups, and finally a chapter on local techniques used in the classification proof. **(a)**

Group Theory. W. R. Scott. New York: Dover, 1987. 479pp. $11.95. ISBN 0-486-65377-3.

An unabridged, corrected reprint of the classic 1964 text. **(a,p)**

Groupes et Algèbres de Lie, Chapitre I: Algebres de Lie. Nicolas Bourbaki. (Éléments de Mathématique) Paris: Diffusion C.C.L.S., 1982. (Available from Scientific & Medical Publications of France, New York, NY) 146pp. $76.00. ISBN 2-903684-09-X.

Reprint of 1971 edition. (See below for translation of chapters 1-3, *Lie Groups and Lie Algebras*.) **(a,s)**

Groupes et Algèbres de Lie, Chapitres II, III. Nicolas Bourbaki. (Éléments de Mathématique) Paris: Diffusion C.C.L.S., 1982. (Available from Scientific & Medical Publications of France, New York, NY) 320pp. $74.00. ISBN 2-903684-08-1.

Reprint of 1972 edition. (See below for translation of chapters 1-3, *Lie Groups and Lie Algebras*.) **(a,s)**

Groupes et Algèbres de Lie, Chapitres 4, 5 et 6. Nicolas Bourbaki. (Éléments de Mathématique) Paris: Masson, 1981. (Available from Scientific & Medical Publications of France, New York, NY) 288pp. $74.00. ISBN 2-225-76076-4.

A rich and useful volume, treating material which has relevance well beyond the theory of Lie groups and algebras. **(a,s)**

Groupes et Algèbres de Lie, Chapitres VII, VIII. Nouvelle Tirage. Nicolas Bourbaki. (Elements de Mathematique) Paris: Masson, 1990. (Available from Scientific & Medical Publications of France, New York, NY) 271pp. $72.00. ISBN 2-225-81911-4. Reprint of 1975 edition.

This book serves as a convenient reference and source of stimulating exercises to those familiar with the basic features of semisimple Lie algebras. **(a,s)**

Groupes et Algèbres de Lie, Chapitre 9: Groupes de Lie Réels Compacts. Nicolas Bourbaki. (Éléments de Mathématique) Paris: Masson, 1982. (Available from Scientific & Medical Publications of France, New York, NY) 138pp. $ 45.00. ISBN 2-225-76461-1.

Careful organization of a multifaceted subject, the focus here is on compact groups. Many references to earlier chapters and other works of the author. (See below for translation of chapters 1-3, *Lie Groups and Lie Algebras.*) **(a,s)**

Harmonic Analysis on Semi-Simple Lie Groups. Garth Warner. Die Grundlehren der mathematischen Wissenschaften;188-189. New York: Springer-Verlag, 1972. *I.* xvi + 529pp. $79.00. ISBN 0-387-05468-5. *II.* viii + 491pp. $79.00. ISBN 0-387-05469-3.

Large compilation of various results and techniques in representation theory. While this work is useful for mathematicians working in the field, it is a painstaking task to wade through both volumes. **(a)**

Infinite Abelian Groups. Laszlo Fuchs. Pure and Applied Mathematics;36. San Diego, CA: Academic Press, 1970-1973. *Vol. I.* xi + 290pp. $64.00. ISBN 0-12-269601-8. *Vol. II.* ix + 363pp. $64.00. ISBN 0-12-269602-6.

This two-volume work gives the basic definitions, contstructions, and theorems about abelian groups. Volume I treats the homological aspects; volume II studies structure theory and applications. References. **(a)**

Infinite Abelian Groups. Irving Kaplansky. Ann Arbor, MI: Books on Demand, 1990. v + 91pp. $24.50. ISBN 0-317-08467-4.

Photocopy of 1954 original. This concise book condenses a wealth of material without detriment to the completeness and clarity of the presentation. **(a)**

An Introduction to the Theory of Groups. Third Edition. Joseph J. Rotman. Dubuque, IA: W. C. Brown, 1984. x + 422pp. $58.00. ISBN 0-697-06882-X.

Treatment of amalgams and HNN extensions based on on pushouts and the Seifert-van Kampen theorem. **(a)**

Lie Groups and Lie Algebras, Chapters 1-3. Nicolas Bourbaki. (Elements of Mathematics). New York: Springer-Verlag, 1989. xvii + 450pp. $79.00. ISBN 0-387-13627-4.

English translation of the first part of Bourbaki's *Groupes et Algèbres de Lie*, covering Lie algebras, free Lie algebras, and Lie groups. (See above for chapters 4-6, 9 *Groupes et Algèbres de Lie.*) **(a,s)**

Local Representation Theory. J. L. Alperin. Cambridge Studies in Advanced Mathematics;11. New York: Cambridge University Press, 1986. x + 178pp. $37.50. ISBN 0-521-30660-4.

Good introduction to the modular representation theory of finite groups. **(a)**

Methods of Representation Theory with Applications to Finite Groups and Orders. Charles W. Curtis, Irving Reiner. New York: Wiley, 1987-1990. *Volume I.* xxi + 819pp. $39.95 paperbound. ISBN 0-471-52367-4. *Volume II.* xv + 951pp. $131.95. ISBN 0-471-88871-0.

These volumes replace the old Curtis-Reiner (1981-1987) as the bible of representation theory. They are the most remarkable, thorough, self-contained treatment of the subject. **(a)**

The Santa Cruz Conference on Finite Groups. Ed: Bruce Cooperstein, Geoffrey Mason. Proceedings of Symposia in Pure Mathematics;37. Providence, RI: American Mathematical Society, 1980. xviii + 634pp. $39.60. ISBN 0-821-81440-0.

Ninety papers for a 1979 summer research symposium on group theory. **(a)**

The Theory of Groups. Second Edition. Marshall Hall, Jr. New York: Chelsea, 1976. xiii + 434pp. $14.95. ISBN 0-8284-0288-4.

Reprint of 1968 (10th printing) of the 1959 First edition. **(a)**

Unitary Group Representations in Physics, Probability, and Number Theory. George W. Mackey. Reading, MA: Addison-Wesley, 1989. xiv + 402pp. $22.20. ISBN 0-201-51009-X.

Reprint of author's famous Oxford lecture notes (1978), with a special preface. **(a)**

Vorlesungen über Continuierliche Gruppen. Sophus Lie. New York: Chelsea, 1971. xii + 810pp. $39.95. ISBN 0-8284-0199-3.

Reprint of Lie's 1893 classic introduction to the theory of continuous transformation groups. Errata have been corrected. **(a)**

HISTORY AND BIOGRAPHY

A. M. Turing's ACE Report of 1946 and Other Papers. Ed: B. E. Carpenter, R. W. Doran. Cambridge, MA: MIT Press, 1986. vii + 140pp. $22.00. ISBN 0-262-03114-0.

Includes Turing's original proposal for an Automatic Computing Engine (ACE), his 1947 London Mathematical Society lecture amplifying the proposal, and Michael Woodger's 1958 report on the first practical implementation of ACE. **(a)**

Abrége d'Histoire des Mathématiques 1700-1900. Jean Dieudonné. Paris: Hermann, 1986. x + 517pp. 210.00 FF. ISBN 2-7056-6024-0.

New revised edition of the 1978 original. Describes the development of several important parts of mathematics from 1700 to 1900. Begins with remarks about mathematicians and how they work. Chapter bibliographies. Historical and terminology indexes. **(a)**

Alan Turing: The Enigma. Andrew Hodges. New York: Simon & Schuster, 1983. 587pp. $10.95. ISBN 0-671-49207-1.

A sympathetic, thoroughly documented portrait of Turing, including numerous sections with superb elementary expositions of his mathematics. **(a,p)**

Albert Einstein's Special Theory of Relativity: Emergence (1905) and Early Interpretation (1905-1911). Arthur I. Miller. Reading, MA: Addison-Wesley, 1981. xxviii + 466pp. $20.76 paperbound. ISBN 0-201-04680-6.

the author gives the state of electro-dynamics, optics, and electromagnetic theory at the turn of the century as the historical context for Einstein's paper, analyzes the internal structure of the paper, and examines subsequent developments. Miller's translation of the paper corrects misprints of the original and earlier translations. **(p)**

Alfred North Whitehead: The Man and His Work. Victor Lowe. Baltimore, MD: Johns Hopkins University Press, 1985-1990. *Volume I: 1861-1910.* xi + 351pp. $37.50. ISBN 0-8018-2488-5. *Volume II: 1910-1947.* ix + 389pp. $38.95. ISBN 0-8018-3960-2.

This is the first full biography of Whitehead. Volume I covers his years as a mathematics lecturer at Trinity College. J. B. Schneewind has edited the unfinished manuscript of volume II, which includes, among other things, Whitehead's arrival in America and an explication of his philosophy. **(a,p)**

Archimedes. E. J. Dijksterhuis. Transl: C. Dikshoorn. Princeton, NJ: Princeton University Press, 1987. 457pp. $17.95; $57.50. ISBN 0-691-02400-6; 0-691-08421-1.

A classic study now augmented with Knorr's bibliographic essay "Archimedes after Dijksterhuis: A Guide to Recent Studies," which provides over 200 updated references. **(a,p)**

Aristarchus of Samos: The Ancient Copernicus. Sir Thomas Heath. New York: Dover, 1981. viii + 425pp. $7.00. ISBN 0-486-24188-2.

Reprint of the classic 1913 work describing the history of Greek astronomy from Thales through Aristarchus. **(p)**

Augustin-Louis Cauchy: A Biography. Bruno Belhoste. Transl: Frank Ragland. New York: Springer-Verlag, 1991. xii + 380pp. $79.00. ISBN 0-387-97220-X.

Fascinating and carefully researched account of the professional career of Cauchy. This biography balances historical and mathematical detail and includes a chapter that examines the philosophical rationale of Cauchy's uncompromising approach to politics, religion, and mathematics. **(a)**

Automation of Reasoning. Ed: Jorg Siekmann, Graham Wrightson. New York: Springer-Verlag, 1983. *Volume 1: Classical Papers on Computational Logic 1957-1966.* xii + 525pp. $99.00. ISBN 0-387-12043-2. *Volume 2: Classical Papers on Computational Logic 1967-1970.* xii + 637pp. $99.00. ISBN 0-387-12044-0.

Reprints of papers on the mechanization of reasoning by computers. Comprehensive bibliography of related works in each volume. **(a)**

The Calculating Passion of Ada Byron. Joan Baum. Hamden, CT: Archon Books, 1986. xix + 133pp. $23.50. ISBN 0-208-02119-1.

A warm, informative biography of "the world's first computer 'programmer,'" Lord Byron's daughter Ada, interpreter of Babbage's "Analytical Engine." In 1843 Ada Byron prepared notes on how to use the analytical engine (still unbuilt) to compute Bernoulli numbers. **(a,p)**

A Century of Mathematics in America. Ed: Peter Duren, Richard A. Askey, Uta C. Merzbach. History of Mathematics;1-3. Providence, RI: American Mathematical Society, 1989. *Part I.* viii + 477pp. $57.00. ISBN 0-8218-0124-4. *Part II.* x + 585pp. $70.00. ISBN 0-8218-0130-9. *Part III.* ix + 675pp. $75.00. ISBN 0-8218-0136-8.

Rich collection of papers on research, people, teaching, organizations, politics, and education on the history of mathematics in America. The final volume contains historical papers written in honor of the AMS centennial. **(a)**

Chinese Mathematics: A Concise History. Li Yan, Du Shiran. Transl: John N. Crossley, Anthony W. -C. Lun. New York: Oxford University Press, 1987. xiii + 290pp. $49.95. ISBN 0-19-858181-5.

The first English translation of an important Chinese view of the history of mathematics in China, from Li Shou to the Qing Dynasty, showing that indigenous ancient Chinese mathematics was strongly algebraic. **(a,p)**

Code of the Quipu: A Study in Media, Mathematics, and Culture. Marcia and Robert Ascher. Ann Arbor, MI: University of Michigan Press, 1981. vii + 166pp. $11.95. ISBN 0-472-06325-1.

Fascinating glimpse into the culture of the Incas who used elaborately knotted colored cord, called quipus, to record and transmit numerical information. The quipus reveal a sophisticated base-ten bookkeeping system. **(a,p)**

A Concise History of Mathematics. Fourth Revised Edition. Dirk J. Struik. New York: Dover, 1987. xii + 228pp. $7.95. ISBN 0-486-60255-9.

A new chapter on the mathematics of the first half of the twentieth century has been added to this edition as well as extending the bibliographies. **(a,p)**

A Convergence of Lives: Sofia Kovalevskaia͡, Scientist, Writer, Revolutionary. Ann Hibner Koblitz. Boston: Birkhäuser, 1983. xx + 305pp. $19.95. ISBN 0-8176-3162-3.

Highly readable and popular account of Kovalevskaia͡'s life, emphasizing her scientific and political interests. Extensive bibliography. **(a)**

Descartes on Polyhedra: A Study of the De Solidorum Elementis. P. J. Federico. Sources in the History of Mathematics and the Physical Sciences;4. New York: Springer-Verlag, 1982. ix + 145pp. $66.00. ISBN 0-387-90760-2.

First English translation of Descartes' general treatment of polyhedra. Includes facsimile of the manuscript with transcription, translation, and commentary. **(a)**

The Development of Mathematics in China and Japan. Second Edition. Yoshio Mikami. New York: Chelsea, 1974. x + 389pp. $23.95. ISBN 0-8284-0149-7.

Corrected reprint of 1913 edition. **(a,p)**

Élements d'Histoire des Mathématiques. Nouvelle Tirage. Nicolas Bourbaki. (Éléments de Mathématique) Paris: Masson, 1984. (Available from Scientific & Medical Publications of France, New York, NY) 376pp. $112.00. 2-225-80320-X.

This reprint of 1974 edition adds a table of contents. It is a fascinating volume which assembles the historical notes from Bourbaki's various *Éléments*. Full bibliography and index. Expensive. **(a)**

Elements of Algebra. Leonhard Euler. Transl: John Hewlett. New York: Springer-Verlag, 1984. lx + 593pp. $47.00. ISBN 0-387-96014-7.

Reprint of the 1849 English translation of Euler's 1770 *Treatise of Algebra* along with C. Truesdall's essay on Euler. **(a,p)**

Emmy Noether: A Tribute to her Life and Work. Ed: James W. Brewer, Martha K. Smith. Pure and Applied Mathematics;69. New York: Dekker, 1981. x + 180pp. $17.50. ISBN 0-8247-1550-0.

An informative collection commemorating the one hundredth anniversary of Noether's birth. A lengthy biography by Clark Kimberling is followed by four personal reflections by MacLane, Taussky, van der Waerden, and Alexandroff, five brief surveys of her mathematics, and finally her address to the 1932 ICM. **(a)**

Episodes in the Mathematics of Medieval Islam. J. L. Berggren. New York: Springer-Verlag, 1986. xiv + 197pp. $36.60. ISBN 0-387-96318-9.

A sampling of Islamic sources dealing with algebra, geometry, and trigonometry plus commentary on the historical and Islamic context of these ideas. **(a)**

Essays in the History of Mathematics. Ed: Arthur Schlissel. Memoirs of the American Mathematical Society;298. Providence, RI: American Mathematical Society, 1984. v + 73pp. $15.00. ISBN 0-8218-2298-5.

Six papers from a special session of the 1981 annual AMS meeting, in which mathematicians offer their perspectives on the origins of their fields of speciality. **(a)**

The Exact Science in Antiquity. Second Edition. O. Neugebauer. New York: Dover, 1969. xvi + 240pp. $6.95. ISBN 0-486-22332-9.

Unabridged, slightly corrected printing of 1957 edition. **(a,p)**

Festschrift Heinrich Weber zu seinem siebzigsten Geburtstag am 5, Marz 1912. New York: Chelsea, 1971. 500pp. $25.00. ISBN 0-8284-0246-9.

Unaltered reprint of the 1912 collection of essays in honor of the Weber's seventieth birthday. No biography, but 29 essays by Dedekind, Hilbert, etc. **(a)**

Fire in the Valley: The Making of the Personal Computer. Paul Freiberger, Michael Swaine. Berkeley, CA: Osborne/McGraw-Hill, 1984. xv + 288pp. $9.95. ISBN 0-88134-121-5.

Substantive and lively history of personal computers, based on interviews with the founders. Full of fascinating details and anecdotes. Archival photographs. **(p)**

From Error-Correcting Codes Through Sphere Packings to Simple Groups. Thomas M. Thompson. Carus Mathematical Monographs;21. Washington, DC: Mathematical Association of America, 1983. xiv + 228pp. $27.95. ISBN 0-88385-023-0.

A fascinating monograph tracing the events leading from Hamming and Golay's discovery of error-correcting codes in 1948 through Conway's explication of the related group ".0" two decades later. **(a)**

The Genesis of the Abstract Group Concept. Hans Wussing. Transl: Abe Shenitzer. Cambridge, MA: MIT Press, 1984. 331pp. $42.00. ISBN 0-262-23109-3.

Pioneering work traces the evolution of the abstract group concept from implicit group theoretic modes of thought in geometry and number theory through the concept of a group as a permutation group to the axiomatization of the concept. **(a)**

Geometry in Ancient and Medieval India. First Edition. T. A. Sarasvati Amma. Delhi: Motilal Banarsidass, 1979. (Available from Orient Book Distributors, Livingston, NJ) xi + 279pp. $18.50. ISBN 0-89684-020-4.

A geometrical survey of Sanskrit and Prakrit scientific literature, from 6th-5th century B.C. to 17th century A.D. **(a)**

Glory and Failure. Michael Lindgren. Transl: Craig G. McKay. Cambridge, MA: MIT Press, 1990. 414pp. $45.00. ISBN 0-262-12146-8.

A social and technical analysis of the first attempts to mechanize the production of numerical tables. Superb bibliography of original source material. **(a,p)**

Great Moments in Mathematics (After 1650). Howard Eves. Dolciani Mathematical Expositions;7. Washington, DC: Mathematical Association of America, 1983. xii + 263pp. $18.00. ISBN 0-88385-311-6.

A very readable, non-detailed book of twenty lectures from the modern era covering the discovery of Fourier series, non-Euclidean geometry, non-commutative algebra, transfinite numbers, and abstract spaces. **(a,p)**

Great Moments in Mathematics (Before 1650). Howard Eves. Dolciani Mathematical Expositions;5. Washington, DC: Mathematical Association of America, 1983. xiv + 270pp. $18.00. ISBN 0-88385-310-8.

Twenty lectures, similar to the companion volume above, on ancient mathematics. **(a,p)**

Hermann Weyl 1885-1985. Ed: K. Chandrasekharan. New York: Springer-Verlag, 1986. 119pp. $49.00. ISBN 0-387-16843-5.

Features three lectures on Weyl's contributions to physics, geometry, and Lie groups as well as photographs and other details of the celebration of the centenary of his birth. **(a)**

The Higher Calculus: A History of Real and Complex Analysis from Euler to Weierstrasse. Umberto Bottazzini. New York: Springer-Verlag, 1986. 332pp. $47.00. ISBN 0-387-96302-2.

A well-written historical account of 19th century analysis with emphasis on the foundations of complex analysis. **(a)**

Hilbert-Courant. Constance Reid. New York: Springer-Verlag, 1986. xv + 547pp. $28.00. ISBN 0-387-96256-5.

A superb account of the closely linked intellectual biographies of two giants who helped create modern mathematics. This is the combined paperback edition of Reid's 1970 *Hilbert* and 1976 *Courant in Gottingen and New York*. **(a,p)**

The Historical Development of the Calculus. C. H. Edwards, Jr. New York: Springer-Verlag, 1979. xii + 351pp. $49.00. ISBN 0-387-90436-0.

Careful, comprehensive history of calculus, from Babylonian geometry to Robinson's infinitesimals. **(a,p)**

A History of Algebra: From al-Khwarizmi to Emmy Noether. B. L. van der Waerden. New York: Springer-Verlag, 1985. xi + 271pp. $69.00. ISBN 0-387-13610-X.

A valuable, well-documented resource for historians of mathematics and algebraists. The development of algebraic equations, groups and algebras are treated in separate sections. **(a)**

The History of Combinatorial Group Theory: A Case Study in the History of Ideas. Bruce Chandler, Wilhelm Magnus. Studies in the History of Mathematics and Physical Sciences;9. New York: Springer-Verlag, 1982. viii + 234pp. $89.00. ISBN 0-387-90749-1.

A detailed survey of the development of theory of groups defined by generators and relations, from 1882 to mid-1970's. Comprehensive bibliography. **(a)**

A History of Computing in the Twentieth Century: A Collection of Essays. Ed: N. Metropolis, J. Howlett, Gian-Carlo Rota. San Diego, CA: Academic Press, 1980. xix + 659pp. $65.00. ISBN 0-12-491650-3.

Thirty-eight papers from a 1976 Los Alamos conference celebrating sources of the computing revolution. **(p)**

A History of Greek Mathematics. Sir Thomas Heath. New York: Dover, 1981. *Volume I: From Thales to Euclid*. xv + 446pp. $10.95. ISBN 0-486-24073-8. *Volume II: From Aristarchus to Diophantus*. xi + 586pp. $10.95. ISBN 0-486-24074-6.

Reprint of Heath's 1921 classic, noted for its careful scholarship. It is more or less in chronological order, but with special chapters on the history of certain important problems and the roles of major figures. **(a,p)**

A History of Mathematics. Second Edition. Carl B. Boyer, Uta Merzbach. New York: Wiley, 1989. xviii + 762pp. $57.95. ISBN 0-471-09763-2.

Chapters dealing with the 19th and 20th centuries have been revised and expanded from 1968 original. Revised reference and bibliography. **(a,p)**

A History of Mathematics. Fifth Edition. Florian Cajori. New York: Chelsea, 1991. xi + 524pp. $29.50. ISBN 0-8284-2303-6.

This edition of the classic, first published in 1919, has made a few minor improvements and revised the chapter on Babylonian mathematics in light of new research. **(a,p)**

History of Mathematics. David Eugene Smith. New York: Dover, 1958. *Vol. I: General Survey of the History of Elementary Mathematics.* xxii + 596pp. $10.00. ISBN 0-486-20429-4. *Vol. II: Special Topics of Elementary Mathematics.* xii + 725pp. $10.95. ISBN 0-486-20430-8.

An unaltered and unabridged reprint of the 1951 edition of the 1923-1925 original. **(a,p)**

The History of Mathematics, An Introduction. David M. Burton. Needham Heights, MA: Allyn & Bacon, 1985. ix + 678pp. $40.00. ISBN 0-205-08095-2.

A well-illustrated work for the general reader, with some emphasis on personalities. Numerous references. **(p)**

The History of Modern Mathematics. Ed: David E. Rowe, John McCleary. San Diego, CA: Academic Press, 1989. *Volume I: Ideas and Their Reception.* xvi + 453pp. $45.00. ISBN 0-12-599661-6. *Volume II: Institutions and Applications.* xvi + 325pp. $37.50. ISBN 0-12-599662-4.

Twenty-four interesting essays on the mathematics of the 19th and 20th centuries. Volume I focuses on pure mathematics; volume II on mathematical physics and applied mathematics. **(a)**

A History of Non-Euclidean Geometry: Evolution of the Concept of a Geometric Space. B. A. Rosenfeld. Transl: Abe Shenitzer. Studies in the History of Mathematics and Physical Sciences;12. New York: Springer-Verlag, 1988. xi + 471pp. $89.00. ISBN 0-387-96458-4.

A readable, corrected, and augmented translation of the 1976 Russian edition, discussing the mathematical and philosophical factors underlying the discovery of non-Euclidean geometry. Extensive and updated list of references. **(a)**

The History of Statistics: The Measurement of Uncertainty Before 1900. Stephen M. Stigler. Cambridge, MA: Harvard University Press, 1986. xvi + 410pp. $14.95. ISBN 0-674-40340-1.

An extremely interesting exposition of how modern statistical methods evolved as solutions to intellectual puzzles posed by scientific data. **(a,p)**

The History of the Calculus and Its Conceptual Development. Carl B. Boyer. New York: Dover, 1959. v + 346pp. $7.95. ISBN 0-486-60509-4.

Reprint of the 1949 original, *The Concepts of the Calculus*. It is a rounded history of the concepts of the subject rather than the theorems and applications of calculus. **(a,p)**

A History of the Calculus of Variations during the Nineteenth Century. I. Todhunter. New York: Chelsea, 1962. 532pp. $23.95. ISBN 0-8284-0164-0.

Unaltered reprint of *A History of the Progress of the Calculus of Variations during the Nineteenth Century* originally published in 1861. **(a)**

A History of the Calculus of Variations from the 17th Through the 19th Century. Herman H. Goldstine. Studies in History of Mathematics and Physical Sciences;5. New York: Springer-Verlag, 1980. xviii + 410pp. $98.00. ISBN 0-387-90521-9.

This excellent resource reviews the key papers from Fermat, Newton, and Bernoulli to Hilbert, Kneser, Caratheodory, and Hahn. **(a)**

A History of the Calculus of Variations in the Eighteenth Century. Robert Woodhouse. New York: Chelsea, 1964. ix + 154pp. $14.95. ISBN 0-8284-0177-2.

Textually unaltered reprint of *A Treatise on Isoperimetrical Problems and the Calculus of Variations* originally published in 1810. **(a)**

A History of Vector Analysis: The Evolution of the Idea of a Vectorial System. Michael J. Crowe. New York: Dover, 1985. xv + 270pp. $7.00. ISBN 0-486-64955-5.

Reprint of 1967 edition, bringing to readers "some of the relevant studies of specific areas which have appeared" since the first edition. **(a)**

Huygens and Barrow, Newton and Hook. V. I. Arnol'd. Transl: Eric J. F. Primrose. Boston: Birkhäuser, 1990. 118pp. $18.50. ISBN 0-8176-2383-3.

A penetrating historical probe that reveals contemporary perspectives on deep mathematical problems explored by Newton and his contemporaries. **(a,p)**

I Have a Photographic Memory. Paul R. Halmos. Providence, RI: American Mathematical Society, 1987. vii + 326pp. $55.00. ISBN 0-8218-0115-5.

A more or less chronologically arranged collection of 600 snapshots of mathematicians from Halmos' larger collection and with his chatty, informative commentary. Index of names. **(a)**

I Want to Be a Mathematician: An Automathography in Three Parts. Paul R. Halmos. Washington, DC: Mathematical Association of America, 1985. xvi + 421pp. $18.00. ISBN 0-88385-445-7.

Paperback reprint of unadulterated Halmos comments on teaching, scholarship, writing, and research. **(a,p)**

Ibn al-Haytham's Completion of the Conics. J. P. Hogendijk. Sources in the History of Mathematics and the Physical Sciences;7. New York: Springer-Verlag, 1985. x + 417pp. $159.00. ISBN 0-387-96013-9.

The entire Arabic text, with Englsh translation, of al-Haytham's attempted reconstruction of the lost book VIII of Apollonius' *Conics*. Expensive. **(a)**

International Mathematical Congresses: An Illustrated History, 1893-1986. Donald J. Albers, G. L. Alexanderson, Constance Reid. New York: Springer-Verlag, 1987. 63pp. $39.80. ISBN 0-387-96479-7.

A picture book of 20th century mathematicians. Brief descriptions of each International Congress of Mathematicians, photos of Fields medalists, lists of plenary addresses, etc. **(a)**

Introduction to Analysis of the Infinite. Leonhard Euler. Transl: John D. Blanton. New York: Springer-Verlag, 1988-1990. *Book I.* xv + 327pp. $49.95. ISBN 0-387-96824-5. *Book II.* xii + 504pp. $59.00. ISBN 0-387-97132-7.

The first English translation of Euler's classic work, which bridged what he believed to be a considerable gap between algebra and analysis. Volume 2 is on the theory of curves. **(a)**

Introduction to Classical Mathematics. Helmut Koch. Transl: John Stillwell. Mathematics and Its Applications;70. Norwell, MA: Kluwer Academic, 1991-. *I: From the Quadratic Reciprocity Law to the Uniformization Theorem.* xvii + 453pp. $149.00. ISBN 0-7923-1231-7.

Revised edition of *Einführung in die klassische Mathematik I.* This unique volume provides clear, concise highlights of 19th century mathematics (from Gauss to Riemann) in historical order, but with modern notation and proofs. Expensive. **(a)**

An Introduction to the History of Structural Mechanics. Edoardo Benvenuto. New York: Springer-Verlag, 1991. *Part I: Statics and Resistance of Solids.* xxi + 306pp. $49.00. ISBN 0-387-96227-1. *Part II: Vaulted Structures and Elastic Systems.* xxi + 245pp. $49.00. ISBN 0-387-97187-4.

An outgrowth of a 1981 Italian work by Benvenuto, this English rendition is especially well done. **(a)**

Ein Jahrhundert Mathematik 1980-1990. Ed: Gerd Fischer, Friedrich Hirzebruch, Winfried Scharlau, Willi Toernig. Dokumente zur Geschichte der Mathematik;6. Braunschweig: Friedr. Vieweg, 1990. xii + 830pp. DM 198.00. ISBN 3-528-06326-2.

A Festschrift on the 100th anniversary of the Deutscher Mathematiker Vereinigung. **(a)**

Janos Bolyai Appendix: The Theory of Space. Ed: Ferenc Karteszi. North-Holland Mathematics Studies;138. New York: Elsevier, 1987. 238pp. $100.00. ISBN 0-317-70053-7.

Revision and translation of a memorial volume first published in 1952, containing Bolyai's ideas of non-Euclidean geometry. Karteszi augments the Appendix with beautifully written explanatory notes. Expensive. **(a)**

John Von Neumann and Norbert Wiener: From Mathematics to the Technologies of Life and Death. Steve J. Heims. Cambridge, MA: MIT Press, 1980. xviii + 547pp. $19.95. ISBN 0-262-08105-9.

Fascinating dual biography of two extraordinary mathematicians, focusing on their professional development, the influence of their ideas on 20th century science, and their different efforts to influence science policy and politics after WWII. **(a,p)**

John von Neumann and the Origins of Modern Computing. William Aspray. Cambridge, MA: MIT Press, 1990. xvii + 376pp. $35.00. ISBN 0-262-01121-2.

A revealing portrait, based on archival material, of the many dimensions of von Neumann's pioneering work in computing. Text shows close ties to applied mathematics. **(a)**

Jordanus de Nemore: De Numeris Datis. Barnabas Bernard. Berkeley: University of California Press, 1981. xii + 212pp. $52.50. ISBN 0-520-04283-2.

The first critical edition (with English translation) of the earliest European treatise on advanced algebra, which was written in the late 13th century by one of the most celebrated natural philosophers of the time. De Numeris Datis is the earliest treatise to make extensive use of letters to represent numbers. **(a)**

Joseph Liouville 1809-1882: Master of Pure and Applied Mathematics. Jesper Lutzen. Studies in the History of Mathematics and Physical Sciences;15. New York: Springer-Verlag, 1990. xix + 884pp. $98.00. ISBN 0-387-97180-7.

A massive, thoroughly researched biography of Liouville in two parts: six chapters devoted to his scientific and professional career and eleven chapters describing his mathematical contributions. Numerous quotations (in English) from Liousville's extensive unpublished notebooks enliven the account. Extensive notes link text to lengthy bibliography. **(a)**

The Last Problem. E. T. Bell. Washington, DC: Mathematical Association of America, 1990. 326pp. $17.50. ISBN 0-88385-451-1.

The editor provides contemporary notes to this reprint of the 1961 original, an entertaining, but idiosyncratic history of the cultures that gave birth to Fermat's Last Theorem. **(a)**

Leonhard Euler 1707-1783. Ed: Emil Fellmann. Boston: Birkhäuser, 1983. 555pp. $52.75. ISBN 3-7643-1343-9.

Twenty-nine papers on Euler's life and work as well as the history of the attempts to publish his collected scholarly writings. This work includes an enormous bibliography of publications about him. **(a)**

Makers of Mathematics. Stuart Hollingdale. New York: Penguin Books, 1989. xiv + 433pp. $10.95. ISBN 0-14-022732-6.

Idiosyncratic history of mathematics told through a series of sketches of the lives and accomplishments of famous men of mathematics. **(a,p)**

The Man Who Knew Infinity: A Life of the Genius Ramanujan. Robert Kanigel. New York: Charles Scribner's, 1991. ix + 438pp. $27.95. ISBN 0-684-19259-4.

Perhaps one of the best scientific biographies ever written, this extraordinarily compelling book is filled with the social, psychological, personal, and mathematical details of Ramanujan and his mentor, Hardy. **(a,p)**

Mathematical Astronomy in Copernicus's De Revolutionibus. N. M. Swerdlow, O. Neugebauer. Studies in the History of Mathematicsand Physical Sciences;10. New York: Springer-Verlag, 1984. $119.00 set. ISBN 0-387-90939-7. *Part 1*. xvi + 537pp. *Part 2*. v + 172pp.

Analysis and exposition of *De Revolutionibus* in relation to the methodology of Ptolemy's *Almagest*. Expensive. **(a)**

The Mathematical Career of Pierre de Fermat (1601-1665). Michael S. Mahoney. Ann Arbor, MI: Books on Demand, 1990. xviii + 419pp. $109.80. ISBN 0-317-08307-4.

Photocopy of 1973 original, an expansion and revision of author's Princeton University thesis. Expensive. **(a)**

Mathematical Thought from Ancient to Modern Times. Second Edition. Morris Kline. New York: Oxford University Press, 1990. *Vol. 1*. xviii + 390 + xxiipp. $15.95 paperbound. ISBN 0-19-506135-7. *Vol. 2*. xx + 391-812 + xxiipp. $14.95 paperbound. ISBN 0-19-506136-5. *Vol. 3*. xvi + 813-1212 + xxiipp. $15.95 paperbound. ISBN 0-19-506137-3.

Unaltered reprint of the 1972 original, a lucid, monumental history of the major themes in mathematics, with half devoted to the 19th and 20th century. A must in every library. **(a,p)**

Mathematical Visions: The Pursuit of Geometry in Victorian England. Joan L. Richards. San Diego, CA: Academic Press, 1988. xiii + 266pp. $34.95. ISBN 0-12-587445-6.

A scholarly account which examines the position of mathematics in English higher education and its role in the development of English mathematics. Extensive list of references. **(a)**

The Mathematical Work of Charles Babbage. J. M. Dubbey. New York: Cambridge University Press, 1978. viii + 235pp. $65.00. ISBN 0-521-21649-4.

Discussion, in great detail, of Babbage's influence on the development of mathematics in England in the 19th century and his contributions to mathematics as distinct from computation. **(a)**

The Mathematical Work of John Wallis. Second Edition. J. F. Scott. New York: Chelsea, 1981. xi + 240pp. $16.50. ISBN 0-8284-0314-7.

Reprint of 1938 London original, this is a thorough and informative analysis of mid-seventeenth century arithmetic, mechanics and algebra. Appendix includes brief biographies of Wallis' contemporaries mentioned in the volume. **(a)**

Mathematics and Its History. John Stillwell. New York: Springer-Verlag, 1989. x + 371pp. $49.80. ISBN 0-387-96981-0.

Written for an advanced undergraduate course, this text aims to give a unified view of mathematics through its history. Excellent references. **(a,p)**

Mathematics in Civilization. Enlarged and Corrected Edition. H. L. Resnikoff, R. O. Wells, Jr. New York: Dover, 1984. viii + 408pp. $11.95. ISBN 0-486-24674-4.

Reprint of the 1973 original, which covered 4,000 years of mathematics, and supplemented with a brief chapter on 20th century mathematics and an appendix of solutions. **(p)**

The Mathematics of Sonya Kovalevskaya. Roger Cooke. New York: Springer-Verlag, 1984. xiii + 234pp. $52.00. ISBN 0-387-96030-9.

Fascinating mathematical biography with extensive quotations from correspondence and a detailed analysis of her major papers. **(a)**

Mathematische Abhandlungen. O. Bolza et al. New York: Chelsea, 1974. viii + 451pp. $25.00. ISBN 0-8284-0275-2.

Unaltered reprint (known informally as "Festschrift Schwarz") of a 1914 collection of papers in honor of H. A. Schwarz. Contributors include Caratheodory, Fejér, Hilbert, Toeplitz, etc. **(a)**

Men of Mathematics. Eric Temple Bell. New York: Simon & Schuster, 1986. xvii + 590pp. $13.95. ISBN 0-671-62818-6.

Reprint of 1937 original. **(a,p)**

Norbert Wiener, 1894-1964. P. R. Masani. Boston: Birkhäuser, 1990. 416pp. $63.00. ISBN 0-8176-2246-2.

A fascinating, well documented reflection on the intellectual and personal character of Wiener, including a biography, mathematics and other topics, and extensive quotations from his correspondence. **(a,p)**

The Origins of Cauchy's Rigorous Calculus. Judith V. Grabiner. Cambridge, MA: MIT Press, 1981. x + 252pp. $42.50. ISBN 0-262-07079-0.

In 1823 Cauchy published *Calcul Infinitesimal*, which marks the beginning of a century's work on the foundations of calculus. Grabiner searches for the roots of this revolution and finds them primarily in the works of Lagrange. Lagrange's proposal to found calculus on algebra provides the context forthe work of Cauchy and Bolzano and helps explain the similarities in their works. An appendix includes translations of key parts of Cauchy's work as well as Grabiner's extensive notes to original sources. **(a)**

The Origins of the Infinitesimal Calculus. Margaret E. Baron. New York: Dover, 1987. viii + 304pp. $8.95. ISBN 0-486-65371-4.

Reprint of 1969 study of the rise of geometric techniques from the Greeks to the early 17th century. **(a,p)**

Pascal's Arithmetical Triangle. A. W. F. Edwards. New York: Oxford University Press, 1987. xii + 174pp. $37.50. ISBN 0-19-520546-4.

A comprehensive history of Pascal's triangle from interpretations of early examples to its development into a powerful tool. **(a)**

Peano: Life and Works of Giuseppe Peano. Hubert C. Kennedy. Studies in the History of Modern Science;4. Dordrecht: Reidel, 1980. (Available from Kluwer Academic, Norwell, MA) xii + 230pp. $63.50;$32.50. ISBN 90-277-1067-8;90-277-1068-6.

First full biography of Peano, describing his life, his development of symbolic logic, his mathematics, and his efforts to promote Interlingua. Appendices. **(a)**

Portraits in Silicon. Robert Slater. Cambridge, MA: MIT Press, 1987. xiv + 374pp. $24.95. ISBN 0-262-19262-4.

Thirty-one fascinating biographies, based on personal interviews, tracing the accomplishments of pioneers of the computer age. **(p)**

The Probabilistic Revolution. Ed: Lorenz Kruger et al. Cambridge, MA: MIT Press, 1987. *Volume 1: Ideas in History*. xv + 449pp. $37.50. ISBN 0-262-11118-7. *Volume 2: Ideas in the Sciencesb. xvii* + 459pp. $37.50. ISBN 0-262-11119-5.

Thirty-five essays investigating the historical roots of the revolution that supplanted determinism with probabilistic models as the paradigm of physical and social science. **(a)**

Quest–An Autobiography. Leopold Infeld. New York: Chelsea, 1980. 361pp. $19.95. ISBN 0-8284-0309-0.

Reprint of the important 1941 work. **(a,p)**

Rabdology. John Napier. Transl: William Frank Richardson. Charles Babbage Institute Reprint Series for History of Computing;15. Cambridge, MA: MIT Press, 1990. xxxvii + 135pp. $40.00. ISBN 0-262-14046-2.

First complete English translation to Napier's 1617 user manual, *Rabdologia*. His brief treatise contains the first written reference to the decimal point. **(a)**

Regiomontanus: His Life and Work. Ernst Zinner. Transl: Ezra Brown. Studies in the History and Philosophy of Mathematics;1. New York: Elsevier, 1990. x + 402pp. $82.00. ISBN 0-444-88792-X.

Extraordinarily detailed account of the scholarly life of this precursor to Copernicus. **(a)**

The Rise of Statistical Thinking 1820-1900. Theodore M. Porter. Princeton, NJ: Princeton University Press, 1986. xii + 333pp. $13.95. ISBN 0-691-02409-X.

A detailed account of the roots of modern statistics in nineteenth century science. **(a)**

Science Awakening. Vol. 1: Egyptian, Babylonian and Greek Mathematics. Fourth Edition. B. L. van der Waerden. Princeton Junction, NJ: Scholar's Bookshelf, 1988. 306pp. $40.00;$14.95. ISBN 0-945726-04-X;0-945726-05-8.

Revised translation of *Ontwakende Wetenschap*. Changes include Huber's discovery which sheds new light on the role of geometry in Babylonian algebra and a rewritten section on Heron's matrices. Illustrated. **(a)**

The Scottish Book: Mathematics from the Scottish Cafe. Ed: R. Daniel Mauldin. Boston: Birkhäuser, 1981. xiii + 268pp. $49.95. ISBN 3-7643-3045-7.

Reproduction of the unique notebook in which the mathematical circle of Banach, Ulam, Steinhaus, Mazur, Kac recorded at their favorite cafe in Lvov, Poland difficult problems in set theory, topology, measure theory, and functional analysis from 1935 to 1941. This is the English translation with occasional commentaries and modern references and introduced by brief anecdotal essays by Ulam, Kac, Zygmund and Erdos. **(a)**

A Sourcebook in Classical Analysis. Ed: Garrett Birkhoff, Uta Merzbach. Cambridge, MA: Harvard University Press, 1973. xii + 470pp. $30.50. ISBN 0-674-82245-5.

Eighty-one brief selections from the 19th Century masters (e.g., Cauchy, Jordan, Gauss, Klein, Lie, etc.), each translated into English and arranged by topic into 13 chapters with introductory interpretive essays. **(a)**

A Source Book in Mathematics, 1200-1800. Ed: D. J. Struik. Princeton, NJ: Princeton University Press, 1986. xiv + 427pp. $21.95;$55.00. ISBN 0-691-02397-2;0-691-08404-1.

Paperback edition of 1969 original. A compendium of mathematical classics from the Latin world from the 13th through the 18th centuries. **(a,p)**

Studies in the History of Mathematics. Ed: Esther R. Phillips. Studies in Mathematics;26. Washington, DC: Mathematical Association of America, 1987. 320pp. $36.50. ISBN 0-88385-128-8.

Ten essays on recent research in the history of mathematics, particularly the period 1870-1950. **(a)**

To Infinity and Beyond: A Cultural History of the Infinite. Eli Maor. Boston: Birkhäuser, 1986. xvi + 275pp. $49.50. ISBN 0-8176-3325-1.

Interesting story of the role of infinity in our culture, touching on mathematics, cosmology, art, and music. A good but not exhaustive bibliography. **(p)**

The Tragicomical History of Thermodynamics 1822-1854. C. Truesdell. Studies in the History of Mathematical and Physical Sciences;4. New York: Springer-Verlag, 1980. ISBN 0-387-90403-4.

A detailed exegesis of the published work of Fourier, Carnot, Joule, Kelvin, Clausius, Rankin and others. **(a)**

The Unreal Life of Oscar Zariski. Carol Parikh. San Diego, CA: Academic Press, 1991. xvii + 264pp. $32.95. ISBN 0-12-545030-3.

Based partially on interviews with Zariski before his death in 1986, this award-winning author brings to life the intellectual ferment and international dimensions of Zariski's career. Complete Zariski bibliography, mathematical survey of his mathematical work, and many photographs. **(a)**

Unrolling Time: Christiaan Huygens and the Mathematization of Nature. Joella G. Yoder. New York: Cambridge University Press, 1988. xi + 238pp. $42.50. ISBN 0-521-34140-X.

A case study of the inter-relationship between mathematics and physics in the work of Huygens (1629-1695). Original drawings supplemented with clearer redrawn versions. **(a)**

Vorlesungen über die Entwicklung der Mathematik im 19. Jahrhundert. Felix Klein. New York: Chelsea, 1956. 2 v. in one (xiii + 385 + x + 208pp.) $12.95. ISBN 0-8284-0074-1.

Reprint of 1926-1927 original. Volume 2 entitled, *Die Grundbegriffe der Invariantentheorie und ihr Eindringen in die mathematische Physik.* **(a)**

Who Got Einstein's Office? Ed Regis. Reading, MA: Addison Wesley, 1987. xv + 316pp. $10.53. ISBN 0-201-12278-2.

A thoroughly researched and well-documented account of the people and ideas at the Institute for Advanced Study which reads like a tale of intellectual intrigue. **(p)**

William Rowan Hamilton: Portrait of a Prodigy. Sean O'Donnell. Dublin: Boole Press, 1983. xvi + 224pp. $24.95. ISBN 0-906783-06-2.

A fascinating, well-documented, and lucid account of the life and accomplishments of Hamilton. **(a)**

Women in Mathematics. Lynn M. Osen. Cambridge, MA: MIT Press, 1974. xii + 185pp. $7.95. ISBN 0-262-65009-6.

Written in an agreeable style, chapters in this book range from Hypatia to the Feminine Mathematique. Bibliography. **(a,p)**

Zermelo's Axiom of Choice: Its Origins, Development, and Influence. Gregory H. Moore. Studies in the History of Mathematics and Physical Sciences;8. New York: Springer-Verlag, 1982. xiv + 410pp. $75.00. ISBN 0-387-90670-3.

Fascinating and detailed history of the controversies surrounding the axiom. **(a)**

100 Years of Mathematics: A Personal Viewpoint. George Temple. New York: Springer-Verlag, 1981. xvi + 316pp. $32.00. ISBN 0-387-91192-8.

A synopsis of key elements in mathematical research in the 19th and 20th centuries. Over 600 references to original papers of the 300 mathematicians described. **(a)**

LINEAR ALGEBRA

Eclectic Linear Algebra. Bobby Fincher. Denton, TX: Spadra Press, 1986. vii + 164pp. $15.00. ISBN 0-937161-34-9.

A concise introduction to elementary linear algebra. **(p)**

Elementary Linear Algebra. Robert S. Johnson, Thomas O. Vinson, Jr. San Diego, CA: Harcourt Brace Jovanovich, 1987. viii + 342pp. $32.00. ISBN 0-15-521082-3.
Highly readable and usefultext. **(p)**

Finite-Dimensional Vector Spaces. Second Edition. Paul R. Halmos. New York: Springer-Verlag, 1987. viii + 200pp. $31.00. ISBN 0-387-90093-4.
Reprint of 1958 original. **(a)**

Infinite Dimensional Lie Algebras. Second Edition. Victor G. Kac. New York: Cambridge University Press, 1985. xvii + 280pp. $32.50. ISBN 0-521-32133-6.
A clear account of Kac-Moody algebras by one of the founders. Exercises. **(a)**

Invariantentheorie. Paul Gordon. New York: Chelsea, 1987. xi + 360pp. $49.95. ISBN 0-8284-0328-7.
Reprint, with only minor changes, of the two volume 1885-1887 edition. **(a)**

Linear Algebra. Second Edition. John B. Fraleigh, Raymond A. Beauregard. Reading, MA: Addison-Wesley, 1990. xiii + 571pp. $39.16. ISBN 0-201-11949-8.
Unique feature of this text is that eigenvalues, eigenvectors, and their applications are introduced much earlier. Text is not intended to be too rigorous. **(p)**

Linear Algebra. Third Edition. Serge Lang. New York: Springer-Verlag, 1987. ix + 285pp. $39.80. ISBN 0-387-96412-6.
Extensively reorganized and rewritten. Covers vector spaces, matrices, etc. **(p)**

Linear Algebra. Revised English Edition. Georgi E. Shilov. Transl/Ed: Richard A. Silverman. New York: Dover, 1977. xi + 387pp. $7.95. ISBN 0-486-63518-X.
Written for students of mathematics and physics, the author brings together ideas from algebra, geometry, and analysis. **(p)**

Linear Algebra: An Introductory Approach. Charles W. Curtis. New York: Springer-Verlag, 1984. x + 337pp. $29.00. ISBN 0-387-90992-3.
This edition shows no major changes from the Allyn & Bacon Third edition. **(a)**

Linear Algebra and Geometry, A Second Course. Irving Kaplansky. New York: Chelsea, 1974. xii + 143pp. $10.95. ISBN 0-8284-0279-5.
Corrected reprint of 1969 text. Refreshing, informal style enriches this concise collection of closely related topics on the interface between linear algebra and classical geometry. **(a,p)**

Linear Algebra with Applications. Third Edition. Steven J. Leon. New York: Macmillan, 1990. xv + 458pp. $36.75. ISBN 0-02-369821-7.
This edition adds a new appendix containing computational exercises to be solved using MATLAB as well as new theorems, applications, and figures. **(a)**

Linear Algebra with Business Applications. John W. Brown, Donald R. Sherbert. New York: McGraw-Hill, 1991. 403 + 49pp. $26.96. ISBN 0-07-056764-6.

This book presents the fundamental techniques and computational algorithms of linear algebra and provides an introduction to the main concepts, but without a lot of formalism. Exercises are in appendix. **(a,p)**

Linear Associative Algebras. Alexander Abian. Elmsford, NY: Pergamon, 1971. vii + 166pp. $37.00. ISBN 0-08-016564-8.

Self-contained introduction to the theory of finite-dimensional linear (associative) algebras. Results are developed in a straight-forward, concrete manner. **(a)**

Matrix Analysis. Roger A. Horn, Charles R. Johnson. New York: Cambridge University Press, 1985. xiii + 561pp. $37.50. ISBN 0-521-30586-1.

For the second course in linear algebra, with an emphasis on applications. **(a)**

Matrix Theory and Applications. Ed: Charles R. Johnson. Proceedings of Symposia in Applied Mathematics;40. Providence, RI: American Mathematical Society, 1990. x + 260pp. $57.00. ISBN 0-8218-0154-6.

Based on the January 1989 AMS Short Course which looked at the ways in which modern matrix theory was stimulated by its interplay with other subjects. **(s)**

Schaum's Outline of Theory and Problems of Matrix Operations. Richard Bronson. New York: McGraw-Hill, 1989. 230pp. $8.95 paperbound. ISBN 0-07-007978-1.

Features an algorithmic approach with emphasis on computationally efficient methods. **(a,p)**

The Theory of Matrices. F. R. Gantmakher. Transl: K. A. Hirsch. New York: Chelsea, 1959. *Vol. 1*. x + 374pp. $19.95. ISBN 0-8284-013104. *Vol. 2*. ix + 276pp. $19.95. ISBN 0-8284-0133-0.

Reprint of translation of 1953 Russian original, *Teorii͡a Matrits,* with some changes in the bibliography for the English-language reader. **(a)**

3000 Solved Problems in Linear Algebra. Seymour Lipschutz. New York: McGraw-Hill, 1989. vii + 480pp. $19.95. ISBN 0-07-038023-6.

This collection covers almost every type of linear algebra problem making it useful as a supplement or refresher. **(a,p)**

LOGIC AND FOUNDATIONS OF MATHEMATICS

Bounded Arithmetic. Samuel R. Buss. Studies in Proof Theory;3. Napoli: Bibliopolis, 1986. (Available from Humanities Press International, Atlantic Highlands, NJ) vi + 221pp. $45.00. ISBN 88-7088-150-4.

This is a slightly revised version of the author's excellent 1985 doctoral dissertation from Princeton University. By paying attention to details and keeping it relatively self-contained, the author has made it accessible to a rather general mathematical audience. **(a)**

Brouwer's Intuitionism. Walter P. van Stigt. Studies in History and Philosophy of Mathematics;2. New York: Elsevier, 1990. xxvi + 530pp. $94.75. ISBN 0-444-88384-3.

A fascinating intellectual biography tracing the roots of Brouwer's philosophical views from his youth through a detour into topology and back into foundational studies. Appendices provide original text and translations of several important unpublished papers and correspondence. **(a)**

Cantorian Set Theory and Limitation of Size. Michael Hallett. Logic Guides;10. New York: Oxford University Press, 1984. xxii + 343pp. $55.00 ISBN 0-19-853179-6.

This comprehensive study corrects a number of common misunderstandings and establishes a new level for historical comprehension of Cantor's monumental contribution to mathematics. **(a)**

Computability: An Introduction to Recursive Function Theory. Nigel Cutland. New York: Cambridge University Press, 1980. x + 251pp. $47.50;$14.95. ISBN 0-521-22384-9;0-521-29465-7.

A well-written introduction to recursion theory, with numerous examples. **(a,p)**

Computability & Unsolvability. Martin Davis. New York: Dover, 1982. xxv + 248pp. $6.50. ISBN 0-486-61471-9.

Reprint of 1958 original with a new preface and new appendix. **(a)**

A Course in Mathematical Logic. I͡U. I. Manin. Transl: Neal Koblitz. Graduate Texts in Mathematics;53. New York: Springer-Verlag, 1977. xiii + 286pp. $39.00 paperbound. ISBN 0-387-90243-0.

While the standard topics from model theory which are closely connected with algebra are omitted, many interesting unorthodox topics are included. Book presents material in a fresh way and proceeds at a high level of mathematical sophistication. **(a)**

Descriptive Set Theory. Yiannis N. Moschovakis. Studies in Logic and the Foundations of Mathematics;100. New York: Elsevier, 1987. xii + 637pp. $54.50. ISBN 0-444-70199-0.

A reprint of 1980 original, this book does a fine job of systematically presenting the main results and methods of the theory, including necessary background material. **(a)**

Elements of Logic and Foundations of Mathematics in Problems. Wiktor Marek, Janusz Onyszkiewicz. Norwell, MA: Kluwer Academic, 1982. viii + 276pp. $19.95 paperbound. ISBN 90-277-2131-9.

Translation of 1972 Polish edition. Problem book covering topics in set theory and mathematical logic. **(a)**

Foundations and Fundamental Concepts of Mathematics. Third Edition. Howard Eves. Boston, MA: PWS-Kent, 1990. xvii + 344pp. $33.00. ISBN 0-534-92163-9.
Slightly revised reprint of the classic work by Eves and Newsom. **(p)**

The Foundations of Arithmetic: A Logico-Mathematical Enquiry into the Concept of Number. Second Revised Edition. G. Frege. Transl: J. L. Austin. Evanston, IL: Northwestern University Press, 1980. xii + xii [superscript e] + xi [superscript e] + 119 + 119 [superscript e]pp. $28.95;$14.95. ISBN 0-8101-0023-1;0-8101-0605-1.
German and English on opposite pages. Reprint of 1953 edition. **(a)**

Foundations of Logic and Mathematics. Rudolf Carnap. Chicago: University of Chicago Press, 1974. iv + 71pp. $4.50. ISBN 0-226-57578-0.
Reprint of 1939 original. **(p)**

Foundations of Set Theory. Second Revised Edition. Abraham A. Fraenkel, Yehoshua Bar-Hillel, Azriel Levey. Studies in Logic and the Foundations of Mathematics;67. New York: Elsevier, 1973. x + 404pp. $36.25. ISBN 0-7204-2270-1.
The chapters on axiomatic set theory and on intuitionism have been extensively revised. Unusually lucid exposition remains essentially non-technical. **(a,p)**

Fundamentals of Contemporary Set Theory. Keith J. Devlin. New York: Springer-Verlag, 1979. viii + 182pp. $24.50. ISBN 0-387-90441-7.
An introduction to the basic topics of pure set theory which every mathematician should know, presented in a conversational style. **(a,p)**

Gödel's Proof. Ernest Nagel, James R. Newman. New York: New York University Press, 1958. 118pp. $6.50. ISBN 0-8147-0325-9.
A detailed expository popular account. **(a,p)**

Harvey Friedman's Research on the Foundations of Mathematics. Ed: L. A. Harrington et al. Studies in Logic and the Foundations of Mathematics;117. New York: Elsevier, 1985. xvi + 408pp. $79.00. ISBN 0-444-87834-3.
A collection of papers, some appearing for the first time, in celebration of Friedman's receipt of the Waterman Award in 1984. Survey papers touch on his contributions to set theory, computational complexity, intuitionism, proof theory, and recursion theory. **(a)**

History and Philosophy of Modern Mathematics. Ed: William Aspray, Philip Kitcher. Minnesota Studies in the Philosophy of Science;XI. Minneapolis, MN: University of Minnesota Press, 1988. viii + 386pp. $35.00;$13.95. ISBN 0-8166-1566-7;0-8166-1567-5.
Fourteen papers from a 1985 conference by mathematicians, historians, and philosophers giving different perspectives on 19th and 20th century mathematics. **(a)**

How to Read and Do Proofs: An Introduction to Mathematical Thought Processes. Second Edition. Daniel Solow. New York: Wiley, 1990. xx + 242pp. $19.95. ISBN 0-471-51004-1.

A down-to-earth exposition which keeps in close touch with the student's perspective while equipping them to understand the procedures and intensions underlying typical condensed proofs in textbooks. This revision includes a new chapter on nested qualifiers and more exercises. **(p)**

Introduction to Axiomatic Set Theory. J. L. Krivine. Transl: David Miller. Dordrecht: Reidel, 1974. (Distributed by Kluwer Academic, Norwell, MA) vi + 100pp. $15.00;$8.00. ISBN 90-277-0169-5;90-277-0411-2.

A concise, highly readable, graduate level text on methods used in constructing relative consistency proofs. No index. **(a)**

Introduction to Combinators and [lambda]-Calculus. J. Roger Hindley, Jonathan P. Seldin. London Mathematical Society Student Texts;1. New York: Cambridge University Press, 1986. v + 360pp. $19.95 paperbound. ISBN 0-521-31839-4.

Both an introduction to the subject and a complete exposition of almost all the new (especially, model-theoretic) results. **(a)**

Introduction to Mathematical Logic. Third Edition. Elliott Mendelson. Pacific Grove, CA: Brooks-Cole, 1987. ix + 341pp. $43.95. ISBN 0-534-06624-0.

Some of the major changes in this edition include more development of recursion theory and proofs of the incompleteness theorems are now based on the diagonalization lemma. Updated bibliography. **(a)**

Introduction to Metamathematics. Stephen Cole Kleene. Bibliotheca Mathematica;1. New York: Elsevier, 1971. x + 550pp. $63.25. ISBN 0-444-10088-1.

Reprint of 1952 original. **(a)**

Introduction to Proof in Abstract Mathematics. Andrew Wohlgemuth. Philadelphia, PA: Saunders, 1990. xii + 366pp. $48.00. ISBN 0-03-026782-X.

Interesting and highly recommended text for transition to abstract mathematics. **(a)**

Introduction to the Foundation of Mathematics. Raymond L. Wilder. Melbourne, FL: Krieger, 1980. xvi + 327pp. $29.50. ISBN 0-89874-170-X.

Reprint of 1965 original. **(a)**

The Language of First-Order Logic. Jon Barwise, John Etchemendy. CSLI Lecture Notes;23. Stanford, CA: Center for the Study of Language and Information, 1990. xiii + 259pp. Computer disk. $27.50 paperbound. ISBN 0-937073-59-8.

"This book was designed for use in a first course of logic, for students interested in the symbolic sciences, and for use with Tarksi's world, a program that runs on the Macintosh computer."—Preface. **(a)**

Leçons sur les Ensembles Analytiques et Leurs Applications. Second Corrected Edition. Nicolas Lusin. New York: Chelsea, 1972. $15.95. ISBN 0-8284-0250-7.

Reprint of 1930 original. Includes a note by Sierpinski and preface by Lebesgue. **(a)**

Logic and Information. Keith Devlin. New York: Cambridge University Press, 1991. xii + 308pp. $34.50. ISBN 0-521-41031-4.

Writing clearly for mathematicians, computer scientists and philosophers, the author tries to restore logic as the science of "reasoning, thinking, and inference." **(a)**

Logic for Computer Scientists. Uwe Schöning. Progress in Computer Science and Applied Logic;8. Boston: Birkhäuser, 1989. 166pp. $25.00. ISBN 0-8176-3453-3.

This book is intended for computer science students, taking an introductory course on theorem proving and logic programming. **(a)**

Logic for Mathematicians. Revised Edition. A. G. Hamilton. New York: Cambridge University Press, 1988. viii + 228pp. $22.95. ISBN 0-521-36865-0.

Well-motivated treatment of fundamental topics aimed at the general mathematics student. Revision includes new sections on Skolemisation and applying well-formed formulas to logic programming. **(a)**

Logic: Techniques for Formal Reasoning. Second Edition. Donald Kalish, Richard Montague, Gary Mar. Ed: Robert J. Fogelin. San Diego, CA: Harcourt Brace Jovanovich, 1980. xvi + 520pp. $21.00. ISBN 0-15-551181-5.

This edition differs from the former by a significant increase in the quantity and quality of exercises and by a more extensive treatment of the invalidity of symbolic arguments obtained by the introduction of models. **(a)**

Mathematical Logic. Revised Edition. W. V. Quine. Cambridge, MA: Harvard University Press, 1976. xii + 346pp. $25.50. ISBN 0-674-55451-5.

Unaltered reprint of 1951 edition, a classic in the field, although it is now dated. **(a)**

Mathematics and Physics. I͡U. I. Manin. Progress in Physics;3. Transl: Ann and Neal Koblitz. Boston, MA: Birkhäuser, 1981. xii + 99pp. $10.00. ISBN 3-7643-3027-9.

A fascinating, subjective reflection on the nature of mathematics and physics, suitable for philosophically-minded audiences versed in both subjects. **(a)**

Mathematics and Plausible Reasoning. G. Pólya. Princeton, NJ: Princeton University Press, 1954. *V. I: Induction and Analogy in Mathematics.* xvi + 280pp. $32.00. ISBN 0-691-08005-4. *Volume II: Patterns of Plausible Inference. Second Edition.* x + 190pp. $32.00. ISBN 0-691-08006-2.

A great number of interesting examples from various parts of mathematics are discussed in order to point out the manner in which mathematical research should be conducted. **(a)**

Mathematics, Form and Function. Saunders Mac Lane. New York: Springer-Verlag, 1986. xi + 476pp. $42.00. ISBN 0-387-96217-4.

An explication of the intrinsic character of mathematics. **(a,p)**

Methods of Mathematics. Fourth Edition. W. V. Quine. Cambridge, MA: Harvard University Press, 1982. x + 333pp. $28.00;$10.95. ISBN 0-674-57175-4;0-674-57175-2.

Major change in this edition is the author's departure from the notations of truth functions and quantifiers of Whitehead and Russell, switching to other symbols for the conditional, the biconditional, and universal quantification. **(a)**

Model-Theoretic Logics. Ed: J. Barwise, S. Feferman. New York: Springer-Verlag, 1985. xviii + 893pp. $140.00. ISBN 0-387-90936-2.

Invaluable resource for research frontiers in abstract model theory. Extensive bibliography. Expensive. **(a)**

Naive Set Theory. Paul R. Halmos. New York: Springer-Verlag, 1974. vii + 104pp. $26.50. ISBN 0-387-90092-6.

Reprint of Van Nostrand classic. **(a)**

Numbers, Sets and Axioms: The Apparatus of Mathematics. A. G. Hamilton. New York: Cambridge University Press, 1982. ix + 255pp. $42.50; $17.95. ISBN 0-521-24509-5; 0-521-287761-8.

Excellent text covering number systems, axiomatic theory, infinite sets, ordinal and cardinal numbers with comments on the subject and philosophical issues. **(a)**

On Numbers and Games. J. H. Conway. L. M. S. Monographs;6. San Diego, CA: Academic Press, 1976. ix + 238pp. $49.00. ISBN 0-12-186350-6.

New, exciting, and highly original theory of real numbers is expounded by Conway in a style that is concise, literate, and delightfully whimsical. **(p)**

Philosophy of Logic. Second Edition. W. V. Quine. Cambridge, MA: Harvard University Press, 1986. x + 109pp. $6.95. ISBN 0-674-66563-5.

A revision of the brief introduction to the philosophy of deductive logic by the pre-eminent person in the field. **(a)**

Proof Theory. Second Edition. Gaisi Takeuti. Studies in Logic and the Foundations of Mathematics;81. New York: Elsevier, 1987. x + 490pp. $90.00. ISBN 0-444-87943-9.

First edition was based on a series of lectures given at the 1968 Buffalo Symposium on Intuitionism and Proof Theory. Among the additions of this present edition is an appendix containing contributions by Kreisel, Pohlers, Simpson, and Feferman. **(a)**

Pursuit of Truth. W. V. Quine. Cambridge, MA: Harvard University Press, 1990. x + 113pp. $17.50. ISBN 0-674-73950-7.

Concise recapitulation of the ideas of evidence, reference, meaning, intention, and truth. **(p)**

Recursive Function Theory and Logic. Ann Yasuhara. San Diego, CA: Academic Press, 1971. xv + 338pp. $79.50. ISBN 0-12-768950-8.

An excellent and wide-ranging introduction. Many good exercises. **(a)**

Recusively Ennumerable Sets and Degrees: A Study of Computable Functions and Computably Generated Sets. Robert I. Soare. New York: Springer-Verlag, 1987. xviii + 437pp. $35.00. ISBN 0-387-15299-7.

"This book represents a kind of progress report over the last 40 years on the programs, ideas, and hopes expressed by E. L. Post in 1944."—Preface. **(a)**

Reflections of Kurt Gödel. Hao Wang. Cambridge, MA: MIT Press, 1987. xxvi + 336pp. $25.00. ISBN 0-262-23127-1.

Written in diary style, this fascinating commentary reveals unknown depth and richness to Gödel's thoughts on mathematics, philosophy of science, and metaphysics. **(a,p)**

Remarks on the Foundations of Mathematics. Revised Edition. Ludwig Wittgenstein. Ed: G. H. von Wright, R. Rhees, G. E. M. Anscombe. Transl: G. E. M. Anscombe. Cambridge, MA: MIT Press, 1983. 444pp. $13.95. ISBN 0-262-73067-7.

Translation of *Bemerkungen ãber die Grundlagen der Mathematik*, the posthumously published writings of Wittgenstein which first appeared in 1956. Additional material included in this edition. Index. **(a)**

Set Theory. Fourth Edition. Felix Hausdorff. Transl: John R. Aumann et al. New York: Chelsea, 1991. 352pp. $27.50. ISBN 0-8284-1119-0.

Translation of 1937 edition of Mengenlehre. **(a)**

Set Theory: An Introduction to Independence Proofs. Kenneth Kunen. Studies in Logic and the Foundations of Mathematics;102. New York: Elsevier, 1980. xvi + 313pp. $68.50. ISBN 0-444-85401-0.

The approach of this introduction to relative consistency proofs in set theory nicely separates the mathematical and combinatorial techniques from the metamathematical difficulties. Many exercises. **(a)**

Set Theory and Its Logic. Revised Edition. W. V. Quine. Cambridge, MA: Belknap Press of Harvard University Press, 1969. xvii + 361pp. $9.95. ISBN 0-674-80207-1.

This book contains a detailed and interesting exposition of introductory chapters of set theory and of several axiom systems. However, it's a bit difficult for the undergraduate. On the other hand, the work is a real treasure for those who want to know the way set theory developed. **(a)**

Set Theory and Metric Spaces. Second Edition. Irving Kaplansky. New York: Chelsea, 1977. xii + 140pp. $10.95. ISBN 0-8284-0298-1.

Essentially a reprint of 1972 original, which arose from course in set theory and metric spaces which Kaplansky and Spanier gave at University of Chicago. **(a)**

Simplified Independence Proofs: Boolean Valued Models of Set Theory. J. Barkley Rosser. Pure and Applied Mathematics;31. San Diego, CA: Academic Press, 1969. xv + 217pp. $59.00. ISBN 0-12-598050-7.

First published account of Scott and Solovay's theory of Boolean-valued models for set theory. **(a)**

Studies in Inductive Logic and Probability. Ed: Rudolf Carnap, Richard C. Jeffrey. Berkeley, CA: University of California Press, 1971-1980. *Volume I.* vi + 264pp. out-of-print ISBN 0-520-01866-4. *Volume II.* 305pp. $38.00. ISBN 00520-03826-6.
Volume 2 is the long delayed and final volume in the set. Articles are numbered successively from volume 1 and an index to both volumes appears at the end of volume 2. **(a)**

Symbolic Logic. Fifth Edition. Irving M. Copi. New York: Macmillan, 1979. xiii + 398pp. $42.00. ISBN 0-02-324980-3.
The major change from the Fourth edition is the addition of a chapter on Zermelo-Fraenkel set theory. **(p)**

Symbolic Logic and Mechanical Theorem Proving. Chin-Liang Chang, Richard Char-Tung Lee. San Diego, CA: Academic Press, 1987. xiii + 331pp. $29.95. ISBN 0-12-170350-9.
Reprint of 1973 original. **(a)**

Théorie des Ensembles. Nouvelle Édition. Nicolas Bourbaki. (Éléments de Mathématique) Paris: Masson, 1990. (Available from Scientific & Medical Publications of France, New York, NY) various pagings. $103.00. ISBN 2-225-81909-2.
Reprint of the 1970 revision of the work, when published in 1954, marked the first time a book systematically developed axiomatic set theory. **(a)**

Theory of Recursive Functions and Effective Computability. Hartley Rogers, Jr. Cambridge, MA: MIT Press, 1987. xxi + 482pp. $16.50. ISBN 0-262-68052-1.
Corrected paperback reprint of 1967 original. Computer science developments confirm the continuing significance of this work: the first comprehensive treatment of the theory of algorithms, partial functions, recursion, and unsolvability. **(s)**

Topoi, the Categorical Analysis of Logic. Robert Goldblatt. Studies in Logic and the Foundations of Mathematics;98. New York: Elsevier, 1979. xv + 486pp. $83.00. ISBN 0-444-85207-7.
Introduction to topos theory and its implications for logic and foundations. **(a)**

The Undecidable: Basic Papers on Undecidable Propositions, Unsolvable Problems and Computable Functions. Ed: Martin Davis. New York: Raven Press, 1965. ii + 440pp. $41.50. ISBN 0-9112-01-4.
This volume contains much of the principal work of the giants in the field of recursive functions prior to 1950. All papers are in English with excellent short introductions by the editor. **(a)**

The Universal Turing Machine: A Half-Century Survey. Ed: Rolf Herken. New York: Oxford University Press, 1988. xiv + 661pp. $115.00. ISBN 0-19-853741-7.

A commemorative volume containing 5 expositions of the context of Turing's work followed by 23 diverse examples of current research. Expensive. **(a)**

Wittgenstein on the Foundations of Mathematics. Crispin Wright. Cambridge, MA: Harvard University Press, 1980. xix + 481pp. $30.00. ISBN 0-674-95385-1.

Systematic account of Wittgenstein's later philosophy of mathematics emphasizing the connections with his philosophy of language. Index, analytical table of contents, and page references to his writings are included. **(a)**

MATHEMATICAL PHYSICS

Integrable Models. Ashok Das. World Scientific Lecture Notes in Physics;30. Teaneck, NJ: World Scientific, 1989. xii + 342pp. $28.00. ISBN 9971-50-911-3.

"The first 8 chapters of the book develop various ideas of classical integrability with the example of the Korteweg-de Vries equations. The methods of zero curvature and quantum inverse scattering are discussed within the context of the nonlinear Schrodinger equation."—Preface. **(a)**

Mathematical Physics. Robert W. Carroll. North-Holland Mathematics Studies;152. New York: Elsevier, 1988. x + 399pp. $94.75. ISBN 0-444-70443-4.

A demanding introduction to mainly PDE-oriented mathematical physics, with a heavy emphasis on scattering and inverse scattering. Extensive bibliography. **(a,s)**

Methods of Mathematical Physics. First English Edition. R. Courant, D. Hilbert. New York: Wiley, 1953-1962. $132.95 set. ISBN 0-471-17990-6. *Vol. I.* xv + 561pp. $69.95; $29.95. ISBN 0-470-17952-X; 0-471-50447-5. *Vol. II.* xxii + 830pp. $85.95; $29.95. ISBN 0-470-17985-6; 0-471-50439-4.

Translation and revision of well-known 1931 German original. **(a,s)**

Principles of Advanced Mathematical Physics. Robert D. Richtmyer. New York: Springer-Verlag, 1978-1981. *Vol. I.* xv + 422pp. $49.00. ISBN 0-387-08873-3. *Vol. II.* ix + 322pp. $59.00. ISBN 0-387-10772-X.

A physicist presents functional analysis with examples from physics, probability theory, differential equations, and quantum mechanics. **(a)**

Topics in Soliton Theory. Robert W. Carroll. North-Holland Mathematics Studies;167. New York: Elsevier, 1991. xii + 428pp. $100.00. ISBN 0-444-88869-1.

"The book [makes] available to nonexperts, students, and scientists... some of the ideas and techniques of "soliton mathematics" (in a broad sense)."—Preface. Extensive references. Expensive. **(a)**

MATHEMATICAL RECREATIONS

Aha! Gotcha: Paradoxes to Puzzle and Delight. Martin Gardner. New York: W. H. Freeman, 1982. vii + 164pp. $10.95 paperbound. ISBN 0-7167-1414-0.

Another delightful book by Gardner, in which he explains, in nontechnical language, why each paradox is paradoxical. References and suggested readings appear at end of book. **(p)**

The Chess Mysteries of Sherlock Holmes. Raymond Smullyan. New York: Alfred A. Knopf, 1979. xiii + 171pp. $10.00. ISBN 0-39450-488-7.

For intellectual puzzle enthusiasts. Fifty captivating chess problems of the type where the object is to deduce certain events in the game's past, each set up and integrated into a whole by a stimulating Holmes Watson dialogue. **(p)**

Chinese Chess. H. T. Lau. Rutland, VT: Charles E. Tuttle, 1985. 248pp. $14.95. ISBN 0-8048-1495-3.

Chinese chess, a variant of the Western game, is explained in this book: pieces and their moves, standard openings, mid-game strategy, and numerous end games. **(p)**

Fifty Challenging Problems in Probability with Solutions. Frederick Mosteller. New York: Dover, 1987. viii + 88pp. $3.95. ISBN 0-486-65355-2.

An unabridged reprint of the 1965 work which demonstrates mathematical thinking at its best. **(p)**

Fractal Music, Hypercards and More. Martin Gardner. New York: W. H. Freeman, 1992. 327pp. $13.95. ISBN 0-7167-2188-0.

Reprint of Gardner's Mathematical Games columns from the 1978-1979 *Scientific American*, supplemented with updates provided by the readers and by papers published by the author after the columns were written. **(p)**

Game, Set, and Math: Enigmas and Conundrums. Ian Stewart. New York: Penguin Books, 1989. viii + 191pp. $9.95 paperbound. ISBN 0-14-013237-6.

Twelve reprints (in English) from the "Mathematical Visions" column in *Pour la Science*, the French translation of *Scientific American.* Mathematical level is more demanding than Gardner, but the style is similar. **(p)**

Handbook of Cubik Math. Alexander H. Frey, Jr., David Singmaster. Hillside, NJ: Enslow, 1982. viii + 193pp. $18.95. ISBN 0-89490-058-7.

A well written elaboration of Singmaster's *Notes on Rubik's Magic Cube* with much the same material in a more accessible form. **(p)**

Hidden Connections, Double Meanings. David Wells. New York: Cambridge University Press, 1988. 164pp. $14.95. ISBN 0-521-31334-1.

A collection of geometric problems that can be solved by "seeing" properly with the mind's eye. **(p)**

The Ins and Outs of Peg Solitaire. John D. Beasley. Recreations in Mathematics; 2. New York: Oxford University Press, 1985. xii + 275pp. $19.95. ISBN 0-19-853203-2.

Details of the rules, notation, equipment, and history of the 33-hole version of peg solitaire. **(p)**

International Mathematical Olympiads, 1978-1985 and Forty Supplementary Problems. Murray S. Klamkin. New Mathematical Library;31. Washington, DC: Mathematical Association of America, 1986. xii + 141pp. $11.50. ISBN 0-88386-631-X.

Problems from recent Olympiads with rewritten and expanded solutions. References and a glossary of uncommon terms. **(p)**

Knotted Doughnuts and Other Mathematical Entertainments. Martin Gardner. New York: W. H. Freeman, 1986. xiii + 278pp. $12.95. ISBN 0-7167-1799-9.

Eleventh collection of reprints from Gardner's *Scientific American* columns. **(p)**

Mathematical Byways in Ayling, Beeling, and Ceiling. Hugh ApSimon. New York: Oxford University Press, 1984. xii + 97pp. $16.95. ISBN 0-19-853201-6.

Collection of intriguing problems involving little formal mathematical knowledge, but considerable ingenuity. **(p)**

Mathematical Carnival. Martin Gardner. Washington, DC: Mathematical Association of America, 1989. 320pp. $17.50. ISBN 0-88385-448-1.

Reprint of 1975 anthology of Gardner's columns with new introduction, preface, and recent updates. **(p)**

The Mathematical Gardner. Ed: David A. Klarner. Pacific Grove, CA: Brooks/Cole, 1981. viii + 382pp. $37.95. ISBN 0-534-98015-5.

Martin Gardner fans offer a 65th birthday collection of games, geometry, packings and tilings, supernatural numbers, codes, fun and games. **(p)**

Mathematical Magic Show. Martin Gardner. Washington, DC: Mathematical Association of America, 1990. 312pp. $17.50. ISBN 0-317-99616-9.

Updated reprint of his 1977 collection published by Knopf. **(p)**

Mathematical Quickies: 270 Stimulating Problems with Solutions. Charles W. Trigg. New York: Dover, 1985. xi + 210pp. $5.95. ISBN 0-486-24949-2.

An unabridged and corrected reproduction of the 1967 edition. **(p)**

Mathematical Recreations and Essays. Thirteenth Edition. W. W. Rouse Ball, H. S. M. Coxeter. New York: Dover, 1987. xvii + 428pp. $8.95. ISBN 0-486-25357-0.

New material in this revision features the role of the computer in extending the list of Mersenne primes and in the resolution of the problem of "squaring the square." **(p)**

Mathematics: Problem Solving Through Recreational Mathematics. Bonnie Averbach, Orin Chein. New York: W. H. Freeman, 1980. xvi + 400pp. $33.95. ISBN 0-7167-1124-9.

Recreational problems, puzzles, and games for fun and for promoting critical thinking. Selected bibliography. **(p)**

More Mathematical Morsels. Ross Honsberger. Dolciani Mathematical Expositions;10. Washington, DC: Mathematical Association of America, 1991. xii + 322pp. $16.00. ISBN 0-88385-313-2.

Delightful collection of expansive solutions for intriguing elementary problems, drawn mostly from *Crux Mathematicorum*. **(p)**

NYSML-ARML Contests, 1973-1985. Second Edition. Ed: Harry D. Ruderman. Reston, VA: National Council of Teachers of Mathematics, 1987. vi + 193pp. $18.00. ISBN 0-87353-247-3.

Chronological arrangment of problems and answers from two prestigious high school mathematics competitions: The New York State Mathematics League and The Atlantic Region Mathematics League. **(p)**

NYSML-ARML Contests, 1983-1988. Gilbert Kessler, Lawrence Zimmerman. Reston, VA: National Council of Teachers of Mathematics, 1989. x + pp. $18.00. ISBN 0-87353-278-3.

See above. **(p)**

Old and New Unsolved Problems in Plane Geometry and Number Theory. Victor Klee, Stan Wagon. Dolciani Mathematical Expositions;11. Washington, DC: Mathematical Association of America, 1991. xv + 333pp. $22.00. ISBN 0-88385-315-9.

Two dozen elementary problems, with analysis, related results, hints, and references, are presented for both amateurs and professionals. **(a,p)**

Penrose Tiles to Trapdoor Ciphers. Martin Gardner. New York: W. H. Freeman, 1989. ix + 311pp. $13.95;$19.95. ISBN 0-7167-1987-8;0-7167-1986-X.

Fractals, surreal numbers, paradoxes, etc. from Gardner's *Scientific American* columns. **(p)**

Polyominoes: A Guide to Puzzles and Problems in Tiling. George E. Martin. Washington, DC: Mathematical Association of America, 1991. ix + 184pp. $21.00. ISBN 0-88385-501-1.

A systematic exploration with many examples and problems of the tiling properties of "polyominoes," shapes formed from "rook-wise" connected squares. **(p)**

Problems for Mathematicians, Young and Old. Paul Halmos. Dolciani Mathematical Expositions;12. Washington, DC: Mathematical Association of America, 1991. xviii + 318pp. $26.00. ISBN 0-88385-320-5.

This appealing book arranges the 165 problems by the mathematical subdiscipline into which they most naturally fit. There is much here for both the veteran professional mathematician as well the budding mathematician in high school. **(a,p)**

Riddles of the Sphinx and Other Mathematical Puzzle Tales. Martin Gardner. New Mathematical Library;32. Washington, DC: Mathematical Association of America, 1987. x + 164pp. $14.50. ISBN 0-88385-632-8.

Thirty-six mathematical riddles embedded in imaginative stories first published in Isaac Asimov's *Science Fiction Magazine*. Great browsing. **(p)**

Rubik's Cube Compendium. Erno Rubik et al. Transl: David Singmaster. Recreations in mathematics;3. New York: Oxford University Press, 1987. xi + 225pp. $28.95. ISBN 0-19-853202-4.

Translation of the 1981 Hungarian volume with Singmaster featuring the numerous Rubik "offspring" that emerged in the intervening years. **(a,p)**

Sliding Piece Puzzles. Edward Hodern. Recreations in Mathematics;4. New York: Oxford University Press, 1986. xiii + 249pp. $29.95. ISBN 0-19-853204-0.

An entertaining collection of more than 250 sliding piece puzzles. **(p)**

The Stanford Mathematics Problem Book: With Hints and Solutions. George Pólya, Jeremy Kilpatrick. Ann Arbor, MI: Books on Demand, 1990. 68pp. $20.00. ISBN 0-317-09309-6.

Photocopy of 1974 original. Intriguing problems with instructive solutions. **(p)**

Time Travel and Other Mathematical Bewilderments. Martin Gardner. New York: W. H. Freeman, 1987. ix + 295pp. $12.95;$17.95. ISBN 0-7167-1925-8;0-7167-1924-X.

Twelfth collection of reprints from Gardner's *Scientific American* columns, covering paradoxes, puzzles, pasttimes, etc. **(p)**

U.S.A. Mathematical Olympiads, 1972-1986. Murray S. Klamkin. New Mathematical Library; 33. Washington, DC: Mathematical Association of America, 1988. xv + 127pp. $14.50. ISBN 0-88385-634-4.

Each of the fifteen contests are followed by solutions, along with further comments, grouped by subject. **(p)**

Wheels, Life, and Other Mathematical Amusements. Martin Gardner. New York: W. H. Freeman, 1983. ix + 261pp. $13.95. ISBN 0-7167-1589-9.

Paperback version of the updated reprints of Gardner's 22 columns from the *Scientific American*. **(p)**

The William Lowell Putnam Mathematical Competition, Problems and Solutions: 1938-1964. A. M. Gleason, R. E. Greenwood, L. M. Kelly. Washington, DC: Mathematical Association of America, 1980. xi +652pp. $45.00. ISBN 0-88385-428-7.

Outstanding collection of problems and solutions from the first twenty-five years of the Putnam Competition. **(a)**

The William Lowell Putnam Mathematical Competition, Problems and Solutions: 1965-1984. Ed: Gerald L. Alexanderson, Leonard F. Klosinski, Loren C. Larson. Washington, DC: Mathematical Association of America, 1985. xii + 147pp. $27.00. ISBN 0-88385-441-4.

A corrected reprinting from the *American Mathematical Monthly* and *Mathematics Magazine* of problems and solutions to the last 20 Putnam exams. **(a)**

MODELLING

Alternate Realities: Mathematical Models of Nature and Man. John L. Casti. New York: Wiley, 1989. xvii + 493pp. $34.95. 0-471-61842-X.

A bold, wide-ranging introduction to the use of modern mathematical tools in the process of modelling natural and human systems. **(a,p)**

A Concrete Approach to Mathematical Modelling. Michael Mesterton-Gibbons. Reading, MA: Addison-Wesley, 1989. xx + 597pp. $39.16. ISBN 0-201-12910-8.

The author takes the student carefully through the model building process: each model is critically developed, analyzed, and assessed. The 12 chapters are centered around either mathematical or modeling techniques. **(a,p)**

Concepts of Mathematical Modeling. Walter J. Meyer. New York: McGraw-Hill, 1984. xi + 440pp. $41.32. ISBN 0-07-041747-4.

Introductory text emphasizes the modeling process. **(p)**

Empirical Model Building. James R. Thompson. New York: Wiley, 1989. xiv + 242pp. $44.95. ISBN 0-471-60105-5.

A broad collection of problem situations suitable for use in undergraduate modelling courses as well as the tools. **(a,p)**

A First Course in Mathematical Modeling. Frank R. Giordano, Maurice D. Weir. Pacific Grove, CA: Brooks-Cole, 1985. xvii + 382pp. $42.95. ISBN 0-53-403367-9.

Highly readable discussion of creative and empirical model construction, model fitting, etc. **(p)**

Mathematical Modelling. J. N. Kapur. New York: Wiley, 1988. xi + 259pp. $29.95. ISBN 0-470-20088-X.

A collection of modeling applications, grouped according to techniques. **(a)**

Mathematical Models and Their Analysis. Frederic Y. M. Wan. New York: Harper & Row, 1989. xvi + 394pp. $47.50. ISBN 0-06-046902-1.

Outstanding text not only on mathematical models, but also on the process of mathematical modelling. Models studied are grouped according to the principle concern of the model. **(a,s)**

NUMBER THEORY

Algebraic Number Fields. Gerald J. Janusz. Pure and Applied Mathematics;55. San Diego, CA: Academic Press, 1973. x + 220pp. $59.50. ISBN 0-12-380250-4.

This self-contained introduction to algebraic number theory and class field theory is ideally suited for a second year graduate course in algebra. **(a)**

Algebraic Number Theory. Serge Lang. Graduate Texts in Mathematics;110. New York: Springer-Verlag, 1986. xiii + 354pp. $29.80. ISBN 0-387-96375-8.

Changes in this edition include corrections, historical updates on Fermat's Last Theorem and the class number problem as well as 2 new short sections. This text remains an excellent modern introduction to the subject. **(a,p)**

Algebraic Number Theory. Second Edition. Ian Stewart, David Tall. New York: Routledge, Chapman & Hall, 1987. xix + 262pp. $25.00 paperbound. ISBN 0-412-29690-X.

Very readable introduction to the subject, assuming some exposure to rings and fields. **(a)**

Algebraic Numbers. Paulo Ribenboim. Pure and Applied Mathematics;27. Ann Arbor, MI: Books on Demand, 1990. xi + 300pp. $78.00. No ISBN.

Photocopy of 1972 original, a fine exposition of algebraic number theory. **(a)**

The Algebraic Theory of Quadratic Forms. T. Y. Lam. Reading, MA: Addison-Wesley, 1973. xi + 344pp. $36.75 paperbound. ISBN 0-8053-5666-5.

Very clearly written book. Included at the end of each chapter are reasonable exercises that definitely enhance further understanding of the subject. **(a)**

Algorithmic Algebraic Number Theory. M. Pohst, H. Zassenhaus. Encyclopedia of Mathematics and Its Applications;30. New York: Cambridge University Press, 1989. xiv + 465pp. $100.00. ISBN 0-521-33060-2.

This introduction to constructive algebraic number theory and some related algorithms assumes a basic algebra background. Great reference for both experts and non-experts. Expensive. **(a)**

Arithmetical Functions. K. Chandrasekharan. Die Grundlehren der mathematischen Wissenschaften;167. New York: Springer-Verlag, 1970. 231pp. $59.00. ISBN 0-387-05114-7.

Treatise on the asymptotic behavior of certain arithmetical functions. **(a)**

The Book of Prime Number Records. Second Edition. Paulo Ribenboim. New York: Springer-Verlag, 1989. xxiii + 479pp. $49.80. ISBN 0-387-97042-8.

More than a book of records, this valiant attempt to keep track of advances on the prime number front is an interesting and delightful account of number theory. It contains recent historical presentations of important problems and includes some proofs. *The Little Book of Big Primes* is an abridged version (see below). **(a)**

Class Field Theory. Emil Artin, John Tate. Reading, MA: Addison-Wesley, 1990. xxxviii + 259pp. $29.95. ISBN 0-201-51011-1.

Photoreproduction of original typewritten text of the 1967 publication of chapters 5-12 (global theory) of the notes from the Princeton seminar on Class Field Theory. A "Special Preface" includes a list of 8 related books published since 1967. **(a)**

A Classical Introduction to Modern Number Theory. Kenneth Ireland, Michael Rosen. Graduate Texts in Mathematics;84. New York: Springer-Verlag, 1982. xiii + 341pp. $45.00. ISBN 0-387-90625-8.

Revised and expanded version of the authors' 1972 *Elements of Number Theory*. Unlike many other texts, the focus is on topics which point to algebraic number theory and arithmetic algebraic number theory. **(a)**

A Classical Invitation to Algebraic Numbers and Class Fields. Harvey Cohn. New York: Springer-Verlag, 1978. xiii + 328pp. $39.80 paperbound. ISBN 0-387-90345-3.

This intriguing book serves as an introduction to classical algebraic number theory, with emphasis on rational (elementary) number theory, and with the overall motivation and goal being class field theory. Two appendices by Olga Taussky. **(a)**

A Concise Introduction to the Theory of Numbers. Alan Baker. New York: Cambridge University Press, 1984. xiii + 95pp. $18.00. ISBN 0-521-28654-9.

A superb introduction to number theory, thorough and self-contained. **(p)**

Continued Fractions. Third Edition. A. I͡A. Khinchin. Ann Arbor, MI: Books on Demand, 1990. xi + 95pp. $28.00. ISBN 0-317-09464-5.

Photocopy of the 1964 University of Chicago translation of 1961 Russian original. Khinchin confines his attention to continued fractions with positive integral elements. **(a)**

Continued Fractions. C. D. Olds. New Mathematical Library;9. Washington, DC: Mathematical Association of America, 1963. 162pp. $12.95. ISBN 0-88385-609-3.

Author gives the beautiful linking of quadratic irrationals with periodic continued fractions. **(a)**

A Course in Arithmetic. J. -P. Serre. Graduate Texts in Mathematics;7. New York: Springer-Verlag, 1973. viii + 115pp. $39.00 paperbound. ISBN 0-387-90040-3.

Translation of 1970 *Cours d'Arithmetique*. **(a)**

A Course in Number Theory and Cryptography. Neal Koblitz. Graduate Texts in Mathematics;114. New York: Springer-Verlag, 1987. 208pp. $35.00. ISBN 0-387-96576-9.

Good introduction to elementary number theory and algebra underlying public-key cryptosystems. **(a)**

Disquisitiones Arithmeticae. English Edition. Carl Friedrich Gauss. Transl: Arthur A. Clarke. New York: Springer-Verlag, 1986. xx + 472pp. $79.00. ISBN 0-387-96254-9.

A slightly revised reprint to the 1966 translation of the 1801 monumental work. **(a)**

Divisor Theory. Harold M. Edwards. Boston: Birkhäuser, 1990. xiv + 166pp. $29.50. ISBN 0-8176-3448-7.

Covers the general theory, including some material not treated by Kronecker, as well as applications to algebraic number theory and algebraic curves. **(a)**

Divisors. Richard R. Hall, Gerald Tenenbaum. Cambridge Tracts in Mathematics;90. New York: Cambridge University Press, 1988. xvi + 167pp. $42.50. ISBN 0-521-34056-X.

This well-written and readable monograph is concerned with the distribution of the divisors of an integer. Brief appendix on distribution functions. **(a)**

Elementary and Analytic Theory of Algebraic Numbers. Second Edition. Wladyslaw Narkiewicz. New York: Springer-Verlag, 1990. xii + 746pp. $89.50. ISBN 0-387-51250-0.

Updated edition to an important reference: recent results plus updated chapter notes and extensive bibliography. **(a)**

Elementary Number Theory. Second Edition. Edmund Landau. Transl: Jacob E. Goodman. New York: Chelsea, 1988. 256pp. $18.95. ISBN 0-8284-0125-X.

A classic, with added exercises by Bateman and Kohlbecker. **(a,p)**

Elementary Theory of Numbers. W. Sierpinski. North-Holland Mathematical Library;31. New York: Elsevier, 1988. xii + 513pp. $102.75. ISBN 0-444-86662-0.

Few changes, other than references, to the 1964 original. This book considers mainly Diophantine and divisibility problems whose solutions lie within the ring of integers. Expensive. **(a)**

Elements of Quaternions. Third Edition. William R. Hamilton. New York: Chelsea, 1969. $99.50 set. ISBN 0-8284-0219-1. *Volume I.* xxxiii + 594pp. *Volume II.* liv + 502pp.

Reprint of 1901 original. **(a)**

Elliptic Curves. Dale Husemoller. Graduate Texts in Mathematics;111. New York: Springer-Verlag, 1986. xv + 350pp. $49.00. ISBN 0-387-96371-5.

Excellent introduction to one of the hottest topics in number theory. **(a)**

Elliptic Functions. Second Edition. Serge Lang. Graduate Texts in Mathematics;112. New York: Springer-Verlag, 1987. xi + 326pp. $39.80. ISBN 0-387-96508-4.

Reprint of 1973 original, an excellent introduction to the subject, covering its classical origins to modern topics. **(a)**

Exercises in Number Theory. D. P. Parent. New York: Springer-Verlag, 1984. x + 541pp. $56.00. ISBN 0-387-96063-5.

Translation of 1978 original, which collected 200 problems from a group of French number theorists. **(a)**

From Fermat to Minkowski: Lectures on the Theory of Numbers and Its Historical Development. Winfried Scharlau, Hans Opolka. New York: Springer-Verlag, 1985. xi + 184pp. $38.00. ISBN 0-387-90942-7.

Introduction to number theory including the theory of binary quadratic forms and zeta-functions. **(a)**

Handbuch der Lehre von der Verteilung der Primzahlen. Third Edition. Edmund Landau. New York: Chelsea, 1974. xviii + 1001pp. $39.50. ISBN 0-8284-0096-2.
English text on the classic, with an appendix by Paul T. Bateman. (**a**)

The Higher Arithmetic: An Introduction to the Theory of Numbers. Fifth Edition. H. Davenport. New York: Dover, 1983. 172pp. $6.75. ISBN 0-486-24452-0.
Reprint of the 1960 classic. (**a**)

Higher Mathematics from an Elementary Point of View. H. Rademacher. Ed: D. Goldfeld. Boston: Birkhäuser, 1983. 128pp. $39.95. ISBN 3-7643-3064-3.
Lightly edited exposition of a 1947 series of lectures, ranging from prime numbers through inversions. (**a**)

History of the Theory of Numbers. Leonard E. Dickson. New York: Chelsea, 1971. $65.00 set. ISBN 0-8284-0086-5. *Vol. I: Divisibility and Primality.* xii + 486pp. *Vol. II: Diophantine Analysis.* xxv + 803pp. *Vol. III: Quadratic and Higher Forms.* v + 313pp.
Unaltered reprinting of the 1919-1923 Carnegie Institution of Washington original. (**a**)

Introduction to Analytic Number Theory. Tom M. Apostol. New York: Springer-Verlag, 1986. xii + 338pp. $42.00. ISBN 0-387-90163-9.
Reprint of 1976 edition, this beautiful introduction to the topic includes a good historical introduction, excellent problems, bibliography, etc. (**a**)

Introduction to Cyclotomic Fields. Lawrence C. Washington. Graduate Texts in Mathematics;83. New York: Springer-Verlag, 1982. xi + 389pp. $55.00. ISBN 0-387-90622-3.
A beautiful survey of the subject, highlighting p-adic L-functions, Stickelberger's theorems, measures and distributions, and Iwasawa theory. (**a**)

An Introduction to Diophantine Approximation. J. W. S. Cassels. New York: Hafner, 1972. x + 168pp. $9.95. ISBN 0-02-842650-9.
Revised reprint of 1957 edition, a self-contained modulo of some integration theory and algebraic number theory. (**a**)

Introduction to Elliptic Curves and Modular Forms. Neal Koblitz. Graduate Texts in Mathematics;97. New York: Springer-Verlag, 1984. viii + 248pp. $45.00. ISBN 0-387-96029-5.
Covers doubly periodic functions and arithmetic on elliptic curves, classical functions of Riemann, etc., modular forms of integer weight, and those of half integer weight. Motivated throughout by the congruent number problem. Numerous exercises. (**a**)

Introduction to Number Theory. Loo-Keng Hua. Transl: Peter Shiu. New York: Springer-Verlag, 1982. xviii + 572pp. $69.00. ISBN 0-387-10818-1.
Translation and update of Hua's monumental 1956 text. (**a**)

An Introduction to the Theory of Numbers. Fifth Edition. G. H. Hardy, E. M. Wright. New York: Oxford University Press, 1989. xvi + 426pp. $32.95 paperbound. ISBN 0-19-853171-0.

Reprint of this important and successful book that first appeared more than 40 years ago and is now a well-know classic. However, it could use a complete subject index. **(a)**

An Introduction to the Theory of Numbers. Fifth Edition. Ivan Niven, Herbert S. Zuckerman, Hugh L. Montgomery. New York: Wiley, 1991. xiii + 529pp. $52.95. ISBN 0-471-62546-9.

A classic re-born: this edition reflects some the the new developments in number theory of the past decade. Extensive notes tie each section to historical and bibliographic sources. **(a)**

Invitation to Number Theory. Oystein Ore. New Mathematical Library;20. Washington, DC: Mathematical Association of America, 1967. viii + 129pp. $12.95. ISBN 0-88385-620-4.

Discusses and gives examples of various number theories and how they function within the science of mathematics. **(a,p)**

Invitation to Number Theory with Pascal. Donald D. Spencer. Ormond Beach, FL: Camelot, 1989. 320pp. $19.95. ISBN 0-89218-126-5.

Highly readable introduction to both number theory and Pascal. **(a)**

Irrational Numbers. Ivan Niven. Carus Mathematical Monograph;11. Washington, DC: Mathematical Association of America, 1956. 164pp. $22.50. ISBN 0-99395-011-7.

Among the topics treated in this exposition on the properties of irrational numbers are algebraic irrationals, transcendental, and normal numbers as well as the solution of Hilbert's seventh problem. **(a)**

Lattice Theory. Third Edition, Reprinted with Corrections. Garrett Birkhoff. Colloquium Publications;25. Providence, RI: American Mathematical Society, 1984. vi + 418pp. $50.00. ISBN 0-8218-1025-1.

Sixth printing of 1940 original. "The purpose of this edition is... to make the deeper ideas of lattice theory accessible to mathematicians generally, to portray its structure, and to indicate some of its most interesting applications. Unlike previous editions, this one contains only a very incomplete bibliography."—Preface. **(a,s)**

Lectures on Number Theory. Adolf Hurwitz, Nikolaos Kritikos. Transl: William C. Schulz. New York: Springer-Verlag, 1986. xiv + 273pp. $29.00. ISBN 0-387-96236-0.

A solid introduction to elementary number theory in an amazingly clear and readible presentation. **(a)**

The Little Book of Big Primes. Paulo Ribenboim. New York: Springer-Verlag, 1991. xvii + 237pp. $29.50. ISBN 0-387-97508-X.

A chatty abridged (omitting highly complex proofs) version of Ribenboim's *Book of Prime Number Records* (see p. 256). Excellent bibliography, appendices, indexes. **(a)**

Logical Number Theory I: An Introduction. Craig Smorynski. New York: Springer-Verlag, 1991. x + 405pp. $49.00. ISBN 0-387-52236-0.

An idiosyncratic introduction to logic via number theory and vice versa, with historical and philosophical digressions. **(a)**

Modular Functions and Dirichlet Series in Number Theory. Second Edition. Tom M. Apostol. Graduate Texts in Mathematics;41. New York: Springer-Verlag, 1990. x + 204pp. $49.80. ISBN 0-387-97127-0.

Concerned primarily with elliptic and modular functions and their number theoretic applications. Updated bibliography. **(a)**

The Number Systems: Foundations of Algebra and Analysis. Second Edition. Solomon Feferman. New York: Chelsea, 1989. xii + 418pp. $23.95. ISBN 0-8284-0333-3.

Reprint of 1964 original, one of the first undergraduate texts written by a prominent American mathematician. **(a)**

Number Theory. Z. I. Borevich, I. R. Shafarevich. Transl: Newcomb Greenleaf. Pure and Applied Mathematics;20. San Diego, CA: Academic Press, 1986. x + 435pp. $53.00. ISBN 0-12-117851-X.

Reprint of 1966 edition of the beautifully written classic. Topics and treatment are so current that it is hard to believe it was written 20 years ago. **(a)**

Number Theory. Ed: K. Győry, G. Halász. New York: Elsevier, 1990. $179.50 set. ISBN 0-444-70476-0. *Volume I: Elementary and Analytic*. 573pp. *Volume II: Diophantine and Algebraic*. 495pp.

Most, but not all, of the papers presented at the July 1987 Janos Bolyai Mathematical Society Conference on Number Theory in Budapest. These give a comprehensive picture of current research in the field. Expensive. **(a)**

Number Theory. Helmut Hasse. Transl: Horst Gunter Zimmer. Die Grundlehren der mathematischen Wissenschaften;229. New York: Springer-Verlag, 1980. xvii + 638pp. $49.00. ISBN 0-387-08275-1.

Corrected and enlarged translation of the Third edition of this classic. **(a,s)**

Number Theory and Its Applications in China. Ed: Wang Yuan, Yang Chungchun, Pan Chegbiao. Contemporary Mathematics;77. Providence, RI: American Mathematical Society, 1988. $24.00. ISBN 0-8218-5084-9.

Nine survey articles and three current research articles describe the research accomplishment of Chinese number theorists from 1949 to 1979. **(a)**

Number Theory and Its History. Oystein Ore. New York: Dover, 1988. x + 370pp. $8.95. ISBN 0-486-65620-9. Reprint of the elementary 1948 classic.

Introduces a selection of important results in the context of historical exposition. (**a**)

Number Theory in Science and Communication: With Applications in Cryptography, Physics, Digital Information, Computing, and Self-Similarity. Second Enlarged Edition, Corrected Printing. M. R. Schroeder. Springer Series in Information Sciences;7. New York: Springer-Verlag, 1990. xix + 374pp. $41.50. ISBN 0-387-15800-6.

This text explores the use of number theory in a broad range of applications. A few new topics have been added to the 1984 edition. Treatment is a bit superficial in some sections. (**a**)

Number Theory: Proceedings of a Symposium, May 4, 1989, Washington, D.C. Washington, DC: National Research Council, 1990. vii + 51pp. Complimentary. No ISBN.

Three superb expositions, written for the general audience, of key ideas of contemporary number theory presented in May 1989 by Lenstra, Mazur, and Odlyzko. (**a**)

Number Theory, Trace Formulas and Discrete Groups. Ed: Karl Egil Aubert, Enrico Bombieri, Dorian Goldfeld. San Diego, CA: Academic Press, 1989. xx + 510pp. $89.00. ISBN 0-12-067570-6.

Eight survey articles and twenty-one research summaries from a 1987 Oslo symposium honoring Atle Selberg's seventieth birthday. Papers focus on automorphic forms, trace formulae, prime numbers, and Riemann zeta function. (**a**)

P-Adic Numbers, P-Adic Analysis, and Zeta-Functions. Second Edition. Neal Koblitz. Graduate Texts in Mathematics;58. New York: Springer-Verlag, 1984. xii + 150pp. $42.00. ISBN 0-387-96017-1.

This edition has changed with the inclusion of material on the Iwasawa logarithm and the p-adic gamma-function. (**a**)

Pi and the AGM: A Study in Analytic Number Theory and Computational Complexity. Jonathan M. and Peter B. Borwein. New York: Wiley, 1987. xv + 414pp. $69.95. ISBN 0-471-83138-7.

AGM (arithmetic-geometric mean) allows one to obtain 5000 digits of pi faster than the familiar arctangent formulas. The hows and whys necessitate a study of elliptic integrals and theta functions. (**a**)

Primes of the Form $x^2 + ny^2$: Fermat, Class Field Theory, and Complex Multiplication. David A. Cox. New York: Wiley, 1989. xi + 351pp. $49.95. ISBN 0-471-50654-0.

The answer to the question in the title leads the reader on a delightful excursion through classical and modern algebraic number theory. Many rich exercises plus interesting historical remarks. (**a**)

Quadratic Forms and Hecke Operators. Anatoliĭ N. Andrianov. Grundlehren der mathematischen Wissenschaften;286. New York: Springer-Verlag, 1987. xii + 374pp. $109.00. ISBN 0-387-15294-6.

A culmination of 20 years work by Andrianov and his collaborators on the theory of theta-series and the Hecke algebra in the setting of Siegel modular forms. Expensive. (**a**)

Ramanujan: Twelve Lectures on Subjects Suggested by His Life and Work. Third Corrected Edition. G. H. Hardy. New York: Chelsea, 1978. 236pp. $15.95. ISBN 0-8284-0136-5.

In eleven chapters Hardy weighs the evidence of Ramanujan's independence, analyzes his proofs, and clearly points out the flaws of the reasonings which he took for proofs. (**a**)

Representations of Integers as Sums of Squares. Emil Grosswald. New York: Springer-Verlag, 1985. xv + 810pp. $69.80. ISBN 0-387-96126-7.

Enthusiastic treatment of a classical problem in number theory: counting lattice points on an n-dimensional sphere or ellipsoid. Extensive references. (**a**)

The Selberg Trace Formula and Related Topics. Ed: Dennis A. Hejhal, Peter Sarnak, Audrey Ann Terras. Contemporary Mathematics;53. Providence, RI: American Mathematical Society, 1986. ix + 554pp. $50.00. ISBN 0-8218-5058-X.

Proceedings of a joint 1984 AMS-IMS-SIAM summer conference which explore Selberg's ideas for mathematical physics, classical analytic number theory, modern adelic number theory, modern group representations, classical harmonic analysis, etc. (**a**)

Studies in Pure Mathematics: To the Memory of Paul Turán. Ed: Paul Erdős. Boston, MA: Birkhäuser, 1983. 773pp. $86.95.

This volume includes over sixty papers written by Turán's friends, students, and collaborators. (**a**)

Theorie der Congruenzen: Elemente der Zahlentheorie. Second German Edition. P. L. Chebyshev. Transl: Hermann Schapira. New York: Chelsea, 1972. xvii + 366pp. $17.95. ISBN 0-8284-0254-X.

Reprint of 1889 German edition, with errata corrected, tables rearranged, and other minor changes. (**a**)

The Theory of Algebraic Numbers. Second Edition. Harry Pollard, Harold G. Diamond. Carus Mathematical Monographs;9. Washington, DC: Mathematical Association of America, 1975. 162pp. $21.00. ISBN 0-88385-018-4.

Makes available and understandable the elementary parts of classical algebraic number theory. An often-used text for introductory courses. (**a**)

The Theory of the Riemann Zeta-Function. Second Edition. E. C. Titchmarsh. Ed: D. R. Heath-Brown. New York: Oxford University Press, 1986. x + 412pp. $52.00 paperbound. ISBN 0-19-853369-1.

Editor has updated and added end-of-the chapter notes to the 1951 classic treatise. Updated references. (**a**)

Topics from the Theory of Numbers. Second Edition. Emil Grosswald. Boston, MA: Birkhäuser, 1984. xv + 333pp. $39.95. ISBN 3-7643-3044-9.

This edition has a thorough revision of analytic and algebraic number theory while keeping the outstanding introduction to elementary number theory essentially the same. **(a)**

Transcendental Number Theory. Alan Baker. New York: Cambridge University Press, 1990. x + 165pp. $17.95 paperbound. ISBN 0-521-2-461-5.

This paperback reissue updates material in this survey of the highlights of transcendental number theory. **(a)**

Transcendental Numbers. Andrei Borisovich Shidlovskiĭ. Transl: Neal Koblitz. Studies in Mathematics;12. Hawthorne, NY: Walter de Gruyter, 1989. xix + 466pp. $89.50. ISBN 0-89925-437-3.

Definitive exposition focusing on the transcendental nature of values of E-functions. Extensive bibliography. **(a)**

Uniform Distribution of Sequences of Integers in Residue Classes. Wladyslaw Narkiewicz. New York: Springer-Verlag, 1984. viii + 125pp. $15.10. ISBN 0-387-13872-2.

Survey of current knowledge about uniform distribution of integers into residue classes. References. **(a)**

Unsolved Problems in Number Theory. Richard K. Guy. Unsolved Problems in Intuitive Mathematics;1. New York: Springer-Verlag, 1981. xviii + 161pp. $29.00. ISBN 0-387-90593-6.

This work covers primes, divisibility, sequences, Diophantine equations, and additive number theory. Problems are introduced with definitions, examples, and references. **(a)**

Untersuchungen über höhere Arithmetik. Second Edition. Carl Friedrich Gauss. Ed: H. Maser. New York: Chelsea, 1981. xiii + 695pp. $39.50. ISBN 0-8284-0191-8.

Second, unaltered reprint of 1965 German translation of *Disquisitiones Arithmeticae.* **(a)**

Vorlesungen über die Theorie der algebraischen Zahlen. Erich Hecke. New York: Chelsea, 1970. 274pp. $12.95. ISBN 0-8284-0046-6.

Reprint of 1923 original with errata corrected and full index added. **(a)**

Vorlesungen über Zahlentheorie. Edmund George Hermann Landau. New York: Chelsea, 1969. 3v. in 1. $21.95. ISBN 0-8284-0032-6.

Reprint originally published in 1927 in 3 volumes. The present work contains all 3 except for the first section of Volume One, which is separately published under the title *Elementare Zahlentheorie* (see below). **(a)**

Vorlesungen über Zahlentheorie: Aus der elementaren Zahlentheorie. Edmund George Hermann Landau. New York: Chelsea, 1950. vii + 180pp. $12.00. ISBN 0-8284-0026-1.

Reprint of the first section of 3 volume work (see above). **(a)**

13 Lectures on Fermat's Last Theorem. Paulo Ribenboim. New York: Springer-Verlag, 1979. xvi + 302pp. $49.00. ISBN 0-387-90432-8.

The problems and analogues of Fermat's Last Theorem are presented in a very readable way, giving a good feel of mathematics. Excellent list of references. **(a)**

NUMERICAL ANALYSIS

The Algebraic Eigenvalue Problem. J. H. Wilkinson. New York: Oxford University Press, 1988. xviii + 662pp. $39.95. ISBN 0-19-853418-3.

A paperback edition of the classic reference on numerical linear algebra. **(a)**

Analysis of Numerical Methods. Eugene Isaacson, Herbert Bishop Keller. New York: Wiley, 1966. xv + 541pp. $72.95. ISBN 0-471-42865-5.

Emphasizes the mathematical foundations upon which known methods are based and from which others might be derived and concentrates on computational error and stability. **(a)**

Basic Numerical Mathematics. John Todd. International Series in Numerical Mathematics;14,22. Boston: Birkhäuser, 1977-1979. *Vol. 1: Numerical Analysis.* 253pp. $52.00. ISBN 0-12-692401-5. (Available from Academic Press, San Diego, CA) *Vol. 2: Numerical Algebra.* 216pp. out-of-print. ISBN 3-7643-0811-7.

Volume 1 is a tersely written introduction to numerical analysis, with detailed solutions of selected exercises. **(a)**

Computational Techniques for Ordinary Differential Equations. Ed: I. Gladwell, D. K. Sayres. San Diego, CA: Academic Press, 1980. xi + 303pp. $72.00. ISBN 0-12-285780-1.

Papers presented by invited speakers at a 1978 University of Manchester conference. Discussion of both initial value and boundary value problems. **(a)**

Elementary Numerical Analysis: An Algorithmic Approach. Third Edition. S. D. Conte, Carl de Boor. New York: McGraw-Hill, 1980. xii + 432pp. $42.20. ISBN 0-07-012447-7.

New topics are discussed in this revision of a successful text. **(a)**

Fast Fourier Transform and Convolution Algorithms. Second Corrected and Updated Edition. Henri J. Nussbaumer. Springer Series in Information Sciences;2. New York: Springer-Verlag, 1982. xii + 276pp. $45.00. ISBN 0-387-11825-X.

This book presents in a unified way the various fast algorithms that are used for the implementation of digital filters and the evaluation of discrete Fourier transforms. **(a,s)**

Introduction to Matrix Computations. G. W. Stewart. San Diego, CA: Academic Press, 1973. xiii + 441pp. $49.95. ISBN 0-12-670350-7.

The author introduces the better algorithms for solving linear equations, linear least squares problems, and eigenvalue problems as well as applications. **(a)**

Introduction to Numerical Analysis. J. Stoer, R. Bulirsch. Transl: R. Bartels, W. Gautschi, C. Witzgall. New York: Springer-Verlag, 1980. ix + 609pp. $39.00. ISBN 0-387-90420-4.

Very substantial introduction, including all standard topics. Many exercises. **(a,p)**

Introduction to Numerical Computations. Second Edition. James S. Vandergraft. San Diego, CA: Academic Press, 1983. xiv + 372pp. $42.95. ISBN 0-12-711356-8.

Well-written text. This edition also includes Newton formulas for interpolation and adaptive methods for integration. List of special symbols. **(a)**

An Introduction to the Approximation of Functions. Theodore J. Rivlin. New York: Dover, 1981. viii + 150pp. $4.95. ISBN 0-486-64069-8.

Reprint of 1969 original includes uniform, least squares and least-first-power approximations. **(a)**

Iterative Methods for the Solution of Equations. J. F. Traub. New York: Chelsea, 1982. xviii + 310pp. $15.95. ISBN 0-8284-0312-0.

Reprint of 1964 edition. Dated, but well done treatment of the foundations of a general theory of iteration algorithms for the numerical solution of real equations and systems of real equations with emphasis on algorithm efficiency. **(a)**

Iterative Solution of Large Linear Systems. David M. Young. San Diego, CA: Academic Press, 1971. xxiv + 570pp. $96.00. ISBN 0-12-773050-8.

An immense and well-organized treatment of the iterative solution of large linear systems with sparse matrices as well as a thorough study of the method of successive over-relaxation and its invariants. Bibliography. **(a)**

Methods of Numerical Integration. Second Edition. Philip F. Davis, Philip Rabinowitz. San Diego, CA: Academic Press, 1984. xiv + 612pp. $75.00. ISBN 0-12-206360-0.

Updated and expanded edition with major changes on the evaluation of multi-dimensional integrals and automatic integration. **(a)**

A New Approach to Scientific Computation. Ed: Ulrich W. Kulisch, Willard L. Miranker. Notes & Reports in Computer Science and Applied Mathematics;7. San Diego, CA: Academic Press, 1983. xv +384pp. $70.00. ISBN 0-12-428660-7.

Twelve papers on the development and use of computer arithmetic for intervals of reals, complex numbers, vectors, and arrays. **(a)**

Numerical Methods and Software. David Kahaner, Cleve Moler, Stephen Nash. New York: Prentice-Hall, 1989. xii + 495pp. $55.00. ISBN 0-13-627258-4.

This volume discusses many of the topics most important to scientists and engineers, in a style intermediate between a cookbook and a textbook. **(a,s)**

Numerical Methods for Partial Differential Equations. Second Edition. W. F. Ames. San Diego, CA: Academic Press, 1977. (Available from Van Nostrand Reinhold, NY) xiv + 365pp. $36.95 paperbound. ISBN 0-442-30737-3.

While this book does give an excellent introduction to finite difference methods for partial differential equations and provides an extensive review of the literature, references and much of the original material have not been updated from the sixties. **(a)**

Numerical Recipes: The Art of Scientific Computing. William H. Press et al. New York: Cambridge University Press, 1986. xx + 818pp. $49.50. ISBN 0-521-38330-7.

A compendium of useful numerical techniques, with algorithms clearly explained. **(a,s)**

Numerical Solution of Elliptic Problems. Garrett Birkhoff, Robert E. Lynch. Philadelphia, PA: Society for Industrial and Applied Mathematics, 1984. xi + 319pp. $36.50. ISBN 0-89871-197-5.

The use and analysis of difference, iterative, integral and finite methods of solving elliptic boundary value problems computationally, especially in two-dimensions. **(a,s)**

Numerical Solution of Partial Differential Equations in Science and Engineering. Leon Lapidus, George F. Pinder. New York:Wiley, 1982. 677pp. $76.95. ISBN 0-471-09866-3.

Each type of partial differential equation is treated in a separate chapter. Numerical methods used include finite differences, finite elements, collocation, and boundary elements. Extensive bibliography. **(s)**

Schaum's Outline of Theory and Problems of Numerical Analysis. Second Edition. Frances Scheid. New York: McGraw-Hill, 1989. 471pp. $10.95. ISBN 0-07-055221-5.

Organized and very concise treatment of undergraduate numerical analysis with hundreds of solved and unsolved problems. **(a,p)**

Solution of Partial Differential Equations on Vector and Parallel Computers. James M. Ortega, Robert G. Voigt. Philadelphia, PA: Society for Industrial and Applied Mathematics, 1985. iii +96pp. $14.50. ISBN 0-89871-055-3.

A state-of-the-art survey, discussing both direct and iterative methods for elliptic equations. **(a,s)**

Spline Functions: Basic Theory. Larry L. Schumaker. New York: Wiley, 1981. xiv + 550pp. $79.05. ISBN 0-471-76475-2.

Comprehensive introduction to the theory of polynomial, generalized, and mutlidimensional splines emphasizing their algebraic, analytic, and approximation-theoretic properties. **(a)**

2000 Solved Problems in Numerical Analysis. Francis Scheid. New York: McGraw-Hill, 1990. v + 698pp. $19.95. ISBN 0-07-055233-9.

Two thousand solved problems in numerical analysis listed under different topics, with a brief summary at the beginning of each chapter. **(a,p)**

OPERATIONS RESEARCH AND OPTIMIZATION

Algorithms, Graphs, and Computers. Richard Bellman, Kenneth L. Cooke, Jo Ann Lockett. Mathematics in Science and Engineering;62. San Diego, CA: Academic Press, 1970. xv + 246pp. $56.00. ISBN 0-12-084840-6.

For students and laymen, this elementary text uses dynamic programming and graph theory to present mathematical and computer methods related to large systems from an operations research point of view. **(a,p)**

Combinatorial Optimization: Algorithms and Complexity. Christos H. Papadimitriou, Kenneth Steiglitz. New York: Prentice-Hall, 1982. xvi + 496pp. $67.00. ISBN 0-13-152462-3.

The goal of this book is to bring together in one volume the important ideas of computational complexity developed by computer scientists over the past 15 years, and the foundations of mathematical programming developed by the operations research community."—Preface. **(a,s)**

Discrete Optimization. R. Gary Parker, Ronald L. Rardin. San Diego, CA: Academic Press, 1988. xi + 472pp. $75.00. ISBN 0-12-545075-3.

Presents matroid and linear programming-based strategies for polynomial-time solvable problems. **(a)**

Dynamic Programming: Models and Applications. Eric V. Denardo. New York: Prentice-Hall, 1982. xii + 227pp. $61.73. ISBN 0-13-221507-1.

An account of key ideas of dynamic programming and their uses in operations research. Topics include shortest route problem, resource allocation, production and inventory control. **(a)**

Fundamentals of Queueing Theory. Second Edition. Donald Gross, Carl M. Harris. New York: Wiley, 1985. xiii + 587pp. $64.95. ISBN 0-471-89067-7.

A thorough and thoughtful revision of an already good book, adding new material on queueing networks and approximations. **(a,s)**

Geometric Algorithms and Combinatorial Optimization. Martin Grotschel, Laslo Lovasz, Alexander Schrijver. Algorithms and Combinatorics;2. New York: Springer-Verlag, 1988. xii + 362pp. $69.00. ISBN 0-387-13624-X.

Develops two geometric algorithms and uses them to prove the polynomial time solvability of several problems in combinatorial optimization. **(a,s)**

A Guide to Simulation. Second Edition. Paul Bratley, Bennett L. Fox, Linus E. Schrage. New York: Springer-Verlag, 1987. xx + 397pp. $55.00. ISBN 0-387-96467-3.

Comprehensive introduction to discrete event simulation. **(a,s)**

Introduction to Optimization Methods. P. R. Adby, M. A. H. Dempster. New York: Routledge, Chapman & Hall, 1974. x + 204pp. $18.95. ISBN 0-412-11040-7.

This book introduces non-linear methods of optimization, covering the classical techniques as well as some of the more recent methods of unconstrained and constrained optimization. **(a)**

Linear and Combinatorial Optimization in Ordered Algebraic Structures. U. Zimmermann. Annals of Discrete Mathematics;10. New York: Elsevier, 1981. ix + 380pp. $113.00. ISBN 0-444-86153-X.

Results and methods for linear and combinatorial optimization problems where linear functions over the reals are replaced by functions over algebraic structures. Bibliography. Expensive. **(a)**

Markov Decision Processes. Paul Thie. Arlington, MA: COMAP, 1983. vii + 95pp. $8.95. ISBN 0-912843-04-7.

Introduction to this useful method of operations research, with interesting applications throughout the text. **(s)**

Maxima and Minima Without Calculus. Ivan Niven. Dolciani Mathematical Expositions;6. Washington, DC: Mathematical Association of America, 1982. xv + 303pp. $32.50. ISBN 0-88385-306-X.

A compendium of algebraic, geometric, and inequality methods for optimization problems, excluding methods of calculus, linear programming, and game theory. **(a)**

Methods of Mathematical Economics: Linear and Nonlinear Programming, Fixed-Point Theorems. Joel Franklin. New York: Springer-Verlag, 1980. x + 297pp. $45.00. ISBN 0-387-90481-6.

A delightfully written text emphasizing linear programming models as well as techniques for solving them. **(p)**

Network Flows and Monotropic Optimization. R. T. Rockafellar. New York: Wiley, 1984. xiii + 616pp. $72.95. ISBN 0-471-88078-7.

Presents an introduction to the theory of network programming problems and generalizations of these results to a larger class of monotropic programming problems. **(a)**

Numerical Methods for Unconstrained Optimization. Ed: W. Murray. San Diego, CA: Academic Press, 1972. xi + 144pp. $74.00. ISBN 0-12-512250-0.

Eight good expository papers from a 1971 joint IMA/NPL conference. References. Index. **(a)**

Numerical Methods for Unconstrained Optimization and Nonlinear Equations. J. E. Dennis, Jr., Robert B. Schnabel. New York: Prentice-Hall, 1983. xiii + 378pp. $68.00. ISBN 0-13-627216-9.

A survey of methods for small- to medium-size problems, excluding methods not based on Newton's method. **(a)**

Optimization by Vector Space Methods. David G. Luenberger. New York: Wiley, 1969. xvii + 326pp. $69.95. ISBN 0-471-55359-X.

"Primary objective of the book is to demonstrate that a rather large segment of the field of optimization can be effectively unified by a few geometric principles of linear vector space theory"—Preface. **(a)**

Optimization Techniques: An Introduction. L. R. Foulds. New York: Springer-Verlag, 1981. xi + 502pp. $54.00. ISBN 0-387-90586-3.

Introduces some of the most important optimization techniques at an undergraduate level. **(p)**

Optimization-Theory and Applications: Problems with Ordinary Differential Equations. Lamberto Cesari. Applications in Mathematics;17. New York: Springer-Verlag, 1983. xiv + 542pp. $98.00. ISBN 0-387-90676-2.

An encyclopedic treatment dedicated to the proposition of both the classical and modern theories. Chapter-by-chapter bibliographical comments and large number of illustrative examples are invaluable. Extensive bibliography. **(a,s)**

Practical Methods of Optimization. Second Edition. R. Fletcher. New York: Wiley, 1987. xiv + 436pp. $69.95. ISBN 0-471-91547-5.

Text emphasizes understanding the considerations involved in choosing practical algorithms to handle some standard problems. Second edition combines into one the material in volumes 1-2 of First edition as well as including new material on network programming, optimality conditions for constrained composite non-smooth optimization, etc. **(a,s)**

Schaum's Outline of Theory and Problems of Operations Research. Richard Bronson. New York: McGraw-Hill, 1982. 328pp. $10.05. ISBN 0-07-007977-3.

This book is divided into 2 parts: mathematical programming and probabilistic methods. Each chapter is divided into 3 sections: methodology, worked problems, problems with answers to test mastery of material. **(a)**

The Single Server Queue. Revised Edition. J. W. Cohen. Applied Mathematics and Mechanics;8. New York: Elsevier, 1982. xiv + 694pp. $85.25. ISBN 0-444-85452-5.

Massive, definitive treatment of single server queues: the basic ideas of stochastic processes needed, the model and variety of methods used in analysis, and variants, etc. Extensive bibliography. **(a)**

The Traveling Salesman Problem: A Guided Tour of Combinatorial Optimization. Ed: E. L. Lawler et al. New York: Wiley, 1985. x + 465pp. $94.95. ISBN 0471-90413-9.

A survey-of-the-art collection of papers on various aspect of the traveling salesman problem. **(a)**

PARTIAL DIFFERENTIAL EQUATIONS

The Boundary Value Problems of Mathematical Physics. O. A. Ladyzhenskaia͡. Transl: Jack Lowater. Applied Mathematical Sciences;49. New York: Springer-Verlag, 1985. xxx + 322pp. $79.00. ISBN 0-387-90989-3.

Translation of 1973 Russian original, a systematic exposition of problems of solvability of linear boundary and initial boundary value problems from partial differential equations with variable coefficients. **(a,s)**

Elliptic Partial Differential Equations of Second Order. Second Edition. David Gilbarg, Neil S. Trudinger. Die Grundlehren der mathematischen Wissenschaften;224. New York: Springer-Verlag,1983. xiii + 513pp. $45.50. ISBN 0-387-13025-X.

Covers the classical and modern points of view of the theory of the linear and non-linear elliptic Dirichlet problem. **(a)**

The Existence of Multi-Dimensional Shock Fronts. Andrew Majda. Memoirs of the American Mathematical Society;281. Providence, RI: American Mathematical Society, 1983. 93pp. $13.00. ISBN 0-8218-2281-0.

The author proves the short-time existence of solutions with shock discontinuities of arbitrary length on an initial smooth surface with arbitrary geometry. **(a)**

A First Course in Partial Differential Equations with Complex Variables and Transform Methods. H. F. Weinberger. New York: Wiley, 1965. ix + 446pp. $69.95. ISBN 0-471-00623-8.

Author successfully presents Fourier series, boundary value problems, integral transforms in a framework embodying the general properties of partial differential equations and their solutions. **(a)**

First-Order Partial Differential Equations. Hyun Ku Rhee, Rutherford Aris, Neal R. Amundson. New York: Prentice-Hall, 1986-1989. *Volume I: Theory and Application of Single Equations.* xiii + 543pp. $56.00. ISBN 0-13-319195-8. *Volume II: Theory and Application of Hyperbolic Systems of Quasi Linear Equations.* xii + 548pp. $80.00. ISBN 0-13-319237-7.

A good set of books on partial differential equations intended for engineering students. **(a)**

Introduction to Partial Differential Equations and Hilbert Space Methods. Karl E. Gustaphson. New York: Wiley, 1980. xv + 270pp. $19.95. ISBN 0-471-04089-4.

Spare, but well-footnoted text is divided into 2 parts, each with enough material for a semester course. **(a)**

Introduction to Partial Differential Equations with Applications. E. C. Zachmanoglou, Dale W. Thoe. New York: Dover, 1986. x + 405pp. $9.95. ISBN 0-486-65251-3.

Unabridged, corrected reprint of 1976 original. **(a,p)**

Introduction to the Theory of Nonlinear Elliptic Equations. Jindrich Necas. New York: Wiley, 1986. 163pp. $49.95. ISBN 0-471-90894-0.

Reprint of 1983 Teubner lecture notes on boundary value problems for nonlinear, second order, elliptical partial differential equations. References. **(a)**

Linear Partial Differential Equations with Constant Coefficients; Existence, Approximation, and Regularity of Solutions. François Treves. Mathematics and Its Applications;6. New York: Gordon and Breach, 1966. x + 534pp. $165.00. ISBN 0-677-01190-3.

Very expensive but very readable account of the theory of constant coefficient partial differential equations. Expensive. **(a)**

Navier-Stokes Equations: Theory and Numerical Analysis. Third Edition. Roger Temam. Studies in Mathematics and Its Applications;2. New York: Elsevier, 1984. xii + 526pp. $118.00;$45.00. ISBN 0-444-87558-1;0-444-87559-X.

"The present book stands at the boundary between computational fluid dynamics and mathematical analysis to which CFD is firmly tied. This new edition contains a revision and an updating of the bibliography."—Preface. Expensive. **(a)**

Partial Differential Equations. Second Edition. P. R. Garabedian. New York: Chelsea, 1986. xii + 672pp. $29.50. ISBN 0-8284-0325-2.

Reprint of 1964 original, written for scientists, engineers, and mathematicians. **(a)**

Partial Differential Equations. Fourth Edition. Fritz John. Applied Mathematical Sciences;1. New York: Springer-Verlag, 1982. x + 249pp. $32.00. ISBN 0-387-90609-6.

Corrected new edition, with new proofs and examples and more problems. **(a)**

Partial Differential Equations: An Introduction. Bernard Epstein. Melbourne, FL: Krieger, 1975. x + 273pp. $24.50. ISBN 0-88275-330-4.

Reprint of 1962 original. **(p)**

Partial Differential Equations of Applied Mathematics. Second Edition. Erich Zauderer. New York: Wiley, 1989. xiii + 779pp. $69.95. ISBN 0-471-87517-1.

This book serves as both a text for beginning graduate students in applied mathematics and a useful reference work for engineers and scientists. **(a,s)**

Pseudodifferential Operators and Spectral Theory. M. A. Shubin. Transl: Stig I. Andersson. New York: Springer-Verlag, 1987. x +278pp. $59.00. ISBN 0-387-13621-5.

An introduction of pseudodifferential operators which applies PDO's to the spectral theory of elliptic operators. Good bibliography. **(a)**

Schaum's Outline of Theory and Problems of Partial Differential Equations. Paul Duchateau, David W. Zachmann. New York: McGraw-Hill, 1986. 241pp. $9.95. ISBN 0-07-017897-6.

Both a reference and supplemental or primary text aimed at providing an understanding of the importance of partial differential equations among the topics of applied mathematics. Problems range from elementary and pratical to fairly advanced and theoretical. **(a,p)**

Shock Waves and Reaction-Diffusion Equations. Joel Smoller. Grundlehren der mathematischen Wissenschaften;258. New York: Springer-Verlag, 1983. xxi + 581pp. $66.00. ISBN 0-387-90752-1.

Comprehensive introduction to partial differential equations, emphasizing, but not limited to, the nonlinear topics of the title. **(a)**

Studies in Partial Differential Equations. Ed: Walter Littman. Studies in Mathematics;23. Washington, DC: Mathematical Association of America, 1982. xiii + 268pp. $30.00. ISBN 0-.

Five articles on current work in the context of the 200-year history of the subject and intended for the non-specialist. **(p)**

PRECALCULUS

Algebra and Trigonometry. Third Edition. Walter Fleming, Dale Varberg. New York: Prentice-Hall, 1988. xv + 604pp. $54.00. ISBN 0-13-021338-1.

Good text in two other versions (see below). **(a,p)**

Algebra For College Students. Third Edition. Jerome E. Kaufmann. Boston: PWS-Kent, 1989. xv + 673pp. $44.95. ISBN 0-534-91580-9.

Intermediate algebra text with some college algebra topics, emphasizing drill. This edition has a major revision of exercise sets. **(a,p)**

College Algebra. Fourth Edition. Raymond A. Barnett. New York: McGraw-Hill, 1989. xxi + 517pp. $45.05. ISBN 0-07-003930-5.

One of a series of precalculus books, this one contains less analytic geometry, a less concise review of intermediate algebra, no trigonometry. Includes a chapter on probability. **(a,p)**

College Algebra. Third Edition. Walter Fleming, Dale Varberg. New York: Prentice-Hall, 1988. xv + 526pp. $52.00. ISBN 0-13-141656-1.

Contains 9 chapters of their *Algebra and Trigonometry* (see above), plus chapters on probability and mathematics of finance. **(a,p)**

College Algebra and Trigonometry. Third Edition. Robert Ellis, Denny Gulick. San Diego, CA: Harcourt Brace Jovanovich, 1989. vii + 632pp. $36.00. ISBN 0-15-507932-8.

Changes from earlier editions include the earlier introduction of linear equations and complex numbers, the elimination of discussion of the use of tables, and a new section on compound and continuous interest. **(a,p)**

College Algebra with Trigonometry. Paul K. Rees, Fred W. Sparks, Charles Sparks Rees. New York: McGraw-Hill, 1991. xx + 724pp. $42.85. ISBN 0-07-051737-1.

Thoroughly readable, with applications from a wide variety of disciplines. **(a,p)**

Faces of Mathematics: An Introductory Course for College Students. Second Edition. A. Wayne Roberts, Dale E. Varberg. New York: Harper & Row, 1982. xiii + 492pp. $40.50. ISBN 0-06-045471-7.

Imaginative text for liberal arts students, featuring problems and their solutions. **(a,p)**

Intermediate Algebra. Third Edition. R. David Gustafson, Peter D. Fisk. Pacific Grove, CA: Brooks-Cole, 1991. 672pp. $40.95. ISBN 0-534-16464-1.

This edition makes more of a connection between topics in order to develop students' abilities to integrate and conceptualize. **(p)**

Modern Mathematics. Seventh Edition. Ruric E. Wheeler. Pacific Grove, CA: Brooks-Cole, 1988. xiv + 637pp. $40.95. ISBN 0-534-08580-6.

Mathematics for liberal arts students and elementary school teachers. This edition rewritten in response to recommendations of CUPM of the Mathematical Association of America and the NCTM. **(p)**

Plane Trigonometry. Second Edition. Walter Fleming, Dale Varberg. New York: Prentice-Hall, 1980. xiii + 317pp. $52.00. ISBN 0-13-679051-8.

Contains 4 chapters from their *Algebra and Trigonometry* (see above), plus a chapter on conic sections. **(a,p)**

Schaum's Outline of Theory and Problems of Technical Mathematics. Paul Calter. New York: McGraw-Hill, 1979. xviii + 460pp. $11.95. ISBN 0-07-009651-1.

Outline of the theory and numerous practice problems of algebra (mainly), trigonometry, plane and solid geometry, and the analytic geometry of the straight line. **(a,p)**

Schaum's Outline of Theory and Problems of Trigonometry. Second Edition. Frank Ayres, Jr., Robert E. Moyer. New York: McGraw-Hill, 1990. 224pp. $10.50. ISBN 0-07-002659-9.

"This edition focuses entirely on plane trigonometry, deemphasizes the use of logarithms, includes the use of a calculator, [etc.]."—Preface. **(a,p)**

PROBABILITY

Brownian Motion and Classical Potential Theory. Sidney C. Port, Charles J. Stone. San Diego, CA: Academic Press, 1978. xii + 236pp. $69.00. ISBN 0-12-561850-6.

A clearly written development of the intimate relationship between Brownian motion and potential theory. Extensive bibliography. **(a)**

Brownian Motion and Diffusion. David Freedman. New York: Springer-Verlag, 1983. xii + 231pp. $42.00. ISBN 0-387-90805-6.

Reprint of 1971 original. Presents Brownian motion as a probability on a Borel sigma-field, emphasizing variation, reflection and invariance principles. **(a)**

Brownian Motion and Martingales in Analysis. Richard Durrett. Pacific Grove, CA: Brooks-Cole, 1984. $49.95. ISBN 0-534-03065-3.

Introduces Brownian motion and the stochastic integral and then gives many applications to proving many results in classical analysis. **(a)**

Classical Potential Theory and Its Probabilistic Counterpart. J. L. Doob. Grundlehren der mathematische Wissenschaften;262. New York: Springer-Verlag, 1984. xxv + 846pp. $98.00. ISBN 0-387-90881-1.

Develops the correspondence between potential theory and probability theory. **(a)**

A Course in Probability Theory. Second Edition. Kai Lai Chung. Probability and Mathematical Statistics;21. San Diego, CA: Academic Press, 1974. xii + 365pp. $54.50. ISBN 0-12-174650-X.

Mathematical treatment presuming elementary real variables and some measure theory. **(a)**

The Craft of Probabilistic Modelling: A Collection of Personal Accounts. Ed: J. Gani. New York: Springer-Verlag, 1986. xiv + 313pp. $49.00. ISBN 0-387-96277-8.

Nineteen essays by pioneers in the art of mathematical modeling based on probabilistic methods. **(a)**

Dependence in Probability and Statistics: A Survey of Recent Results. Ed: Ernst Eberlein, Murad S. Taqqu. Progress in Probability and Statistics;11. Boston: Birkhäuser, 1986. xi + 473pp. $46.75. ISBN 0-8176-3323-5.

Twenty-three papers from a 1985 conference concerned with limiting results, assuming various dependence structures. **(a)**

The Doctrine of Chances; or, A Method of Calculating the Probabilities of Events in Play. Third Edition. Abraham de Moivre. New York: Chelsea, 1967. xi + 368pp. $22.50. ISBN 0-8284-0200-0.

"Fuller, clearer, and more correct" reprint of the 1756 original. Includes biography of de Moivre. **(a)**

Elementary Probability Theory with Stochastic Processes. Third Edition. Kai Lai Chung. New York: Springer-Verlag, 1979. xvi + 325pp. $39.95. ISBN 0-387-90362-3.

New edition includes minor improvements as well as photographs of important probabilists. **(a,p)**

Ergodic Theorems. Ulrich Krengel. Studies in Mathematics;6. Hawthorne, NY: Walter de Gruyter, 1985. viii + 357pp. $69.95. ISBN 3-11-008478-3.

Topics covered include classical ergodic theorems, subadditive processes, etc., that can be read independently of each other. (**a**)

Essai d'Analyse sur les Jeux de Hazard. Third Edition. Pierre Remond de Montmort. New York: Chelsea, 1980. xlii + 414pp. $39.50. ISBN 0-8284-0307-4.

Reprint of 1713 edition, which analyzed games of chance popular in 18th century Europe plus Montmort's correspondence with Nicolas Bernoulli. (**a**)

Foundations of Probability. Alfred Renyi. Oakland, CA: Holden-Day, 1970. xvi + 366pp. $44.00. ISBN 0-8162-7114-3.

Very interesting and highly original presentation of modern probability theory. (**a**)

Foundations of the Theory of Probability. Second English Edition. A. N. Kolmogorov. Transl: Nathan Morrison. New York: Chelsea, 1956. vii + 84pp. $9.95. ISBN 0-8284-0023-7.

English translation of 1933 German original, *Grundbegriffe der Wahrscheinlichkeitsrechnung*. Added bibliography by A. T. Bharucha-Reid. (**a,p**)

Introduction to Probability. J. Laurie Snell. New York: McGraw-Hill, 1988. xix + 517pp. $39.95. ISBN 0-685-13441-5.

Several features distinguish this text: its extensive use of compute programs, its inclusion of interesting historical remarks, andits presentation of discrete and continuous probability in a separate and parallel manner. (**a,p,s**)

Introduction to Probability Theory. Kiyosi Ito. New York: Cambridge University Press, 1984. ix + 213pp. $49.50;$18.95. ISBN 0-521-26418-9;0-521-26960-1.

This book contains some of the material that constitutes a standard first-year graduate course in probability. (**p**)

An Introduction to Probability Theory. P. A. P. Moran. New York: Oxford University Press, 1984. 542pp. $35.00. ISBN 0-19-853242-3.

Corrected reprint of 1968 original, a measure-theoretic probability theory text. (**p**)

Introduction to Probability Models. Fourth Edition. Sheldon M. Ross. San Diego, CA: Academic Press, 1989. xiv + 544pp. $44.50. ISBN 0-12-598464-2.

This edition adds material on Markov chains, minimal theory, and simulation. (**a**)

An Introduction to Probability Theory and Its Applications. William Feller. New York: Wiley, 1968-1971. *Vol. I.* Third Edition. xviii + 509pp. $74.95. ISBN 0-471-25768-7. *Vol. II.* Second Edition. xxiv + 669pp. $74.95. ISBN 0-471-25709-5.

While these latest editions of this classic improve the treatment of many topics, misprints and other errors continue. (**a**)

Lady Luck: The Theory of Probability. Warren Weaver. New York: Dover, 1982. 392pp. $6.95. ISBN 0-486-24342-7.

Reprint of 1963 classic, a witty and non-technical introduction to probability. (**p**)

Lectures from Markov Processes to Brownian Motion. Kai Lai Chung. Grundlehren der mathematischen Wissenschaften;249. New York: Springer-Verlag, 1982. viii + 239pp. $54.00. ISBN 0-387-90618-5.

Impressive work, beginning with Markov processes and extended through martingales, Feller and Hunt processes to Brownian motion, ending with areas of future study. **(a)**

Letters on Probability. Alfred Renyi. Transl: Laszlo Vekerdi. Detroit, MI: Wayne State University Press, 1972. v + 86pp. $18.95. ISBN 0-8143-1465-1.

Four fictitious letters from Pascal to Fermat elaborate the philosophical foundations of probability. This is a very imaginative little book. **(a)**

Markov Chains. David Freedman. New York: Springer-Verlag, 1983. xiv + 382pp. $49.80. ISBN 0-387-90808-0.

Reprint of 1971 edition. Contains numerous examples, but no exercises. **(a)**

Markov Random Fields and Their Applications. Ross Kindermann, J. Laurie Snell. Contemporary Mathematics;1. Providence, RI: American Mathematical Society, 1980. ix + 142pp. $17.80. ISBN 0-8218-5001-6.

Introductory survey of the basic theory is followed by applications to economics and voting behavior. **(a)**

Martingale Limit Theory and Its Application. P. Hall, C. C. Heyde. San Diego, CA: Academic Press, 1980. xii + 308pp. $69.00. ISBN 0-12-319350-8.

Authors use martingale limit arguments to investigate generalizations of the law of large numbers, the central limit theorem and the law of the iterated logarithm. The authors' intriguing thesis is that martingale limit theorems should have a central position in probability theory. **(a)**

Mathematics of Kalman-Bucy Filtering. Second Edition. P. A. Ruymgaart, T. T. Soong. Springer Series in Information Sciences;14. New York: Springer-Verlag, 1988. xii + 170pp. $39.00. ISBN 0-387-18781-2.

Excellent exposition covers difference equations and processes, with mean-square calculus in Hilbert spaces. **(a,s)**

Monte Carlo Methods. Volume I: Basics. Malvin H. Kalos, Paula A. Whitlock. New York: Wiley, 1986. $39.95. ISBN 0-471-89839-2.

A unified treatment, focusing on random walks and variance reduction techniques. **(a,s)**

Paradoxes in Probability Theory and Mathematical Statistics. Gabor J. Szekely. Norwell, MA: Kluwer Academic, 1986. xii + 250pp. $89.00. ISBN 90-277-1899-7.

A fascinating exposition of paradoxical results in probability, statistics, random processes and foundations of probability, with historical detail on the origin and resolution of each paradox. **(a)**

Probabilities and Potential. Claude Dellacherie, Paul-Andre Meyer. North-Holland Mathematics Studies;29. New York: Elsevier, 1978. 189pp. $64.00. ISBN 0-7204-0701-X.

This volume contains little enough probability and no potential theory whatsover." However, the book contains those preliminaries on measure and stochastic processes that will make a discussion of the title topics possible later. Extensive bibliography. **(a)**

Probability: An Introduction. Samuel Goldberg. New York: Dover, 1986. xiv + 322pp. $8.95. ISBN 0-486-65252-1.

Reprint of 1960 original. **(a)**

Probability and Statistics for Engineers and Scientists. Fourth Edition. Ronald E. Walpole, Raymond H. Myers. New York: Macmillan, 1989. xiv + 765pp. $45.00. ISBN 0-02-424210-1.

This edition retains the fundamental structure of basic probability and statistical inference, but provides the student with a view of modern data analysis techniques: a new chapter on statistical quality control plus new material on a variety of topics. **(a)**

Probability and Stochastic Processes: With a View Toward Applications. Second Edition. Leo Breiman. Redwood City, CA: Scientific Press, 1986. xii + 324pp. $40.00. ISBN 0-89426-076-6.

This edition is primarily a reissue of the well-received First edition. The many interesting and thought-provoking exercises throughout form an essential part of book. **(a)**

Probability Inequalities in Multivariate Distributions. Y. L. Tong. San Diego, CA: Academic Press, 1980. xiii + 239pp. $57.00. ISBN 0-12-69450-6.

Comprehensive, systematic, and self-contained treatment, including applications. Extensive bibliography. **(a)**

Probability, Statistics and Truth. Second Revised English Edition. Richard von Mises. New York: Dover, 1981. xii + 244pp. $5.95. ISBN 0-486-24214-5.

Reprint of 1957 English edition, a pioneering classic in the foundations of probability theory. **(a)**

Probability Theory and Elements of Measure Theory. Heinz Bauer. Transl: Lisa Rosenblatt. San Diego, CA: Academic Press, 1972. xiii + 427pp. $149.00. ISBN 0-12-082820-0.

Very attractive book for a one year graduate course which truly integrates measure and probability. Expensive. **(a)**

Random Walks and Electric Networks. Peter G. Doyle, J. Laurie Snell. Carus Mathematical Monograph;22. Washington, DC: Mathematical Association of America, 1984. xiii + 159pp. $28.95. ISBN 0-88385-024-9.

A delightful excursion among various interpretations of random walks all leading to several proofs of Polya's 1921 theorem. **(a)**

Theory and Applications of Stochastic Differential Equations. Zeev Schuss. New York: Wiley, 1980. xiii + 321pp. $59.95. ISBN 0-471-04394-X.

Points of origin, basic theory and wide range of applications of stochastic differential equations which contain white noise are presented here. Extensive references. **(a,s)**

Theory of Probability: A Critical Introductory Treatment. Bruno de Finetti. Transl: Antonio Machi, Adrian Smith. New York: Wiley, 1990. *Vol. 1.* xix + 300pp. $29.95. ISBN 0-471-92611-6. *Vol. 2.* xviii + 375pp. $29.955. ISBN 0-471-92612-4.

These volumes make available to English-language readers important writings of this pioneer of subjective probability. **(a)**

The Theory of Probability: And the Elements of Statistics. Fifth Edition. B. V. Gnedenko. Transl: B. D. Seckler. New York: Chelsea, 1989. 529pp. $23.95. ISBN 0-8284-2132-1.

Translation of Fourth (1965) edition of *Kurs Teoriĭ Veroi͡atnosteĭ*. **(a)**

PROGRAMMING

Applied Mathematical Programming. Stephen P. Bradley, Arnoldo C. Hax, Thomas L. Magnanti. Reading, MA: Addison-Wesley, 1977. xvi + 716pp + 19pp. $41.56. ISBN 0-201-00464-X.

"This book presents in a unifying manner the methodology of mathematical programming, an overview of mathematical programming in practice, and a series of selected projects reporting on actual implementation of mathematical programming models in business and the public sector."—Preface. **(a,s)**

Elementary Linear Programming with Applications. International Edition. Bernard Kolman, Robert E. Beck. San Diego, CA: Academic Press, 1980. xiii + 399pp. $49.00. ISBN 0-12-417860-X.

Clearly written text on linear programming methods. **(a,p)**

Elements of Programming Style. Second Edition. Brian W. Kernighan, P. J. Plauger. New York: McGraw-Hill, 1978. xii + 168pp. $29.65 paperbound. ISBN 0-07-034207-5.

An extensive revision of the 1974 classic, the programmer's "Strunk and White." Pithy rules. **(a,p)**

Foundations of Logic Programming. J. W. Lloyd. New York: Springer-Verlag, 1984. x + 124pp. $59.00. ISBN 0-387-13299-6.

The mathematical theory for using logic as a programming language. **(a)**

Introduction to Logic Programming. Christopher John Hogger. APIC Studies in Data Processing;21. San Diego, CA: Academic Press, 1984. xiii + 278pp. $85.00. ISBN 0-12-352090-8.

Comprehensive introduction to the use of symbolic logic as a programming language. **(a)**

Introduction to Mathematical Programming: Applications and Algorithms. Wayne L. Winston. Boston: PWS-Kent, 1991. xi + 794pp. ISBN 0-534-92520-0.

This text is aimed at undergraduate majors in business and operations research as well as MBA students. As such, it is mathematically sound, but not demanding in terms of preparation. **(p)**

Introduction to the Theory of Programming Languages. Bertrand Meyer. New York: Prentice-Hall, 1990. xvi + 447pp. $30.00. ISBN 0-13-498510-9.

A modern, formal treatment of mathematical models of programming languages. **(a)**

Linear Programming. Vasek Chvatal. New York: W. H. Freeman, 1983. xiii + 478pp. $30.95;$29.95. ISBN 0-7167-1195-8, 0-7167-1587-2.

Innovative, attractive introduction to the subject with a sound presentation of basic theory, applications, and advanced techniques. Excellent historical notes and bibliography. **(a,p)**

Linear Programming and Applications, A Course Text. International Student Edition. Will McLewin. London: Input-Output, 1982. (Distributed by State Mutual Book and Periodical, New York) xiv + 216pp. $33.00. ISBN 0-904870-12-X.

Develops the simplex method, duality, and revised simplex methods using both tableau and matrix operations approaches, then covers various applications including Shor-Khachian ellipsoid method, network flows, and game theory. **(a)**

Linear Programming and Extensions. G. B. Dantzig. Princeton, NJ: Princeton University Press, 1963. xvi + 625pp. $75.00. ISBN 0-691-08000-3.

This book, prepared as a Rand Corporation research report for the U.S. Air Force, has remained a basic work on linear programming. **(a)**

The Mathematics of Nonlinear Programming. Anthony L. Peressini, Francis E. Sullivan, J. J. Uhl, Jr. New York: Springer-Verlag, 1988. x + 273pp. $42.00. ISBN 0-387-96614-5.

This text focuses on geometry, penalty functions, Lagrange multipliers, and the simplex method in constrained optimization problems. **(a)**

The Programmer's Book of Rules. George Ledin, Jr., Victor Ledin. New York: Van Nostrand Reinhold, 1979. 248pp. $19.95 paperbound. ISBN 0-317-64257-X.

Two-hundred-fifty do's and don'ts, with brief explanations and examples and references to the literature of programming methodology. **(p)**

Programming Principles, An Introduction. John Motil. Dubuque, IA: William C. Brown, 1984. 461pp. $50.75. ISBN 0-697-06845-5.

An unusual introductory text with structured algorithm development primary and programming separate and secondary. **(a)**

Programming Languages: A Grand Tour. Third Edition. Ed: Ellis Horowitz. New York: W. H. Freeman, 1987. ix + 512pp. $43.95. ISBN 0-88175-142-1.

Thirty source articles, many of which are classics, on programming language design issues. This updated edition is a broader and less bulky anthology. **(a)**

Recursion Via Pascal. J. S. Rohl. Computer Science Texts;19. New York: Cambridge University Press, 1984. x + 192pp. $27.95. ISBN 0-521-26934-2.

A carefully organized study of recursion in procedural (rather than functional) programming. **(a)**

Theory of Linear and Integer Programming. Alexander Schrijver. New York: Wiley, 1986. xi + 471pp. $94.95. ISBN 0-471-90854-1.

A highly theoretical treatment of linear and integer programming suitable for graduate work, which discusses complexity theory, lattices and linear Diophantine equations, etc. **(a)**

REAL ANALYSIS

Advanced Calculus: An Introduction to Analysis. Third Edition. Watson Fulks. New York: Wiley, 1978. xiv + 731pp. $69.95. ISBN 0-471-02195-4.

Revisions include separation of the continuity and differentiation of functions of one real variable and modernization of the definition of the derivative of a vector function of a vector variable. **(a)**

Brownian Motion. T. Hida. Transl: T. P. Speed. Applications of Mathematics;11. New York: Springer-Verlag, 1980. xvi + 325pp. $72.00. ISBN 0-387-90439-5.

This interesting book is the result of the author's investigation of the development of Wiener measure on function spaces. **(a)**

Elementary Analysis: The Theory of Calculus. Kenneth A. Ross. New York: Springer-Verlag, 1980. viii + 264pp. $26.40. ISBN 0-387-90459-X.

Bridges the gap between elementary calculus and an in-depth-study of real analysis. **(a,p)**

The Elements of Real Analysis. Second Edition. Robert G. Bartle. xv + 480pp. New York: Wiley, 1976. $64.95. ISBN 0-471-05464-X.

This edition greatly expands the treatment of the differentiation and integration of functions of several variables while consolidating the discussion of functions of one variable. **(a)**

The Fourier Integral and Certain of Its Applications. Norbert Wiener. New York: Cambridge University Press, 1988. xvii + 201pp. $17.95. ISBN 0-521-35884-1.

Third printing of a true classic, first published in 1933, in which Wiener gave his own unique view centered on Tauberian theorems which served to introduce the idea of Banach algebras and abstract harmonic analysis. **(a,s)**

The Generalized Riemann Integral. Robert M. McLeod. Carus Mathematical Monograph;20. Washington, DC: Mathematical Association of America, 1980. xiii + 275pp. $24.95. ISBN 0-88385-021-4.

Continuous building of ideas from what is known to the monotone and dominated convergence theorems and other results commonly reached only through the Lebesgue theory. Proofs are there as well as beautiful organization to get the main ideas across. **(a)**

Intermediate Real Analysis. Emanuel Fischer. New York: Springer-Verlag, 1983. xiv + 770pp. $42.00. ISBN 0-387-90721-1.

Extraordinarily complete treatment of the first course in real analysis. **(a)**

An Introduction to Nonstandard Real Analysis. Albert E. Hurd, Peter A. Loeb. Pure and Applied Mathematics;118. San Diego, CA: Academic Press, 1985. xii + 232pp. $44.95. ISBN 0-12-362440-1.

This text introduces nonstandard analysis and its applications in calculus, topology, and integration theory. **(a)**

Introduction to Real Analysis. Robert G. Bartle, Donald R. Sherbert. New York: Wiley, 1982. xii + 370pp. $56.95. ISBN 0-471-05944-7.

Appropriate for an intermediate course, this text covers limits, continuity, the Riemann integral, convergence of functions, some topological ideas, all in one real-variable, all with examples. **(a)**

Introduction to the Theory of Fourier Integrals. Third Edition. E. C. Titchmarsh. New York: Chelsea, 1986. x + 394pp. $23.95. ISBN 0-8284-0324-4.

Essentially unaltered reprint of 1948 edition of this classic introduction on the theory of Fourier integrals. **(a)**

Introductory Real Analysis. Revised English Edition. A. N. Kolmogorov, S. V. Fomin. Transl/Ed: Richard A. Silverman. New York: Chelsea, 1975. xii + 403pp. $8.95. ISBN 0-8284-0023-7.

Corrected reprinting of 1970 English edition of *Elementy Funktsiĭ i Funkt͡sional'nogo Analiza.* **(a,p)**

A Primer of Real Functions. Third Edition. Ralph P. Boas, Jr. Carus Mathematical Monographs;13. Washington, DC: Mathematical Association of America, 1981. xi + 232pp. $23.00. ISBN 0-088385-022-2.

Reprint of the 1972 classic on theory of sets and real functions, with some additional references. **(a)**

Probability and Measure. Second Edition. Patrick Billingsly. New York: Wiley, 1986. xii + 622pp. $64.95. ISBN 0-471-80478-9.

A revision of the 1979 edition and expanded to include a central limit theorem for martingales and more on the asymptotic behavior of Markov chains. **(a)**

Real Analysis. Third Edition. H. L. Royden. New York: Macmillan, 1988. xvii + 444pp. $48.00. ISBN 0-02-404151-3.

Author corrects his previous treatment of Baire and Borel measures in locally compact spaces and adds a new chapter on invariant measures. **(a)**

Real Analysis and Probability. Richard M. Dudley. Pacific Grove, CA: Brooks-Cole, 1989. xi + 436pp. $59.95. ISBN 0-534-10050-3.

Assumes analysis at level of Rudin's classic, while probability is short and intense. Exercises and bibliographies follow each chapter. **(a)**

Real Analysis: Modern Techniques and Their Applications. Gerald B. Folland. New York: Wiley, 1984. xiv + 350pp. $61.95. ISBN 0-471-80958-6.

Fairly self-contained treatment, from measure and integration to Fourier analysis, probability and integration on groups and manifolds, with chapter notes, exercises, and references. Not your ordinary textbook in real analysis, but a fascinating journey through various fields of analysis. **(a)**

Real Variables. Alberto Torchinsky. Reading, MA: Addison-Wesley, 1988. xii + 403pp. $43.25. ISBN 0-201-15675-X.

Covers standard topics on Lebesgue measure and integration (in Rn) as well as some of the basics of functional analysis. **(a)**

Regular Variation. N. H. Bingham, C. M. Goldie, J. L. Teugels. Encyclopedia of Mathematics and Its Applications;27. New York: Cambridge University Press, 1989. xix + 491pp. $35.16 paperbound. ISBN 0-521-37943-1.

Comprehensive account of the theory and applications of regular variation. Extensive bibliography. **(a)**

Toeplitz Forms and Their Applications. Second Edition. Ulf Grenander, Gabor Szegö. New York: Chelsea, 1984. ix + 245pp. $18.95. ISBN 0-8284-0321-X.

Reprint of 1955 edition. **(a)**

SOFTWARE SYSTEMS

A Beginner's Book of TeX. Raymond Seroul, Silvio Levy, Dominique Foata. New York: Springer-Verlag, 1991. xii + 282pp. $29.95. ISBN 0-387-97562-4.

An introductory book to help TeX users get quickly started on their own. It contains a careful explanation of all fundamental concepts and commands plus a wealth of commented examples and "tricks," based on the authors long experience with TeX. Last third of the book is devoted to a dictionary-index, summarizing all the material in the text and going into greater detail in many areas. **(a,p)**

Computer Modern Typefaces. Donald E. Knuth. Reading, MA: Addison-Wesley, 1986. xv + 588pp. $33.56. ISBN 0-201-13446-2.

"Most of the pages are filled with somewhat cryptic code in the Metafont language, accompanied by annotated diagrams that illustrate what the Metafont system might draw when presented with the given code. Samples of text show the Computer Modern typefaces in use."—Preface. **(a)**

A Gentle Introduction to TeX: a Manual for Self Study. Michael Doob. (TeXniques;12) Providence, RI: TeX User's Group, 1984. 88pp. $15.00. No ISBN.

This manual starts from the very beginning and moves toward more complicated situations. No previous knowledge of TeX is assumed. Exercises. **(a,p)**

Guide to Expert Systems. Donald A. Waterman. Reading, MA: Addison-Wesley, 1985. xviii + 419pp. $28.76. ISBN 0-201-08313-2.

Written for readers without a background in artificial intelligence. **(p)**

The Joy of TeX: A Gourmet Guide to Typesetting with the AMS-TeX Macro Package. Second Edition. M. D. Spivak. Providence, RI: American Mathematical Society, 1990. xxii + 309pp. $40.00. ISBN 0-8218-2997-1.

The definitive guide to AMS-TeX, the typesetting system designed by the AMS as the standard for their books and journals. Probably a better introduction to TeX for novices than *The TeXbook* (see below) since it does not attempt to explain everything. This edition reflects the changes introduced in Version 2.1 of the AMS-TeX macro package. **(a,p,s)**

LaTeX, A Document Preparation System. Leslie Lamport. Reading, MA: Addison-Wesley, 1986. xiv + 242pp. $23.96 paperbound. ISBN 0-201-15790-X.

This is a definitive user's guide and reference manual to LaTeX (a macro package for TeX that adds commands for automatic numbering of sections, theorems and references). It introduces readers to LaTeX, shows them how to get started, and gradually introduces them to more advanced techniques. **(a)**

METAFONT: The Program. Donald E. Knuth. Reading, MA: Addison-Wesley, 1986. xv + 560pp. $33.56. ISBN 0-201-13438-1.

"The algorithms of Metafont span a wide range of ideas, from random number generation and trigonometry to cobinatorial optimization; therefore this program epitomizes many kinds of software all at once."—Preface. **(a)**

The METAFONTbook. Donald E. Knuth. Reading, MA: Addison-Wesley, 1986. xi + 361pp. $29.56 paperbound. ISBN 0-201-13444-6.

This book enables readers, with only minimal computer science or word processing experience, to master the basic as well as the more advanced features of METAFONT programming. **(a)**

TeX by Example: A Beginner's Guide. Arvind Borde. San Diego, CA: Academic Press, 1992. xiv + 169pp. $19.95. ISBN 0-12-117650-9.

Bibliographical references. **(a,p)**

TeX by Topic: A TeXnician's Reference. Victor Eijkhout. Reading, MA: Addison-Wesley, 1992. vii + 307pp. $26.95 softbound. ISBN 0-201-56882-9.

Not an introductory volumem, this book is a systematic, complete reference manual. The universe of TeX is presented in about 40 separate chapters, beginning with a list of control sequences, then a treatment of theory, and concluding with remarks and examples. **(a)**

TeX for the Beginner. Wynter Snow. Reading, MA: Addison-Wesley, 1992. xii + 377 + 23pp. $26.95. ISBN 0-201-54799-6.

This book is a practical, hands-on tutorial that shows the reader how to use TeX. It also provides a sound conceptual bases for learning more about TeX. LaTeX notes describe what you do differently from TeX. **(a)**

TeX for the Impatient. Paul W. Abrahams, Karl Berry, Kathryn A. Hargreaves. Reading, MA: Addison-Wesley, 1990. xix + 357pp. $20.76. ISBN 0-201-51375-7.

A very useful supplement (but not a substitute) to The TeXbook. Features include a variety of brief illustrative examples followed by an alphabetic glossary of TeX concepts then by discussions of the various commands. **(a)**

TeX: The Program. Donald E. Knuth. Reading, MA: Addison-Wesley, 1986. xv + 594pp. $45.25. ISBN 0-201-13437-3.

The complete text of the TeX program, written in WEB. The program consists of 1377 consecutively numbered single topic sections, each of which contains commentary. **(a)**

The TeXbook. Nineteenth Printing, Revised. Donald E. Knuth. Reading, MA: Addison-Wesley, 1991. ix + 483pp. $29.95. ISBN 0-201-13448-9.

Latest printing of the witty TeX bible. It is not intended for the person who wants to learn the system quickly as it provides layer upon layer of cleverly-illustrated examples of TeX behavior. **(a)**

STATISTICS

Advanced Theory of Statistics. Fourth Edition. Maurice Kendall, Alan Stuart. New York: Hafner, 1977-1983. *Volume 1: Distribution Theory.* xii + 472pp. o.p. ISBN 0-02-847630-1. *Volume 2: Inference and Relationship.* x + 748pp. o.p. ISBN 0-02-847820-7. *Volume 3: Design and Analysis, and Time-Series.* x + 780pp. $95.00. ISBN 0-02-847640-9.

"There have been major revisions of the text of this edition and many minor improvements and extensions, including new exercises, based on recent research" —Preface. (Fifth edition entitled *Kendall's Advanced Theory of Statistics* (see p. 291)) **(a)**

Akaike Information Criterion Statistics. Y. Sakamoto, M. Ishiguro, G. Kitagawa. Norwell, MA: Kluwer Academic, 1986. xix + 290pp. $142.00. ISBN 90-277-2253-6.

Translation of 1933 original: a unified new approach to elementary statistical analysis in which the goodness of an assumed model is measured by the Akaike Information Criterion. Expensive. (**a**)

Analysis of Experiments with Missing Data. Yadolah Dodge. New York: Wiley, 1985. xvii + 499pp. $59.95. ISBN 0-471-88736-6.

Presents recently developed theories and techniques for analyzing designed experiments with missing data. (**a**)

Analysis of Messy Data. George A. Milliken, Dallas E. Johnson. New York: Van Nostrand Reinhold, 1984-1989. *Volume 1: Designed Experiments*. xiv + 473pp. $45.00. ISBN 0-534-02713-X. *Volume 2: Nonreplicated Experiments*. viii + 199pp. $58.95. ISBN 0-442-24408-8.

These volumes cover each topic from a practical viewpoint with formulas included for those readers who would like to program the techniques on their personal computers. Several tables not available in many other books appear here. Appendices, references, index. (**a,s**)

Analysis of Ordinal Categorical Data. Alan Agresti. New York: Wiley, 1984. ix + 287pp. $57.95. ISBN 0-471-89055-3.

Treats methods of analyzing cross-classification tables having ordered, rather than just nominal, categories for at least one of the classifications. Good references. (**a**)

Analysis of Survival Data. D. R. Cox, D. Oakes. New York: Routledge Chapman & Hall, 1984. ix + 201pp. $35.00. ISBN 0-412-24490-X.

Thorough introduction to the analysis of survival data, including parametric and nonparametric analysis techniques for both univariate data sets and data sets with explanatory variables. (**a**)

Applications, Basics, and Computing of Exploratory Data Analysis. Paul F. Vellemen, David C. Hoaglin. Boston, MA: Duxbury Press, 1981. xxi + 354pp. $18.50. ISBN 0-87872-273-4.

Presents nine techniques of exploratory data analysis, illustrates their applications to real data, and provides BASIC and Fortran programs for their implementation. (**a**)

Applied Linear Regression. Sanford Weisberg. New York: Wiley, 1980. xii + 283pp. $44.95. ISBN 0-471-04419-9.

Well-written and concise presentation of many of the techniques in regression analysis, illustrated with real data. Good set of references. (*a,p*)

Applied Multivariate Analysis: Using Bayesian and Frequentist Methods of Inference. Second Edition. S. James Press. Melbourne, FL: Krieger, 1982. xix + 600pp. $39.50. ISBN 0-88275-976-0.

Revision of author's 1972 text, with updated content, references, and computer program listings. (**a**)

Applied Multivariate Statistical Analysis. Second Edition. Richard A. Johnson, Dean W. Wichern. New York: Prentice-Hall, 1988. xvi + 607pp. $60.00. ISBN 0-13-041146-9.

Changes in this edition include the addition of a chapter on canonical correlation analysis and expansion of the chapter on factor analysis. **(p)**

Applied Regression Analysis: A Research Tool. John O. Rawlings. Belmont, CA: Wadsworth, 1988. xviii + 553pp. $57.95. ISBN 0-534-09246-2.

Non-theoretical treatment beginning with a review of least squares and ending with collinearity, response surface modeling, analysis of unbalanced data, etc. **(a,p)**

Approximation Theorems of Mathematical Statistics. Robert J. Serfling. New York: Wiley, 1980. xiv + 371pp. $69.95. ISBN 0-471-02403-1.

Covers a broad range of limit theorems useful in mathematical statistics. **(a)**

Asymptotic Techniques for Use in Statistics. O. E. Barndorff-Nielsen, D. R. Cox. New York: Routledge, Chapman & Hall, 1989. x + 252pp. $49.95. ISBN 0-412-31400-2.

A survey of asymptotic approximation methods for the distribution of means of random variables. **(a)**

The Asymptotic Theory of Extreme Order Statistics. Second Edition. Janos Galambos. Melbourne, FL: Robert E. Krieger, 1987. xv + 414pp. $49.50. ISBN 0-89874-957-3.

Rigorous presentation covering all known asymptotic models. Extensive bibliography. Each chapter has a literature survey. **(a)**

Asymptotic Theory of Statistical Inference. B. L. S. Prakasa Rao. New York: Wiley, 1987. xiv + 438pp. $64.95. ISBN 0-471-84335-0.

Explores the relationship between recent advances in probability theory and stochastic processes and the asymptotic theory of statistical inference. **(a)**

Asymptotics in Statistics: Some Basic Concepts. Lucien Le Cam, Grace Lo Yang. New York: Springer-Verlag, 1990. viii + 180pp. $29.00. ISBN 0-387-97372-9.

Updated rewriting of a set of lectures, in French, which LeCam gave in 1968 to include a discussion of the locally asymptotically mixed normal case. **(a)**

Bandit Problems: Sequential Allocation of Experiments. Donald A. Berry, Bert Fristedt. New York: Routledge, Chapman & Hall, 1985. viii + 275pp. $29.50. ISBN 0-412-24810-7.

Comprehensive treatment of the problem in statistical decision theory of making sequential selections from two or more stochastic processes in order to maximize an expected payoff. **(a)**

Basic Concepts of Probability and Statistics. Second Edition. J. L. Hodges, Jr., E. L. Lehmann. Oakland, CA: Holden-Day, 1970. 454pp. $36.95. ISBN 0-8162-4004-3.

Introduces modern theories of statistics in a simple context. **(p)**

Bayesian Reliability Analysis. Harry F. Martz, Ray A. Waller. Melbourne, FL: R. E. Krieger, 1991. xix + 745pp. $64.95. ISBN 0-89464-395-9.

Reprint of 1982 original, a comprehensive presentation of Bayesian methods for reliability analysis. References. **(a)**

Bayesian Statistics 2. Ed: J. M. Bernardo et al. New York: Elsevier, 1985. 778pp. $143.75. ISBN 0-444-87746-0.

Proceedings of the Second Valencia International Meeting on Bayesian Statistics, held in Spain in September 1983. Expensive. **(a)**

Bayesian Statistics 3. Ed: J. M. Bernardo et al. New York: Oxford University Press, 1988. xi + 805pp. $125.00. ISBN 0-19-852220-7.

Proceedings of the Third Valencia International Meeting on Bayesian Statistics, held in Spain in June 1987. Expensive. **(a)**

Biometry: The Principles and Practice of Statistics in Biological Research. Second Edition. Robert R. Sokal, F. James Rohlf. New York: W. H. Freeman, 1981. xviii + 859pp. $44.95. ISBN 0-7167-1254-7.

A non-mathematical, non-theoretical treatment with biological examples of elementary statistics, analysis of variance, etc. **(p)**

Categorical Data Analysis. Alan Agresti. New York: Wiley, 1990. xv + 558pp. $52.95. ISBN 0-471-85301-1.

This book is divided into four main parts, each with very good problem sets. Extensive references. **(a)**

A Celebration of Statistics: The ISI Centenary Volume. Ed: Anthony C. Atkinson, Stephen E. Fienberg. New York: Springer-Verlag, 1985. xv + 606pp. $52.00. ISBN 0-387-96111-9.

A collection of 25 refereed papers on a variety of topics, including prediction and antropy, etc. **(a)**

Classification and Regression Trees. Leo Breiman et al. Belmont, CA: Wadsworth, 1984. x + 358pp. $29.95. ISBN 0-534-98053-8.

Expository discussion and theoretical framework of the use of binary trees as a nonparametric data analysis method. Good set of references. **(a)**

Comparative Statistical Inference. Second Edition. Vic Barnett. New York: Wiley, 1982. xvi + 325pp. $74.95. ISBN 0-471-10076-5.

Thoroughly revised, the purpose is still "to develop the various principles and methods of inference and decision making." **(a,p)**

Counting for Something: Statistical Principles and Personalities. William S. Peters. New York: Springer-Verlag, 1986. xviii + 275pp. $39.00. ISBN 0-387-96364-2.

An elementary introduction to descriptive statistics along with a thoughtful discussion of the history of probability and statistics. **(p)**

Discrimination and Classification. D. J. Hand. New York: Wiley, 1981. x + 218pp. $89.95. ISBN 0-471-28048-8.

Treats statistical discriminant analysis, pattern recognition, and cluster analysis. **(a)**

Drawing Inferences from Self-Selected Samples. Ed: Howard Wainer. New York: Springer-Verlag, 1986. xiii + 301pp. $27.00. ISBN 0-387-96379-0.

Four papers and discussion from a conference on the topic. **(a)**

Dynamic Graphics for Statistics. Ed: William S. Cleveland, Marylyn E. McGill. Belmont, CA: Wadsworth, 1988. xii + 424pp. $52.95. ISBN 0-534-09144-X.

A collection of 16 papers on the analysis of multidimensional data using dynamic graphical methods. **(a)**

Elements of Statistical Analysis. Hans W. Gottinger. Hawthorne, NY: Walter de Gruyter, 1980. 244pp. $35.25. ISBN 3-11-007169-X.

This text covers the foundations of probability, Bayesian estimation, etc. with most topics requiring no calculus. **(p)**

Elements of Statistical Computing. Volume 1: Statistical Computing. Ronald A. Thisted. New York: Chapman & Hall, 1988. xx + 427pp. $45.00. ISBN 0-412-01371-1.

First 6 chapters of a projected 24 chapter treatise describing the current state-of-the-art in statistical computing. **(a,s)**

Estimation of Variance Components and Applications. C. R. Rao, J. Kleffe. North-Holland Series in Statistics and Probability;3. New York: Elsevier, 1988. xiii + 370pp. $82.00. ISBN 0-444-70023-4.

Covers essentially all known results on estimation of variance components through MINQE, IMINQE, and MML methods. References. **(a)**

Experimental Designs. Second Edition. William G. Cochran, Gertrude M. Cox. New York: Wiley, 1957. 611pp. $74.50. ISBN 0-471-16203-5.

A classic. Includes chapters on factorial experiments and incomplete block designs. **(a,p)**

Exploring Data Tables, Trends, and Shapes. Ed: David C. Hoaglin, Frederick Mosteller, John W. Tukey. New York: Wiley, 1985. xxii + 527pp. $57.95. ISBN 0-471-09776-4.

An important presentation of advanced robust and exploratory techniques. **(a)**

The Fascination of Statistics. Ed: Richard J. Brook et al. Popular Statistics;4. New York: Dekker, 1986. xi + 433pp. $39.75. ISBN 0-8247-7329-2.

Brief descriptions for the lay audience of the use of statistical methods in real applications. **(p)**

Finite Mixture Distributions. B. S. Everitt, D. J. Hand. New York: Routledge, Chapman & Hall, 1981. xi + 143pp. $42.50. ISBN 0-412-22420-8.

Deals with the problem of parameter estimation, primarily by maximum likelihood or the method of moments, when the number of components is known, with some discussion of how to estimate this number. Good set of references. **(a)**

Fitting Equations to Data: Computer Analysis of Multifactor Data. Second Edition. Cuthbert Daniel, Fred S. Wood. New York: Wiley, 1980. xviii + 458pp. $69.95. ISBN 0-471-05370-8.

Changes in this revision include extended use of "component and component-plus-residuals plots" as tools in the point by point study of the data. **(a)**

Generalized Linear Models. Second Edition. P. McCullagh, J. A. Nelder. New York: Chapman & Hall, 1989. xiii + 261pp. $49.95. ISBN 0-412-31760-5.

A unified study of a class of models that generalizes classical linear models. This edition has revised and expanded log-linear and related models, inerse linear and related models, and model checking to name a few topics. Good set of references. **(a)**

Generalized Multivariate Analysis. Fang Kai-Tai, Zhang Yao-Ting. New York: Springer-Verlag, 1990. xi + 220pp. $59.00. ISBN 0-387-17651-9.

Summarizes the theory developed during the past 15 years of multivariate analysis based on elliptically contoured distributions. Good set of references. **(a)**

Goodness-of-Fit Statistics for Discrete Multivariate Data. Timothy R. C. Read, Noel A. C. Cressie. New York: Springer-Verlag, 1988. xi + 211pp. $44.00. ISBN 0-387-96682-X.

Thorough treatment is unified by using the authors' power-divergence family of statistics. Good bibliography. **(a)**

Graphical Exploratory Data Analysis. S. H. C. du Toit, A. G. W. Steyn, R. H. Stumpf. New York: Springer-Verlag, 1986. ix + 314pp. $29.80. ISBN 0-387-96313-8.

Survey of the best known and most widely used methods of analyzing and portraying data graphically. Includes working computer programs. **(a)**

Introduction to Mathematical Statistics. Fifth Edition. Paul G. Hoel. New York: Wiley, 1984. x + 435pp. $54.95. ISBN 0-471-89045-6.

In this edition, the chapter on sampling theory has been rewritten and many new problems have been added. **(a)**

Introduction to Mathematical Statistics. Fourth Edition. Robert V. Hogg, Allen T. Craig. New York: Macmillan, 1978. x + 438pp. $45.00. ISBN 0-02-355710-9.

This revision of a well-known text mainly changes the order of topic presentation. **(a,p)**

Introduction to Statistical Analysis. Fourth Edition. Wilfrid J. Dixon, Frank J. Massey, Jr. New York: McGraw-Hill, 1983. xv + 678pp. $43.08. ISBN 0-07-017073-8.

Changes in this edition include additional material on multiple regression, multiple comparisons, chi-square tests and their power, and expanded treatment of orthogonal contrasts. **(a,p)**

Introduction to Statistics: The Nonparametric Way. Gottfried E. Noether. New York: Springer-Verlag, 1991. xii + 414pp. $49.50. ISBN 0-387-97284-6.

This text provides an alternative to the standard pre-calculus statistics course: the classical normal-theory procedures are included after each topic has been more thoroughly discussed using a nonparametric approach. **(a,p)**

Introductory Statistics for Business and Economics. Third Edition. Thomas H. and Ronald J. Wonnacott. New York: Wiley, 1984. xxii + 746pp. $28.95. ISBN 0-471-09716-0.

New edition of a careful text that only assumes high school algebra. **(p)**

The Inverse Gaussian Distribution: Theory, Methodology, and Applications. Raj S. Chhikara, J. Leroy Folks. Statistics;95. New York: Dekker, 1989. viii + 213pp. $89.75. ISBN 0-8247-7997-5.

Summarizes known properties of the inverse Gaussian distribution and its uses in theoretical and applied statistics. Extensive bibliography. **(a)**

Kendall's Advanced Theory of Statistics. Fifth Edition. Alan Stuart, J. Keith Ord. New York: Oxford University Press, 1987- . *Volume 1: Distribution Theory.* xvi + 604pp. $95.00. ISBN 0-19-520561-8. *Volume 2: Classical Inference and Relationship.* xi + 719pp. $125.00. ISBN 0-19-520909-5.

"In this new edition the structure of the work is being re-shaped to take account of the development of new areas in the theory of statistics and the relative decline of others; indeed, the process of re-shaping began in the Fourth edition of Volume 3. The text has been completely revised, considerable amounts of new material incorporated into some parts of it, and much new detail elsewhere. In consequence, the number of text pages has increased by more than a quarter."—Preface. (Fourth Edition entitled *Advanced Theory of Statistics* (see p. 285)). Expensive. **(a,s)**

Mathematical Methods of Statistics. Harald Cramer. Princeton Mathematical Series;9. Princeton, NJ: Princeton University Press, 1966. xvi + 575pp. $65.00. ISBN 0-691-08004-6.

This work has been a standard reference and text for many years. First published in Sweden in 1945, the author presents mathematical statistics as a part of general probability theory. **(a,p)**

Mathematical Statistics. Fourth Edition. John E. Freund, Ronald E. Walpole. New York: Prentice-Hall, 1987. xv + 608pp. $59.00. ISBN 0-13-562075-9.

The major changes in this edition include introducing Jacobians in connection with functions of several random variables, improving the formulation of the hypotheses used in the analysis of contingency tables, expanding the material on multiples regression using matrix notation, and including tables for some of the most popular nonparametric tests. **(a)**

Mathematical Statistics: A Decision Theoretic Approach. Thomas S. Ferguson. San Diego, CA: Academic Press, 1967. xi + 396pp. $49.95. ISBN 0-12-253750-5.

This book is designed to make Wald's decision theory accessible to first year mathematics graduate students and mainly includes those topics which can be justified from a decision-theoretic point-of-view, with a chapter on confidence sets. **(a)**

Mathematical Statistics and Data Analysis. John A. Rice. Belmont, CA: Wadsworth, 1988. xx + 594pp. $59.95. ISBN 0-534-08247-5.
Emphasizes data analysis, real problems, and computer use. **(a,p)**

Mathematical Statistics, Basic Ideas and Selected Topics. Peter J. Bickel, Kjell A. Doksum. Oakland, CA: Holden-Day, 1977. xv + 493pp. $49.95. ISBN 0-8162-0784-4.
This excellent book covers modern theoretical statistics at the senior-graduate level. Appendix reviews basic probability theory. **(a,p)**

Measurement Error Models. Wayne A. Fuller. New York: Wiley, 1987. xxiii + 440pp. $57.95. ISBN 0-471-86187-1.
Presentation of both theory and applications of models in which the explanatory variables are measured with error. Bibliography. **(a)**

Multiple Imputation for Nonresponse in Surveys. Donald B. Rubin. New York: Wiley, 1987. xxix + 258pp. $43.95. ISBN 0-471-08705-X.
Gives a method for handling nonresponses in sample surveys. **(a)**

Multivariate Analysis. Second Edition. Maurice Kendall. New York: Macmillan, 1980. iii + 210pp. $39.95. ISBN 0-19-520573-1.
An interesting mix of theory and practice with an emphasis on what to do. It serves the general reader particularly well. **(p)**

Multivariate Descriptive Statistical Analysis: Correspondence Analysis and Related Techniques for Large Matrices. Ludovic Lebart, Alain Morineau, Kenneth M. Warwick. Transl: Elisabeth Moraillon Berry. New York: Wiley, 1984. xvi + 231pp. $53.95. ISBN 0-471-86743-8.
An updated and augmented translation of 1977 French original, describing statistical techniques that are particularly suited to the description and analysis of large data sets. Good references. **(a)**

Multivariate Observations. G. A. F. Seber. New York: Wiley, 1984.xx + 686pp. $79.95. ISBN 0-471-88104-X.
Presentation of both modern data-oriented techniques in multivariate analysis and classical methods supported by comments about robustness and general practical applicability. Extensive references. **(a)**

Multivariate Statistical Simulation. Mark E. Johnson. New York: Wiley, 1987. ix + 230pp. $44.95. ISBN 0-471-82290-6.
Focuses on computer generation of continuous multivariate probability distributions. Good set of references. **(a)**

News & Numbers. Victor Cohn. Ames, IA: Iowa State University Press, 1989. xii + 178pp. $9.95. ISBN 0-8138-1442-1.
A journalist's guide to statistics, emphasizing those of health, environment and science. **(p)**

Nonlinear Regression. G. A. F. Seber, C. J. Wild. New York: Wiley, 1989. xx + 768pp. $64.95. ISBN 0-471-61760-1.
Extensive, wide-ranging, predominantly theoretical treatment of nonlinear model fitting. Good references. **(a)**

Order Restricted Statistical Inference. F. T. Wright, R. L. Dykstra. New York: Wiley, 1988. xix + 521pp. $69.95. ISBN 0-471-91787-7.
Comprehensive treatment of statistical inference under inequality restraints. Extensive references. **(a)**

Perspectives on Contemporary Statistics. Ed: David C. Hoaglin, David S. Moore. MAA Notes;21. Washington, DC: Mathematical Association of America, 1992. xiii + 175pp. $20.00. ISBN 0-88385-075-3.
Focusing on current practice, these nine essays explain the topics central to the teaching of statistics to beginners. **(a,p)**

Plots, Transformations, and Regression: An Introduction to Graphical Methods of Diagnostic Regression Analysis. A. C. Atkinson. New York: Oxford University Press, 1985. xiii + 282pp. $32.50. ISBN 0-19-853359-4.
A valuable book to use while carrying out regression analysis with a computer package. **(a)**

Probability and Statistical Inference. Third Edition. Robert V. Hogg, Eliot A. Tanis. New York: Macmillan, 1988. ix + 658pp. $46.50. ISBN 0-02-355810-5.
This edition includes more material on exploratory data analysis, more graphs, and more real applications. **(a)**

Probability, Statistical Optics, and Data Testing: A Problem Solving Approach. Second Edition. B. Roy Frieden. Series in Information Science;10. New York: Springer-Verlag, 1991. xx + 443pp. $59.00. ISBN 0-387-53310-9.
An intensive introduction to the power and applicability of probabilistic and statistical modelling. **(a)**

Randomization Tests. Eugene S. Edgington. Statistics;31. New York: Dekker, 1980. xii + 287pp. $59.75. ISBN 0-8247-6878-7.
Treats the use of randomization tests to determine the significance of results when only nonrandom samples are available. **(a)**

Recent Developments in Clustering and Data Analysis. Ed: Edwin Diday et al. San Diego, CA: Academic Press, 1988. xv + 452pp. $85.00. ISBN 0-12-215485-1.
Thirty five papers comprise the proceedings of the March 1987 Japanese-French Scientific Seminar held in Tokyo. **(a)**

Regression Analysis: Theory, Methods, and Applications. Ashish Sen, Muni Srivastava. New York: Springer-Verlag, 1990. xv + 347pp. $39.00. ISBN 0-387-97211-0.

Up-to-date account of the theory and methods of regression analysis, assuming a background in mathematical statistics. Includes a computer disk. **(a)**

Regression Analysis With Applications. G. Barrie Wetherill et al. New York: Chapman & Hall, 1986. xi + 311pp. $35.00. ISBN 0-412-27490-6.

This volumes focuses on the problems that may occur when writing a regression program containing validity checks and diagnostics. **(a)**

Regression Diagnostics: Identifying Influential Data and Sources of Collinearity. David A. Belsley, Edwin Kuh, Roy E. Welsch. New York: Wiley, 1980. xv + 292pp. $57.95. ISBN 0-471-05856-4.

Detailed treatment emphasizing diagnosis of potential data problems rather than inference or curve fitting. Good bibliography. **(a)**

Residuals and Influence in Regression. R. Dennis Cook, Sanford Weisberg. New York: Routledge, Chapman & Hall, 1982. x + 230pp. $42.50. ISBN 0-412-24280-X.

Well-written, comprehensive monograph on diagnostic methods using residuals and then a variety of methods for the study of influence. Good set of references. **(a)**

Robust Statistics: The Approach Based on Influence Functions. Frank R. Hampel et al. New York: Wiley, 1986. xxi + 502pp. $49.95. ISBN 0-471-82921-8.

Robust estimation in theory and use from the point-of-view of influence functions. **(a)**

Robustness of Statistical Methods and Nonparametric Statistics. Ed: Dieter Rasch, Moti Lal Tiku. Norwell, MA: Kluwer Academic, 1985. 172pp. $97.50. ISBN 90-277-2076-2.

Most of the invited and contributed papers from a 1983 conference. **(a)**

S: An Interactive Environment for Data Analysis and Graphics. Richard A. Becker, John M. Chambers. Belmont, CA: Wadsworth, 1984. xii + 550pp. $37.95. ISBN 0-534-03313-X.

Appealing introduction to Bell Lab's UNIX-based statistical package S, an interactive system emphasizing graphical and exploratory methods. **(a)**

Sampling Techniques. Third Edition. William G. Cochran. New York: Wiley, 1977. xvi + 428pp. $66.95. ISBN 0-471-16240-X.

This book is very useful for students who wish to obtain practical knowledge of sample survey techniques. **(p)**

Say It With Figures. Sixth Edition. Hans Zeisel. New York: Harper & Row, 1985. xvi + 272pp. $7.95. ISBN 0-06-131994-5.

This latest edition of a 1947 classic adds a very helpful chapter on regression analysis. **(p)**

Schaum's Outline of Theory and Problems of Statistics. Second Edition. Murray R. Spiegel. New York: McGraw-Hill, 1988. 504pp. $10.95 paperbound. ISBN 0-07-060234-4.

"It has been designed for use either as a supplement to all current standard texts or as a textbook for a formal course in statistics. It should also be of considerable value as a book of reference."—Preface. Book has been greatly enlarged from First edition. **(a,p)**

Scientific Reasoning: The Bayesian Approach. Colin Howson, Peter Urbach. Peru, IL: Open Court, 1989. xiii + 312pp. $16.95;$36.95. ISBN 0-8126-9085-0;0-8126-9084-2.

A clearly written work for the non-specialist, explaining and defending Bayesian methods. **(a,p)**

A Second Course in Business Statistics: Regression Analysis. William Mendenhall, James T. McClave. Santa Clara, CA: Dellen, 1981. xiii + 637pp. $38.25. ISBN 0-89517-027-2.

Briefly reviews the basic ideas of statistics with several case studies from business. **(a)**

Sensitivity Analysis in Linear Regression. Samprit Chatterjee, Ali S. Hadi. New York: Wiley, 1988. xiv + 315pp. $39.95. ISBN 0-471-82216-7.

Good summary of known results and procedures of the factors that determine a least squares fit and the sensitivity of the fit. **(a)**

Some Theory of Sampling. William Edwards Deming. New York: Dover, 1966. xvii + 602pp. $15.95. ISBN 0-486-64684-X.

Corrected, unabridged reprint of the original 1950 classic that introduces the practical nature of statistical sampling. **(p)**

Statistical Analysis with Missing Data. Roderick J. A. Little, Donald B. Rudin. New York: Wiley, 1987. xiv + 278pp. $44.95. ISBN 0-471-80254-9.

Comprehensive work, divided into two applied statistician. Part I reviews historical approaches and Part II presents a systematic approach using likelihood-based methods. **(a)**

Statistical Computation. J. H. Maindonald. New York: Wiley, 1984. xviii + 370pp. $59.95. ISBN 0-471-86452-8.

This book encourages the use of statistical software with emphasis on multiple linear regression and related topics. **(a)**

Statistical Computing. William J. Kennedy, Jr., James E. Gentle. Statistics;33. New York: Dekker, 1980. xi + 591pp. $49.75. ISBN 0-8247-6898-1.

Discusses numerical methods and algorithms, both general and statistical, and techniques for implementing them. Extensive references. **(a)**

Statistical Concepts and Methods. Gouri K. Bhattacharyya, Richard A. Johnson. New York: Wiley, 1977. xv + 639pp. $64.95. ISBN 0-471-07204-4.

While emphasizing assumptions, this text provides many real examples and extensive coverage of the standard topics as well as some nonstandards ones (e.g., sample surveys). **(a)**

The Statistical Exorcist: Dispelling Statistics Anxiety. Myles Hollander, Frank Proschan. Popular Statistics;3. New York: Dekker, 1984. xi + 247pp. $39.75. ISBN 0-8247-7225-3.

Twenty-six vignettes, followed by related exercises, introduce the everyday use of statistics in decision-making, sampling, analyzing data, and estimating probabilities. **(p)**

Statistical Independence in Probability, Analysis, and Number Theory. Mark Kac. Carus Mathematical Monographs;12. Washington, DC: Mathematical Association of America, 1969. 112pp. $12.50. ISBN 0-88385-012-5.

Reprint of Kac's very popular work that uses techniques of classical analysis to link random numbers, the normal law, etc. This splendid book ranges from the primitive idea of statistical independence to diverse applications. **(a)**

Statistical Inference Based on Ranks. Thomas P. Hettmansperger. New York: Wiley, 1984. xvii + 323pp. $37.50. ISBN 0-471-88474-X.

Develops a coherent and unified set of statistical methods (based on ranks) for carrying out inferences in various experimental situations. References. **(a)**

Statistical Inference for Spatial Processes. B. D. Ripley. New York: Cambridge University Press, 1988. viii + 148pp. $37.50. 0-521-35234-7.

Prize winning essay dealing with some of the distinctive features and special problems of statistical inference on spatial stochastic processes. **(a)**

Statistical Inference for Stochastic Processes. Ishwar V. Basawa, B. L. S. Prakasa Rao. San Diego, CA: Academic Press, 1980. xiv + 435pp. $151.00. ISBN 0-12-080250-3.

Discussion of standard models, theory of inference for general processes in discrete and continuous time, and results on Bayesian, non-parametric, and sequential procedures. Good set of references. Expensive. **(a)**

Statistical Inference in Linear Models: Statistical Methods in Model Building. Ed: Helga and Olaf Bunke. Transl: John Bibby, Michal Basch. New York: Wiley, 1986-1989. *Volume I: Statistical Inference in Linear Models.* (Available from Books on Demand, Ann Arbor) 614pp. $165.80. ISBN 0-357-3395-5. *Volume II: Nonlinear Regression, Functional Relations, and Robust Methods.* ISBN 0-471-91239-5.

Presents a comprehensive, unified and rigorous development of the theory of the linear model in the areas of estimation, testing, confidence regions, Bayesian methods, and optimal design. **(a)**

Statistical Information and Likelihood: A Collection of Critical Essays by Dr. D. Basu. Ed: J. K. Ghosh. New York: Springer-Verlag, 1988. xviii + 365pp. $49.00. ISBN 0-387-96751-6.

Interesting collection of 19 articles by Basu on the foundations of statistical inference. **(a)**

Statistical Methods. Eighth Edition. George W. Snedecor, William G. Cochran. Ames, IA: Iowa State University Press, 1989. xx + 503pp. $44.95. ISBN 0-8138-1561-6.

"The present edition was prepared by several members of the Department of Statistics... Much of the material remains as previously published. A significant change in this edition occurs in the notation used to describe the operations of multiple regression. Matrix algebra replaces the original summation operators, and a short appendix on matrix algebra is included."—Preface. **(a,p)**

The Statistical Theory of Linear Systems. E. J. Hannan, Manfred Deistler. New York: Wiley, 1988. xiii + 380pp. $52.95. ISBN 0-471-80777-X.

Integrates recent developments in the theory and applications of system indentification. References. **(a)**

Statistics. David Freeman et al. New York: W. W. Norton, 1978. xv + 589pp. $44.95. ISBN 0-393-09076-0.

This excellent introductory "non-mathematical" text present many fascinating applications. **(p)**

Statistics: A Guide to the Unknown. Third Edition. Ed: Judith M. Tanur et al. Belmont, CA: Wadsworth, 1989. xxv + 284pp. $18.95. ISBN 0-534-09492-9.

Forty-one short essays illustrate the role of statistics in the biological, political, social, and physical worlds. Twelve new essays have been added to this edition and all original essays have been updated by their authors. **(p)**

Statistics for Experimenters: An Introduction to Design, Data Analysis, and Model Building. George E. P. Box, William G. and J. Stuart Hunter. New York: Wiley, 1978. xviii + 653pp. $59.95. ISBN 0-471-09315-7.

"An introduction to the philosophy of experimentation and the part statistics plays." This book uses no calculus and introduces statistical ideas as they are needed. **(a)**

Statistics: Theory and Methods. Donald A. Berry, Bernard W. Lindgren. Pacific Grove, CA: Brooks-Cole, 1990. xii + 763pp. $48.95. ISBN 0-534-09942-4.

Calculus-based text covering the usual topics, beginning with probability. **(p)**

Survival Analysis. Rupert G. Miller, Jr., Gail Gong, Alvaro Muñoz. New York: Wiley, 1981. xi + 238pp. $39.95. ISBN 0-471-09434-X.

Deals with the analysis of positive-valued random variables based on censored observations with emphasis on recent nonparametric approaches, with applications in medical research. References. **(a)**

Theory of Games and Statistical Decisions. David Blackwell, M. A. Girshick. New York: Wiley, 1954. 355pp. $9.95. ISBN 0-486-63831-6.

This book is intended primarily for first-year graduate students. It restricts the formal development to discrete distributions and continuous ones are used only for the purpose of illustration. **(a)**

Theory of Point Estimation. E. L. Lehmann. New York: Wiley, 1983. xii + 506pp. $51.95. ISBN 0-471-05849-1.

Comprehensive treatment of point estimation in Euclidean sample spaces, covering both exact and large-sample theory. Extensive references. **(a)**

Time Series Analysis: Forecasting and Control. Revised Edition. George E. P. Box, Gwilym M. Jenkins. Oakland, CA: Holden-Day, 1976. xxi + 575pp. $59.95. ISBN 0-8162-1104-3.

No significant change in the contents of this edition of a standard reference. **(a,s)**

Time Series Analysis of Irregularly Observed Data. Ed: Emanuel Parzen. Lecture Notes in Statistics;25. New York: Springer-Verlag, 1984. 363pp. $33.00. ISBN 0-387-96040-6.

Sixteen papers from a 1983 Texas A&M University symposium. **(a)**

Understanding Robust and Exploratory Data Analysis. Ed: David C. Hoaglin, Frederick Mosteller, John W. Tukey. New York: Wiley, 1983. xvi + 447pp. $59.95. ISBN 0-471-09777-2.

Treats the rationale and development of the techniques of exploratory data analysis and of robust and resistant techniques. References. **(a)**

The Visual Display of Quantitative Information. Edward R. Tufte. Cheshire, CT: Graphics Press, 1983. 197pp. $36.00. ISBN 0-9613921-0-X.

A beautiful, brilliant book, synthesizing statistics with the history of graphic design. **(a,p,s)**

STOCHASTIC PROCESSES

Basic Stochastic Processes: The Mark Kac Lectures. Reza Iranpour, Paul Chacon. New York: Macmillan, 1988. xiv + 258pp. $48.00. ISBN 0-02-359820-4.

A short course in stochastic processes based on Kac's lectures. No bibliography. **(a)**

Brownian Motion and Stochastic Calculus. Ionnis Karatzas, Steven E. Shreve. Graduate Texts in Mathematics;113. New York: Springer-Verlag, 1988. xxiv + 470pp. $68.00. ISBN 0-387-96535-1.

This is a valuable book for every graduate student studying stochastic processes and for those interested in pure and applied probability. **(a)**

Extreme Values, Regular Variation, and Point Processes. Sidney L. Resnick. Applied Probability;4. New York: Springer-Verlag, 1987. xii + 320pp. $69.00. ISBN 0-387-96481-9.

Well-written work for the point process approach. Extensive bibliography. **(a)**

A First Course in Stochastic Processes. Second Edition. Samuel Karlin, Howard M. Taylor. San Diego, CA: Academic Press, 1975. xvi + 557pp. $49.50. ISBN 0-12-398552-8.

"We have supplied, in new chapters, broad introductory discussions of several classes of stochastic processes not dealt with in the first edition"—Preface. **(a)**

Introduction to Stochastic Differential Equations. Thomas C. Gard. Pure and Applied Mathematics; 114. New York: Dekker, 1988. xi + 234pp. $65.00. ISBN 0-8247-7776-X.

A basic account, emphasizing the role of Ito's formula in stochastic analysis. **(a)**

Introduction to Stochastic Processes. E. Cinlar. New York: Prentice-Hall, 1975. x + 402pp. $53.00. ISBN 0-13-498089-1.

Well-written, this book is intended for those who are interested in the practical applications and consequences of the theory. **(a,p)**

Introduction to Stochastic Processes. Paul G. Hoel et al. Prospect Heights, IL: Waveland Press, 1987. x + 203pp. $19.95 paperbound. ISBN 0-88133-267-4.

Reprint of 1972 original. This is a well-written text with many illustrative examples and good problem sets. **(a,p)**

Introduction to the Theory of Coverage Processes. Peter Hall. New York: Wiley, 1988. xviii + 408pp. $49.95. ISBN 0-471-85702 -5.

Coverage process is a countable collection of sets in Rn whose centers form a (spatial) point process. Theory and techniques of this topic are clearly presented in detail. **(a)**

An Introduction to the Theory of Point Processes. D. J. Daley, D. Vere-Jones. New York: Springer-Verlag, 1988. xxi + 702pp. $79.00. ISBN 0-387-96666-8.

A well-organized work on the theory of point processes. Extensive bibliography. **(a)**

Limit Theorems for Stochastic Processes. Jean Jacod, Albert N. Shiryaev. New York: Springer-Verlag, 1987. xvii + 601pp. $99.00. ISBN 0-387-17882-1.

Limit theorems are developed to provide a systematic approach to convergence in probability law for processes that are semi-martingales. Extensive bibliography. **(a)**

Markov Processes: Characterization and Convergence. Stewart N. Ethier, Thomas G. Kurtz. New York: Wiley, 1986. x + 534pp. $64.95. ISBN 0-471-08186-8.

Comprehensive treatment of the idea that weak convergence is tied to the characterization of a limit process. Very good bibliography. **(a)**

Point Processes and Queues: Martingale Dynamics. Pierre Bremaud. New York: Springer-Verlag, 1981. xix + 354pp. $59.00. ISBN 0-387-90536-7.

A radically different approach to point processes, i.e., a martingale approach which allows for a unified treatment of dynamical point process systems along the line of Wiener-driven systems. Extensive bibliography. **(a)**

A Second Course in Stochastic Processes. Samuel Karlin, Howard M. Taylor. San Diego, CA: Academic Press, 1981. xviii + 542pp. $54.50. ISBN 0-12-398650-8.

Much material is condensed into this book. Several applications of each stochastic process are developed. **(a)**

Stochastic Processes. J. L. Doob. New York: Wiley, 1990. vii + 654pp. $32.95. ISBN 0-471-52369-0.

Paperbound reprint of 1953 original edition, which was the first systematic survey of stochastic processes. **(a)**

Stochastic Processes. Sheldon M. Ross. New York: Wiley, 1983. viii + 309pp. $64.95. ISBN 0-471-09942-2.

In this well-organized text, Poisson and Markov processes are developed using the sample path approach, then random walks and martingales, and finishing with stochastic order relations. **(a)**

The Theory of Optimal Stopping. Y. S. Chow, Herbert Robbins, David Siegmund. New York: Dover, 1991. xii + 141pp. $6.95. ISBN 0-486-66650-6.

Reprint of 1971 original. Numerous examples and exercises. Good, but dated, bibliography. **(a)**

TOPOLOGY

Basic Topology. M. A. Armstrong. New York: Springer-Verlag, 1983. xii + 251pp. $19.80. ISBN 0-387-90839-0.

Reprint, with minor corrections, of 1979 text. An elegant introduction, emphasizing geometric insights in a wide range of contexts. **(a,p,s)**

Counterexamples in Topology. Second Edition. Lynn Arthur Steen, J. Arthur Seebach, Jr. New York: Springer-Verlag, 1978. xi + 244ppp. $35.50. ISBN 0-387-90312-7.

A useful book which contains definitions about topological and metric spaces, examples, reference charts, exercises, with notes, bibliography, and index. **(a)**

A Course in Simple-Homotopy Theory. M. M. Cohen. Graduate Texts in Mathematics;10. New York: Springer-Verlag, 1982. x + 114pp. $32.00. ISBN 0-387-90056-X.

An excellent exposition of Whitehead's attempt to develop a geometric theory of homotopy equivalences, and the role of that theory in geometric topology. **(a)**

Differential Topology. Victor Guillemin, Alan Pollack. New York: Prentice-Hall, 1974. xvi + 222pp. $59.00. ISBN 0-13-212605-2.

This book gives a very good elementary, intuitive introduction to differential topology. **(a,s)**

Dimension Theory. Revised Edition. Witold Hurewicz, Henry Wallman. Princeton Mathematical Series;4. Princeton, NJ: Princeton University Press, 1974. vii + 165pp. $32.50. ISBN 0-691-07947-1.

Reprint of 1941 original. Interesting reading, this book fulfills the authors' aim "to give a connected and simple account of the most essential parts of dimension theory." **(a)**

Dimension Theory. Keio Nagami. Pure and Applied Mathematics;37. San Diego, CA: Academic Press, 1970. xi + 256pp. $81.00. ISBN 0-12-513650-1.

Includes an appendix by Yukihiro Kodama. Nagami's principal concern lies with the properties of the various dimension functions in more general spaces and his book illuminates their relationships and differences. **(a)**

First Concepts of Topology: The Geometry of Mappings of Segments, Curves, Circles, and Disks. W. G. Chinn, N. E. Steenrod. New Mathematical Library;18. Washington, DC: Mathematical Association of America, 1966. viii + 160pp. $12.95. ISBN 0-88385-618-2.

This clearly written book presents topology as mathematicians see it, defining open sets and their coverings, and from them proves the key theorems. **(a)**

General Topology. Nicolas Bourbaki. (Elements of Mathematics). New York: Springer-Verlag, 1989. *Chapters 1-4*. vii + 437pp. $69.00. ISBN 0-387-19374-X. *Chapters 5-10*. iv + 363pp. $59.00. ISBN 0-387-19372-3.

Reprint of Addison-Wesley's 1966 English translation. Many challenging exercises and superb historical notes. (See p. 303 for chapters 5-10, *Topologie Générale*.) **(a)**

General Topology. Revised and Completed Edition. Ryszard Engelking. Berlin, Germany: Heldermann Verlag, 1989. viii + 529pp. $90.30. ISBN 3-88538-006-4.

This new edition of the 1957 classic has been updated to reflect current research, with additions and corrections to the bibliography. **(a)**

General Topology. Stephen Willard. Reading, MA: Addison-Wesley, 1970. xii + 369pp. $43.16. ISBN 0-201-08707-3.

One of the best general topology texts with a smooth, accurate, clear exposition. **(a,p)**

General Topology and Homotopy Theory. I. M. James. New York: Springer-Verlag, 1984. 248pp. $53.00. ISBN 0-387-90970-2.

A reasonably self-contained work, tightly written and elegant, on certain parts of general topology which are particularly relevant to homotopy theory. **(a)**

Lehrbuch der Topologie. Second Edition. H. Seifert, W. Threlfall. New York: Chelsea, 1980. vii + 353pp. $9.95. ISBN 0-8284-0031-8.

Textually unaltered reprint of 1934 original. **(a)**

Modern General Topology. Second Revised Edition. Jun-iti Nagata. North-Holland Mathematical Library;33. New York: Elsevier, 1985. x + 522pp. $133.50. ISBN 0-444-87655-3.

This revision of earlier editions of the classic adds recent developments in general topology, i.e., linearly ordered spaces, cardinal functions, and dyadic spaces. Expensive. **(a)**

On Knots. Louis H. Kauffman. Annals of Mathematics Studies;115. Princeton, NJ: Princeton University Press, 1987. xv + 480pp. $55.00;$22.50. ISBN 0-691-08434-3;0-691-08435-1.

Expanded version of notes from a Winter 1984 seminar held at the University of Zaragoza, Spain. **(a)**

The Shape of Space: How to Visualize Surfaces and Three-Dimensional Manifolds. Jeffrey R. Weeks. Pure and Applied Mathematics;96. New York: Dekker, 1985. x + 324pp. $49.75. ISBN 0-8247-7437-X.

Well-written and illustrated informal introduction to geometric topology of two and three dimensions. Bibliography. **(a,p,s)**

Stone Spaces. Peter T. Johnstone. Cambridge Studies in Advanced Mathematics;3. New York: Cambridge University Press, 1986. xxi + 370pp. $29.95 paperbound. ISBN 0-521-33779-8.

This book is about one of the interfaces between topology and algebra, the one created by M. H. Stone's discovery that the category of Boolean algebras is dual to the category of totally disconnected compact Hausdorff spaces, a.k.a. Stone spaces. **(a)**

Topological Dynamics. Walter Helbig Gottschalk, Gustav Arnold Hedlund. Colloquium Publications;36. Providence, RI: American Mathematical Society, 1955. vii + 151pp. $37.00. ISBN 0-8218-1036-7.

Brief and condensed style makes this work a bit difficult to read. Part I gives definitions and theorems under most general conditions plus a presentation of topological dynamics from a certain point of view. Part 2 deals with some specific examples of flows. **(a)**

A Topological Picture Book. George K. Francis. New York: Springer-Verlag, 1987. xv + 194pp. $39.80. ISBN 0-387-96426-6.

A fascinating collection of hand-drawn pictures supported by "topological stories" gathered from "expert friends" whose work the author has helped illustrate. An excellent resource to strengthen spatial imagination. **(a,p)**

Topologie. Erster Band. Pavel S. Aleksandrov, Heinz Hopf. New York: Chelsea, 1965. xiii + 636pp. $19.50. ISBN 0-8284-0197-7.

Textually unaltered reprint of the 1935 original. **(a)**

Topologie Générale, Chapitres 5 à 10. Nouvelle Èdition. Nicolas Bourbaki. (Éléments de Mathématique) Paris: Diffusion C.C.L.S., 1982. (Available from Scientific & Medical Publications of France, New York, NY) various pagings. $82.00. ISBN 2-903684-05-7.

Reprint of 1971 edition. (See p. 301 for chapters 1-4, *General Topology.*) **(a)**

Topology. Klaus Janich. Transl: Silvio Levy. New York: Springer-Verlag, 1984. ix + 192pp. $38.00. ISBN 0-387-90892-7.

Highly readable and relatively sophisticated text covers all of topology. No problems or exercises, but beautifully presented intuition. **(a,p)**

Topology. New Edition, Revised and Augmented. K. Kuratowski. Transl: J. Jaworowski. San Diego, CA: Academic Press, 1966-1968. *Vol. I.* xx + 560pp. $65.00. ISBN 0-12-429201-1. *Vol. II.* xiv + 608pp. $103.00. ISBN 0-12-429202-X.

This translation of the set-theoretic topologist's "bible" places more emphasis on general topological spaces. Expensive. **(a)**

Topology, A First Course. James R. Munkres. New York: Prentice-Hall, 1975. xvi + 413pp. $56.00. ISBN 0-13-925495-1.

A very substantial book and an outstanding text. **(a)**

Topology from the Differentiable Viewpoint. John W. Milnor. Charlottesville, VA: University Press of Virginia, 1965. ix + 64pp. $5.95. ISBN 0-8139-0181-2.

Based on notes by David Weaver of lectures given at the University of Virgina in December 1963. **(a)**

Variétés Différentielles et Analytiques, Fascicule de Résultats. Nouvelle Tirage. Nicolas Bourbaki. (Éléments de Mathématique) Paris: Diffusion C.C.L.S., 1983. (Available from Scientific & Technical Publications of France, New York, NY) 196pp. $50.00. ISBN 2-903684-10-3.

"This volume collects the basic concepts and main results of the theory of differentiable manifolds (over the field of reals) and of analytic manifolds (over a complete nondiscrete field with valuation)."—Introduction. **(a)**

COLLECTED AND SELECTED WORKS OF INDIVIDUAL MATHEMATICIANS (ARRANGED BY INDIVIDUAL)

The Selected Works of J. Frank Adams. Ed: J. P. May, C. B. Thomas. New York: Cambridge University Press, 1992. *Volume I.* xvi + 536pp. $74.95. ISBN 0-521-41063-0. *Volume II.* xvi + 529pp. $74.95. ISBN 0-521-41065-7.

This selection brings together all of the major research contributions of one of the world's leading topologists. They are organized by subject matter rather than strict chronological order, although most fall within the period 1955-1988. **(a)**

Collected Papers. Lars Valerian Ahlfors. Boston: Birkhäuser, 1982. $145.00 set ISBN 0-8176-3077-5. *Volume 1.* xix + 520pp. ISBN 0-8176-3075-9. *Volume 2.* xix + 515pp. ISBN 0-8176-3076-7.

Volume 1 covers the years 1929-55 and volume 2 the years 1954-79, with bibliographical references. (**a**)

Collected Mathematical Papers. A. Adrian Albert. Ed: Richard E. Block et al. Providence, RI: American Mathematical Society, 1992. $165.00 (set). ISBN 0-8218-0003-5 (set) *Pt. 1: Associative Algebras and Riemann Surfaces.* 743pp. $80.00. 0-8218-0005-1. *Pt. 2: Nonassociative Algebras and Miscellany.* 938pp. $94.00. ISBN 0-8218-0007-8.

This set contains the collected works of Albert, a leading algebraist of the 20th Century. Expensive. (**a**)

The Collected Papers of T. W. Anderson, 1943-1985. Ed: George P. H. Styan. New York: Wiley, 1990. $185.00 (set). ISBN 0-471-62422-5. *Volume 1.* xliii + 825pp. *Volume 2.* v + 827-1680pp.

These two volumes comprise the 109 research papers by Anderson on probability, statistics, econometrics, and matrix theory, published from 1943-1985. Also included are 16 commentaries, a list of "Other publications" (books, book reviews, papers published after 1985, Festschrift, interviews), and a list of his 22 Ph.D. students. Author and subject indexes. Expensive. (**a**)

Collected Papers: Emil Artin. Ed: Serge Lang, John T. Tate. New York: Springer-Verlag, 1988. xvi + 560pp. $59.00. ISBN 0-387-90686-X.

Unaltered reprint of 1965 Addison-Wesley original. (**a**)

Michael Atiyah: Collected Works. Michael Atiyah. New York: Oxford University Press, 1988. *Volume 1: Early Papers: General Papers.* xxiii + 364pp. $59.00. ISBN 0-19-853275-X. *Volume 2: K-Theory.* xxiii + 829pp. $89.00. ISBN 0-19-853276-8. *Volume 3: Index Theory, 1.* xxiii + 593pp. $79.00. ISBN 0-19-853277-6. *Volume 4: Index Theory, 2.* xxiii + 617pp. $79.00. ISBN 0-19-853278-4. *Volume 5: Gauge Theories.* xxiii + 685pp. $89.00. ISBN 0-19-853279-2.

A scholarly record of accomplishment with each section introduced by the author's commentary to document the genesis of his ideas. (**a**)

Collected Papers of Karl Egil Aubert. Ed: Paulo Ribenboim. Queen's Papers in Pure and Applied Mathematics;89. Kingston, ON: Queen's University Printing Services, 1992. xxxviii + 603pp. $50.36. ISSN 0079-8797.

The papers, including his doctoral thesis, of this important Norwegian algebraist are arranged chronologically. Biographical sketch. (**a**)

The Mathematical Works. Isaac Barrow. Ed: W. Whewell. Hildesheim: G. Olms, 1973. xix + 414 + 320pp. $118.50. ISBN 3-487-04788-8.

Reprint of 1860 edition. Preface in English; text in Latin. (**a**)

The Bellman Continuum: A Collection of the Works of Richard E. Bellman. Ed: Robert S. Roth. Teaneck, NJ: World Scientific, 1986. xxii + 868pp. $90.00. ISBN 9971-50-090-6.

Collection of some of the works of Bellman in the field of mathematics, including bits of notes left at the time of his death. After an introductory chapter which briefly summarizes his life, the book considers 9 general topics from Bellman's more than 600 papers and 30 books written over the past 40 years. **(a)**

Die Werke von Jakob Bernoulli. Jakob Bernoulli. Ed: Naturforschenden Gesellschaft in Basel. Boston: Birkhäuser, 1969-1989. *Band 1: Astronomie, Philosophie Naturalis*. xii + 541pp. $92.50. ISBN 3-7643-0028-0. *Band 2: Elementarmathematik*. xiii + 685pp. $234.00. ISBN 3-7643-0028-0. *Band 3: Die Wahrscheinlichkeits-theorie*. x + 585pp. $127.00. ISBN 3-7643-0713-7.

The first 3 volumes of a 4 volume set of the first critical edition of the collected works of Jakob Bernoulli. Includes appendices, bibliography, and indexes. Expensive. **(a)**

The Collected Works of Arne Beurling. Ed: L. Carleson et al. Boston: Birkhäuser, 1989. $89.00. ISBN 0-8176-3412-6 set. *Volume 1: Complex Analysis*. xx + 475pp. *Volume 2: Harmonic Analysis*. xx + 389pp.

Papers and seminar notes by Beurling who did pioneering work in extremal length, inner functions, and spectral theory. Brief foreword assesses his mathematical contributions. **(a)**

The Collected Papers of R. H. Bing. Ed: Sukhjit Singh, Steve Armentrout, Robert J. Daverman. Providence, RI: American Mathematical Society, 1988. $172.00 set. ISBN 0-8218-0117-1. *Volume 1*. xix + 886pp. *Volume 2*. xvii + 759pp.

A splendid celebration of Bing's mathematical life including all of his mathematical papers and abstracts, organized in part with Bing's help. Singh has prepared an autobiographical essay from tapes made by Bing. Expensive. **(a)**

Selected Papers. Errett Bishop. Teaneck, NJ: World Scientific, 1986. xxiv + 414pp. $63.00. ISBN 9971-50-127-9.

Papers on complex and functional analysis and constructive mathematics are arranged chronologically. Vita, introduction by John Wermer, and bibliography of Bishop's papers. **(a)**

Armand Borel: Oeuvres, Collected Papers. Armand Borel. New York: Springer-Verlag, 1983. $269.00 set. ISBN 0-387-12126-9. *Volume I, 1948-1958*. viii + 718pp. *Volume II, 1959-1968*. iv + 790pp. *Volume III, 1969-1982*. iv + 718pp.

The official record of Borel's work, concentrated in topology, algebraic geometry, and especially algebraic groups. Expensive. **(a)**

The Collected Work of George E. P. Box. Ed: George C. Tiao. Belmont, CA: Wadsworth, 1985. *Volume I*. xiv + 657pp. $59.95. ISBN 0-534-03307-5. *Volume II*. xiv + 710pp. $59.95. ISBN 0-534-03308-3.

The set includes a major portion of his articles. **(a)**

Richard Brauer: Collected Papers. Ed: Paul and Warren J. Fong. Mathematicians of Our Time;17,18,19. Cambridge, MA: MIT Press, 1980. *Volume I: Theory of Algebras and Finite Groups*. liv + 615pp. $80.00. ISBN 0-262-02135-8. *Volume II: Finite Groups*. viii + 586pp. $80.00. ISBN 0-262-02148-X. *Volume III: Finite Groups, Lie Groups, Number Theory, Polynomials and Equations, Geometry, and Biography*. x + 689pp. out-of-print. ISBN 0-262-02149-8.

All but 29 of Brauer's 129 papers. Volume I also has a brief recounting by Brauer of his mathematical education, a reprinted professional biography, and a complete list of his publications. **(a)**

Gesammelte Abhandlungen mathematischen und philosophischen Inhalts, mit erläuternden Anmerkungen sowie mit Ergänzungen aus dem Briefwechsel Cantor-Dedekind. Georg Cantor. Ed: Ernst Zermelo. New York: Springer-Verlag, 1990. vii + 486pp. $96.00. ISBN 0-387-09849-6.

Reprint of 1932 edition. **(a)**

Selected Papers, S. Chandrasekhar. Chicago: University of Chicago Press, 1989-1991. *Volume 1: Stellar Structure and Stellar Atmospheres*. vi + 791pp. $29.95; $74.95. ISBN 0-226-10090-1; 0-226-10089-8. *Volume 2: Radiative Transfer and Negative Ion of Hydrogen*. xvi + 622pp. $29.95; $74.95. ISBN 0-226-10093-6; 0-226-10092-8. *Volume 3: Stochastic, Statistical, and Hydromagnetic Problems in Physics and Astronomy*. xiv + 642pp. $29.95; $79.95. ISBN 0-226-10095-2; 0-226-10094-4. *Volume 4: Plasma Physics, Hydrodynamic and Hydromagnetic Stability, and Applications of the Tensor-Virial Theorem*. xiii + 585pp. $29.95; $79.95. ISBN 0-226-10097-9; 0-226-10096-0. *Volume 5: Relativistic Astrophysics*. xx + 587pp. $29.95; $79.95. ISBN 0-226-10099-5; 0-226-10098-7. *Volume 6: The Mathematical Theory of Black Holes and of Colliding Plane Waves*. xix + 739pp. $35.00; $79.95. ISBN 0-226-10101-0; 0-226-10100-2.

Six volume collection. The latest contains 37 papers, mostly from 1975-85 and introduced by Chandrasekhar's colleague and frequent co-author Xanthopoulos. **(a)**

Shiing-shen Chern: Selected Papers. New York: Springer-Verlag, 1978-1989. Volume I. xxxi + 476pp. $59.00. ISBN 0-387-90339-9. *Volume II*. xxx + 444pp. $49.00. ISBN 0-387-96816-4. *Volume III*. xiv + 504pp. $54.00. ISBN 0-387-96817-2. *Volume IV*. xiv + 462pp. $49.00. ISBN 0-387-96820-2.

This set contains Chern's papers arranged chronologically and includes a complete bibliography and recently revised scientific autobiography. **(a)**

Mathematical Papers. William K. Clifford. New York: Chelsea, 1968. lxx + 658pp. $49.50. ISBN 0-8284-0210-8.

Reprint of 1882 edition. Bibliographical references in preface. **(a)**

The Collected Mathematical Papers of Leonard Eugene Dickson. Ed: A. Adrian Albert. New York: Chelsea, 1975-83. $295.00 set (v.1-5). ISBN 0-8284-0273-6. *Volume I*. xvii + 680pp. *Volume II*. 766pp. *Volume III*. 580pp. *Volume IV*. 636pp. *Volume V*. 644pp. *Volume VI*. 714pp. $55.00. ISBN 0-8284-0305-6.

A collection of nearly all of Dickson's research papers and two of his books. Bibliography and index. Expensive. (**a**)

Selected Writings on Computing: A Personal Perspective. Edsger W. Dijkstra. New York: Springer-Verlag, 1982. xvii + 362pp. $55.00. ISBN 0-387-90652-5.

Selections from Dijkstra's scientific correspondence during the intellectual preoccupations of one of the pioneers of computer science. (**a**)

The Dilworth Theorems: Selected Papers of Robert P. Dilworth. Ed: Kenneth P. Bogart, Ralph Freese, Joseph P. S. Kung. Boston: Birkhäuser, 1990. xxvi + 465pp. $59.50. ISBN 0-8176-3434-7.

A collection of Dilworth's important papers on ordered sets and lattice theory. Each section is preceded by background exposition by Dilworth and followed by articles on later influences. Extensive references. (**a**)

Beno Eckmann Selecta. Ed: Max-Albert Knus, Guido Mislin, Urs Stammbach. New York: Springer-Verlag, 1987. xii + 835pp. $129.00. ISBN 0-387-17518-0.

A collection of Eckmann's major papers as well as a complete bibliography and a brief biographical note, on the occasion of his 70th birthday. Expensive. (**a**)

The Collected Papers of Albert Einstein. English Translation. Ed: John Stachel. Transl: Anna Beck. Princeton, NJ: Princeton University Press, 1987-1989. *Volume 1: The Early Years, 1879-1902.* lxvi + 433pp. $75.00. ISBN 0-691-08407-6. (There is also a paperbound edition which is a selective translation of hard-case. xxii + 196pp. $27.50. ISBN 0-691-08475-0.) *Volume 2: The Swiss Years, Writings, 1900-1909.* xxxii + 656pp. $85.00;$27.95. ISBN 0-691-08526-9; 0-691-08549-8.

First two of an anticipated 30 volume set constituting the definitive, comprehensive edition of all of Einstein's papers, letters, and related documents. Each manuscript is untranslated, typeset in a manner that is faithful to the original German; all editor's notes are in English. The accompanying paperback contains typewritten English translations of all original documents in the official edition. (**a**)

Mathematische Werke. Second Edition. Gotthold Eisenstein. New York: Chelsea, 1989. *Band I.* xiii + 502pp. *Band II.* xiii + 434pp. $125.00 set. ISBN 0-8284-1280-X.

First complete collection of Eisenstein's works together with his letters to Hermite, Richelot, Stern and Gauss as well as a youthful autobiography and Biermann's biographical article. Expensive. (**a**)

Calvin C. Elgot: Selected Papers. Calvin C. Elgot. Ed: Stephen L. Bloom. New York: Springer-Verlag, 1982. xxiv + 460pp. $66.00. ISBN 0-387-90698-3.

Photographic reproductions of 13 original papers on the study of computation in the framework of category theory and algebra. (**a**)

Leonhardi Euleri Opera Omnia. Series Prima: Opera Mathematica. Leonhard Euler. Ed: Academiae Scientiarum Naturalium Helveticae. Boston: Birkhäuser, 1911-1956. All volumes $176.50 each. *Vol. I: Vollständige Anleitung zur Algebra.* 651pp. ISBN

3-7643-1400-1. *Vol. II: Commentationes Arithmeticae: 1.* 611pp. ISBN 3-7643-1401-X. *Vol. III: Commentationes Arithmeticae: 2.* 543pp. ISBN 3-7643-1402-8. *Vol. IV: Commentationes Arithmeticae: 3.* 431pp. ISBN 3-7643-1403-6. *Vol. V: Commentationes Arithmeticae: 4.* 374pp. ISBN 3-7643-1404-4. *Vol. VI: Commentationes Algebraicae ad Theoriam Aequationum Pertinentes.* 509pp. ISBN 3-7643-1405-2. *Vol. VII: Commentationes Algebraicae ad Theoriam Combinatorionum et Probabilitatum Pertinentes.* 580pp. ISBN 3-7643-1406-0. *Vol. VIII: Introductio in Analysin Infinatorum: 1.* 392pp. ISBN 3-7643-1407-9. *Vol. IX: Introductio in Analysin Infinatorum: 2.* 676pp. ISBN 3-7643-1408-7. *Vol. X: Institutiones Calculi Differentialis.* 676pp. ISBN 3-7643-1409-5. *Vol. XI: Institutiones Calculi Integralis: 1.* 462pp. ISBN 3-7643-1410-9. *Vol. XII: Institutiones Calculi Integralis: 2.* 542pp. ISBN 3-7643-1411-7. *Vol. XIII: Institutiones Calculi Integralis: 3.* 505pp. ISBN 3-7643-1412-5. *Vol. XIV: Commentationes Analyticae ad Theoriam Serierum Infinitarum Pertinentes: 1.* 617pp. ISBN 3-7643-1413-3. *Vol. XV: Commentationes Analyticae ad Theoriam Serierum Infinitarum Pertinentes: 2.* 2. Auflg. 1990. 722pp. ISBN 3-7643-1414-1. *Vol. XVI/1: Commentationes Analyticae ad Theoriam Serierum Infinitarum Pertinentes: 3.1.* 355pp. ISBN 3-7643-1415-X. *Vol. XVI/2: Commentationes Analyticae ad Theoriam Serierum Infinitarum Pertinentes: 3.2.* 332pp. ISBN 3-7643-1416-8. *Vol. XVII: Commentationes Analyticae ad Theoriam Integralium Pertinentes: 1.* 457pp. ISBN 3-7643-1417-6. *Vol. XVIII: Commentationes Analyticae ad Theoriam Integralium Pertinentes: 2.* 475pp. ISBN 3-7643-1418-4. *Vol. XIX: Commentationes Analyticae ad Threoriam Integralium Pertinentes: 3.* 494pp. ISBN 3-7643-1419-2. *Vol. XX: Commentationes Analyticae ad Theoriam Integralium Ellipticorum Pertinentes: 1.* 371pp. ISBN 3-7643-1420-6. *Vol. XXI: Commentationes Analyticae ad Theoriam Integralium Ellipticorum Pertinentes: 2.* 380pp. ISBN 3-7643-1421-4. *Vol. XXII: Commentationes Analyticae ad Theoriam Aequationum Differentialium Pertinentes: 1.* 420pp. ISBN 3-7643-1422-2. *Vol. XXIII: Commentationes Analyticae ad Theoriam Aequationum Differentialium Pertinentes: 2.* 455pp. ISBN 3-7643-1423-0. *Vol. XXIV: Methodus Inveniendi Lineas Curvas Maximi Minimive Proprietate Gaudentes Sive Solutio Problematis Isoperimetrici Latissimo Sensu Accepti.* 308pp. ISBN 3-7643-1424-9. *Vol. XXV: Commentationes Analyticae ad Calculum Variationum Pertinentes.* 343pp. ISBN 3-7643-1425-7. *Vol. XXVI: Commentationes Geometricae: 1.* 362pp. ISBN 3-7643-1426-5. *Vol. XXVII: Commentationes Geometricae: 2.* 400pp. ISBN 3-7643-1427-3. *Vol. XXVIII: Commentationes Geometricae: 3.* 381pp. ISBN 3-7643-1428-1. *Vol. XXIX: Commentationes Geometricae: 4.* 448pp. ISBN 3-7643-1429-X.

This series of Euler's collected works contains his contributions on mathematics, in Latin, French, and German. Expensive. **(a)**

Collected Papers on Mathematics, Logic, and Philosophy. Gottlob Frege. Ed: Brian McGuinness. Transl: Max Black et al. Cambridge, MA: Basil Blackwell, 1984. viii + 412pp. $65.00. ISBN 0-631-12728-3.

"Gottlob Frege's occasional writings are all collected in the present volume, about half of them now first appearing in English. They form a corpus as great and as important as his 3 systematic works."—Editor's preface. **(a)**

Selecta. Kurt Otto Friedrichs. Ed: Cathleen S. Morawetz. Boston: Birkhäuser, 1986. $200.00 set. ISBN 0-8176-3270-0. *Vol. 1.* 427pp. $102.00. ISBN 0-8176-3268-9. *Vol. 2.* 608pp. $144.00. ISBN 0-8176-3269-7.

This selection contains reproductions of 34 articles and one report. Papers in the volumes are divided into 7 subject groupings together with their commentators-reviewers: Lax, Kato, John, Wasow, Weitzner, Nirenberg, and Isaacson. Biography by Constance Reid. List of degrees and awards received by Freidrichs. Expensive. **(a)**

Izrail M. Gelfand: Collected Papers. Ed: S. G. Gindikin et al. New York: Springer-Verlag, 1987-1989. *Volume I.* vi + 883pp. $149.50. ISBN 0-387-13619-3. *Volume II.* x + 1039pp. $159.00. ISBN 0-387-19035-X. *Volume III.* x + 1075pp. $149.50. ISBN 0-387-19399-5.

Gelfand's 450 research papers from Banach algebras to cell biology are collected into 3 volumes. Complete bibliography. Expensive. **(a)**

Kurt Gödel: Collected Works. Ed: Solomon Feferman et al. New York: Oxford University Press, 1986. *Volume I: Publications 1929-1936.* xvi + 474pp. $45.00. ISBN 0-19-503964-5.

The first of two volumes of published works to be followed by volumes of unpublished manuscripts and lectures. Each article has a note to elucidate it and place it in proper historical context. Includes English translations and a biographical essay. **(a)**

Mathematical Papers of George Green. Ed: N. M. Ferrers. New York: Chelsea, 1970. x + 336pp. $25.00. ISBN 0-8284-0229-9.

Reprint of 1871 original with errata corrected. The curious may amuse themselves by trying to locate something here that looks like Green's theorem. **(a)**

The Collected Works of Philip Hall. Comp: K. W. Gruenberg, J. E. Roseblade. New York: Oxford University Press, 1988. xi + 776pp. $125.00. ISBN 0-19-853254-7.

This beautiful book consists of almost 50 years of publications, arranged chronologically, by one of the greatest mathematicians of this century. Obituary is by Roseblade. Expensive. **(a)**

P.R. Halmos: Selecta. New York: Springer-Verlag, 1983. *Expository Writing.* Ed: Donald E. Sarason, Leonard Gilman. xix + 304pp. $39.00. ISBN 0-387-90756-4. *Research Contributions.* Ed: Donald E. Sarason, Nathaniel A. Friedman. xxviii + 458pp. $59.50. ISBN 0-387-90755-6.

The two volumes present a selection of Halmos' mathematical writings, from research publications to expository and popular writings. **(a)**

The Mathematical Papers of Sir William Rowan Hamilton. New York: Cambridge University Press, 1931-1967. *Volume 1: Geometrical Optics.* Ed: A. W. Conway, J. L. Synge. xxviii + 534pp. out-of-print. No ISBN. *Volume 2: Dynamics.* Ed: A. W.

Conway, A. J. McConnell. xv + 656pp. out-of-print. No ISBN. *Volume 3: Algebra.* Ed: H. Halberstam, R. E. Ingram. xxiv + 672pp. $140.00. ISBN 0-521-05183-5.

First three volumes in a projected four volume set, the last of which will contain Hamilton's contributions to analysis and geometry. Papers which have never appeared in print before are published in all volumes. Historical notes. Expensive. **(a)**

Collected Papers of G. H. Hardy; Including Joint Papers with J. E. Littlewood and Others. Ed: Committee appointed by the London Mathematical Society. New York: Oxford University Press, 1966-1979. *Volume I.* x + 700pp. out-of-print. No ISBN. *Volume II.* vii + 702pp. out-of-print. No ISBN. *Volume III.* viii + 748pp. out-of-print. No ISBN. *Volume IV.* ix + 722pp. out-of-print. No ISBN. *Volume V.* xv + 694pp. out-of-print. No ISBN. *Volume VI.* xii + 854pp. o.p. ISBN 0-19-853340-3. *Volume VII.* xviii + 897pp. $130.00. ISBN 0-19-853347-0.

The complete bibliography of Hardy's mathematical papers, from 1899 to 1949. Titchmarsh's obituary of Hardy and a table showing the arrangement of the papers appears in the first volume. Expensive. **(a)**

Harish-Chandra, Collected Papers. Ed: V. S. Varadarajan. New York: Springer-Verlag, 1984. $269.00 set. ISBN 0-387-90782-3. *Volume I: [1944-1954].* lxxvii + 566pp. *Volume II: [1955-1958].* v + 539pp. *Volume III: [1959-1968].* v + 670pp. *Volume IV: [1970-1983].* v + 461pp.

All of Harish-Chandra's papers, arranged chronologically to emphasize the evolution of his thinking, from theoretical physics to group representations to harmonic analysis on reductive groups. Includes a brief biography, three surveys of his work, and frontispiece photographs in each volume. Expensive. **(a)**

Mathematische Abhandlungen. Helmut Hasse. Ed: Heinrich Wolfgang Leopoldt, Peter Roquette. Hawthorne, NY: de Gruyter, 1975. $467.00 set. ISBN 3-11-005931-2. *Band 1.* xv + 535pp. *Band 2.* xv + 525pp. *Band 3.* x + 532pp.

These 3 volumes contain the selected papers of Hasse, arranged by topics. Volume 1 contains Hasse's refections on the period during which he developed his celebrated local-global principle for quadratic forms. No survey of his work nor a biography is included in this collection. Expensive. **(a)**

Gesammelte Werke. Second Edition. Ludwig Otto Hesse. New York: Chelsea, 1972. xi + 727pp. $39.50. ISBN 0-8284-0261-2.

This corrected reprint of the 1897 volume edited by several of his students omits various textbooks by Hesse. Bibliography. **(a)**

Gesammelte Abhandlungen. Band I: Zahlentheorie. Fourth Edition. David Hilbert. New York: Chelsea, 1981. $19.50. ISBN 0-8284-0195-0.

Reprint of 1932-1935 original. *Band II: Algebra. Invariantentheorie. Geometrie* and *Band III: Analysis. Grundlagen der Mathematik, Physik, Verschiedenes, Lebensgeschichte* are out-of-print. **(a)**

The Collected Mathematical Works of George William Hill. New York: Johnson Reprint, 1965. $265.00 set. ISBN 0-384-23255-8. *Volume One*. xviii + 363pp. *Volume Two*. v + 339pp. *Volume Three*. 577pp. *Volume Four*. vi + 577pp.

Reprint of 1905-1907 Carnegie Institution of Washington original. Expensive. **(a)**

Einar Hille: Classical Analysis and Functional Analysis, Selected Papers. Ed: Robert R. Kallman. Mathematicians of Our Time;11. Cambridge, MA: MIT Press, 1975. xlvii + 708pp. $65.00. ISBN 0-262-08080-X.

Hille's own selection of his important papers that document the transition from classical to functional analysis. **(a)**

Gesammelte Abhandlungen, Collected Papers. Friedrich Hirzebruch. New York: Springer-Verlag, 1987. $199.00 set. ISBN 0-387-18087-7. *Band I, 1951-1962*. viii + 814pp. *Band II, 1963-1987*. iv + 818pp.

Virtually all of Hirzebruch's papers, in chronological order, with added commentary by the author to indicate progress since original publication. Expensive. **(a)**

Pao-Lu Hsu Collected Papers. Ed: Kai Lai Chung. New York: Springer-Verlag, 1983. xii + 589pp. $98.00. ISBN 0-387-90725-4.

English reprints of all of Hsu's published mathematical papers as well as biographical information and short discussions of Hsu's work in inference, multivariate analysis, and probability. **(a)**

Loo-Keng Hua Selected Papers. Ed: H. Halberstam. New York: Springer-Verlag, 1983. xiv + 889pp. $98.00. ISBN 0-387-90744-0.

Selection of about 50 papers, supplemented by topical surveys on number theory, algebra and geometry, and function theory. Brief biography. Complete publication list. **(a)**

Kiyosi Ito: Selected Papers. Ed: Daniel W. Stroock, S. R. S. Varadhan. New York: Springer-Verlag, 1987. xxi + 647pp. $69.00. ISBN 0-387-96326-X.

This collection offers the major part of Ito's work on stochastic integrals and stochastic differential equations. Includes an autobiography. **(a)**

Gesammelte Werke. Second Edition. C. G. J. Jacobi. New York: Chelsea, 1969. $275.00 set (volumes 1-7). ISBN 0-8284-0226-4. *Erster Band*. x + 546pp. *Zweiter Band*. vi + 527pp. *Dritter Band*. viii + 612pp. *Vierter Band*. 541pp. *Funfter Band*. vi + 515pp. *Sechster Band*. viii + 433pp. *Siebenter Band*. viii + 440pp. *Supplementband* (Volume 8). vi + 300pp. $20.00. ISBN 0-8284-0227-2.

Volumes 1-7 of the 1881-1891 edition have been reprinted with minor changes and rearrangements. These were edited by C. W. Borchardt and K. Weierstrasse. Volume 8, consisting of Jacobi's *Vorlesungen über Dynamik*, was first published in 1866 and edited by A. Clebsch and in a Second edition in 1884 edited by K. Weierstrasse. It is this Second edition which has been reprinted, unaltered. Expensive. **(a)**

Nathan Jacobson: Collected Mathematical Papers. Nathan Jacobson. Boston: Birkhäuser, 1989. $225.00 (set). ISBN 0-8176-3362-6. *Volume 1* [*1934-1946*]. xviii + 454pp. $79.00. ISBN 0-8176-3410-X. *Volume 2* [*1947-1965*]. xviii + 556pp. $79.00. ISBN 0-8176-3411-8. *Volume 3* [*1965-1988*]. xviii + 596pp. $79.00. ISBN 0-8176-3446-0.

Complete collection of Jacobson's papers, both research and expository. He begins each volume with an autobiographical account of the specified period. Expensive. (**a**)

Jack Carl Kiefer Collected Papers. Ed: Lawrence D. Brown et al. New York: Springer-Verlag, 1985-1986. $145.00 set. ISBN 0-387-96003-1. *I: Statistical Inference and Probability* [*1951-1963*]. xxxi + 502pp. $42.00. ISBN 0-387-96003-1. *II: Statistical Inference and Probability* [*1964-1984*]. xiii + 590pp. $42.00. ISBN 0-387-96003-1. *III: Design of Experiments*. xxiv + 718pp. $66.00. ISBN 0-387-96004-X. *Supplementary Volume*. vi + 56pp. o.p. ISBN 0-387-96383-9.

Includes all of Kiefer's papers as well as brief commentaries. Short biography. Complete bibliography. (**a**)

Collected Works. Kunihiko Kodaira. Princeton, NJ: Princeton University Press, 1975. $138.00 set. ISBN 0-685-51710-1. *Volume I*. xx + 647pp. $75.00. ISBN 0-691-08158-1. *Volume II*. x + 494pp. $60.00. ISBN 0-691-08163-8. *Volume III*. x + 480pp. $60.00. ISBN 0-691-08164-6.

This set of Kodaira's collected works was issued on the occasion of his 60th birthday and contain 70 articles from the period 1937 through 1971. The first volume includes a brief biography, a detailed survey of his work, and a list of those papers that serve to link Kodaira's main fields of interest. Expensive. (**a**)

Selected Works of A. N. Kolmogorov. Ed: A. N. Shiryaev. Transl: G. Lindquist. Mathematics and Its Applications (Soviet Series);25,26,27. Dordrecht: Kluwer, 1991-1992. *Volume I: Mathematics and Mechanics*. xix + 551pp. $199.00. ISBN 900277-2796-1. *Vol. II: Probability Theory and Mathematical Statistics*. xvi + 597pp. $199.00. ISBN 9-0277-2797-X. *Volume III: Information Theory and the Theory of Algorithms*. xxv + 275pp. $132.00. ISBN 9-0277-2798-8.

The annotated English translation of Kolmogorov's selected works, originally published in Russian in 1985-1987. Expensive. (**a**)

Leopold Kronecker's Werke. Leopold Kronecker. Ed: K. Hensel. New York: Chelsea, 1968. $175.00 set. ISBN 0-8284-0224-8. *Vol. I*. ix + 483pp. *Vol. II*. viii + 540pp. *Vol. III, Part I*. vii + 473pp. *Vol. III, Part II*. iii + 215pp. (Parts I and II bound as one) *Vol. IV*. x + 508pp. *Vol. V*. x + 527pp.

Reprint of 1895-1930 original. Errata have been corrected and half-title pages before each paper have been omitted. Expensive. (**a**)

Oeuvres de Laguerre, Publiées Sous les Auspices de l'Académie des Sciences. Second Edition. Ch. Hermite, H. Poincaré, E. Rouché. New York: Chelsea, 1972. $69.50 set. ISBN 0-8284-0263-9. *Tome 1: Algèbre. Calcul Intégral*. xv + 468pp. *Tome 2: Géométrie*. 711pp.

Unaltered reprint (except for the correction of errata) of 1898-1905 original. (**a**)

Selected Papers. S. Lefschetz. New York: Chelsea, 1971. 639pp. $39.50. ISBN 0-8284-0234-5.

A collection of 18 papers on topology dating from 1921 to 1954 plus a complete Lefschetz bibliography through 1969. **(a)**

G. Lejeune Dirichlet's Werke. Peter Gustav Lejeune-Dirichlet. New York: Chelsea, 1969. 2 v. in 1. (644 + 422pp). $59.50. ISBN 0-8284-0225-6.

Reprint of two volumes originally published in 1889-1897 of his papers, a few letters, and commentary. **(a)**

Collected Works. Stanislaw Lesniewski. Ed: Stanislaw J. Surma, Jan T. Srzednicki, D. I. Barnett. Nijhoff International Philosophy Series;44. Boston: Kluwer, 1992. *Volume I.* xvi + 382pp. *Volume II.* vi + 412pp. $274.00 ISBN 0-7923-1512-X.

First English language version of all of Lesniewski's papers on mathematical logic and the foundations of mathematics, originally published in either Polish or German. Annotated bibliography by V. Frederick Rickey. Index. Expensive. **(a)**

Collected Papers of J. E. Littlewood. New York: Oxford University Press, 1982. *V.I.* xxxviii + 790pp. $175.00. ISBN 0-19-853353-5. *V.II.* xxxviii + 886pp. $135.00. ISBN 0-19-853355-1.

All of Littlewood's papers arranged in five groups. Most papers are followed by a brief commentary by the editors to provide modern perspective. Introductory biographical memoir. Expensive. **(a)**

Charles Loewner, Collected Papers. Ed: Lipman Bers. Boston: Birkhäuser, 1988. xii + 517pp. $79.00. ISBN 0-8176-3377-4.

Loewner's published contributions in the areas of complex analysis and differential geometry have been gathered in this work. **(a)**

Selected Papers: Saunders Mac Lane. Ed: I. Kaplansky. New York: Springer-Verlag, 1979. xiii + 556pp. $65.00. ISBN 0-387-90394-1.

Twenty-one papers along with a biographical note by Putnam, a personal tribute by Lyndon, and technical reviews of Mac Lane's work by Kaplansky, Eilenberg, and Kelly. Complete bibliography and list of his Ph.D. students. **(a)**

The Collected Letters of Colin MacLaurin. Ed: Stella Mills. Nantwich, Cheshire: Shiva Publishing, 1982. xx + 496pp. $35.00. ISBN 0-906812-08-9.

One of Scotland's greatest mathematicians, MacLaurin helped to interpret and defend Newton. This volume of letters traces the development of his ideas in exchanges with Newton and Stirling and other 18th century scientists and mathematicians. **(a)**

Percy Alexander MacMahon: Collected Papers. Ed: George E. Andrews. Mathematicians of Our Time;13,24. Cambridge, MA: MIT Press, 1978-1986. *Volume I: Combinatorics.* xxix + 1438pp. $75.00. ISBN 0-262-13121-8. *Volume II: Number Theory, Invariants, and Applications.* xxv + 952pp. $95.00. ISBN 0-262-13214-1.

MacMahon's work on symmetric functions, determinants, number theory, and invariant theory with commentary, references, and summaries. **(a)**

Selected Papers of Demetrios C. Magiros: Applied Mathematics, Nonlinear Mechanics, and Dynamical Systems Analysis. Ed: S. G. Tzafestas. Norwell, MA: Kluwer Academic, 1985. xv + 518pp. $153.00. ISBN 90-277-2003-7.

This book contains 43 of Magiros' 54 papers from 1946 to 1984. An appendix has 2 papers in Russian and at the end of the book is a list of unpublished works. Expensive. **(a)**

Wilhelm Magnus, Collected Papers. Ed: Gilbert Baumslag, Bruce Chandler. New York: Springer-Verlag, 1984. xvi + 726pp. $59.80. ISBN 0-387-90879-X.

Collected works of one of the few mathematicians who has done significant work in two unrelated fields: combinatorial group theory and diffraction problems and related topics in analysis. **(a)**

The Collected Papers of Raymond D. Mindlin. Ed: H. Deresiewicz, M. P. Bieniek, F. L. DiMaggio. New York: Springer-Verlag, 1989. $143.00 set. ISBN 0-387-96933-0. *Volume I.* xxviii + 596pp. *Volume II.* xxviii + 591pp. Copies of 100 of his research papers in applied mechanics from 1934 to 1986. Expensive. **(a)**

Gesammelte Abhandlungen. Hermann Minkowski. New York: Chelsea, 1967. 2 v. in 1. $59.50. ISBN 0-8284-0208-6.

Collected works of Minkowski plus a memorial to Minkowski which Hilbert delivered in 1909. **(a)**

Collected Papers, Marston Morse. Marston Morse. Teaneck, NJ: World Scientific, 1987. $370 set. ISBN 9971-978-94-6. *Volume 1.* xlix + 532pp. *Volume 2.* vi + 598pp. *Volume 3.* vi + 639pp. *Volume 4.* vi + 602pp. *Volume 5.* vi + 578pp. *Volume 6.* vi + 559pp.

The collected works of Marston Morse containing over 50 papers on Morse theory. Expensive. **(a)**

The Mathematical Papers of Isaac Newton. Ed. D. T. Whiteside, M. A. Hoskin. New York: Cambridge University Press, 1967-1981. *V. 1: 1664-1666.* xlvi + 590pp. out-of-print. No ISBN. *V. 2: 1667-1670.* xxii + 520pp. out-of-print. No ISBN. *V. 3: 1670-1673.* xxxvii + 576pp. $155.00. ISBN 0-521-07119-4. *V. 4: 1674-1684.* xxxii + 678pp. $175.00. ISBN 0-521-07740-0. *V. 5: 1683-1684.* xxii + 627pp. $182.50. ISBN 0-521-08262-5. *V. 6: 1684-1691.* xxxiv + 614pp. $182.50. ISBN 0-521-08719-8. *V. 7: 1691-1695.* xlvii + 706pp. $222.50. ISBN 0-521-08720-1. *V. 8: 1697-1722.* lv + 704pp. $240.00. ISBN 0-521-20103-9.

The first edition of all of Newton's extant mathematical notes and papers has been carefully annotated. Each volume has a general introduction as well as introductions and numerous notes to each section, English translations to Newton's Latin pieces, manuscript facsimiles, appendices, detailed analytical table of contents, and index of names. Expensive. **(a,p)**

Joint Statistical Papers of J. Neyman and E. S. Pearson. J. Neyman, E. S. Pearson. Berkeley, CA: University of California Press, 1967. iv + 299pp. $48.00. ISBN 0-520-00991-6. Companion volume to *The Selected Papers of E. S. Pearson* (out-of-print; not listed) and *A Selection of Early Statistical Papers of J. Neyman* (see below).

This selection includes 10 joint papers published from 1928 to 1938 and concludes with a paper by Neyman alone. **(a)**

A Selection of Early Statistical Papers of J. Neyman. J. Neyman. Berkeley, CA: University of California Press, 1967. ix + 429pp. $48.00. ISBN 0-520-00992-4.

This volume of Neyman's selected papers includes 28 contributions (4 of them joint) published or written before the early 1940's. A bibliography lists 144 papers and 12 books and monographs either written or edited by the author. **(a)**

Emmy Noether: Gesammelte Abhandlungen, Collected Papers. N. Jacobson. New York: Springer-Verlag, 1983. viii + 777pp. $89.00. ISBN 0-387-11504-8.

While the collected papers are in German, Paul Alexandroff's memorial address and Nathan Jacobson's introduction are in English. **(a)**

Kiyoshi Oka, Collected Papers. Ed: Hans Grauert, Reinhold Remmert. Die Grundlehren der mathematischen Wissenschaften;265. New York: Springer-Verlag, 1984. xiv + 223pp. $89.00. ISBN 0-387-13240-6.

Ten fundamental papers (1936-1962) and 2 brief notes, followed by commentary, by one of the founders of function theory of several complex variables. **(a)**

Collected Mathematical Papers. Alexander Ostrowski. Boston, MA: Birkhäuser, 1983-1985. $488.20 set. ISBN 0-8176-1512-1. *Volume 1: Determinants, Linear Algebra, Algebraic Equations.* 904pp. o.p. ISBN 3-7643-1506-7. *Volume 2: Multivariate Algebra, Formal Algebra.* 652pp. o.p. ISBN 3-7643-1507-5. *Volume 3: Number Theory, Geometry, Topology, Convergence.* 532pp. o.p. ISBN 3-7643-1508-3. *Volume 4: Real Function Theory, Differential Equations, Differential Transformations.* 632pp. $82.95. ISBN 3-7643-1509-1. *Volume 5: Complex Function Theory.* 544pp. $77.95. ISBN 3-7643-1510-5. *Volume 6: Conformal Mapping, Numerical Analysis, Miscellany.* 718pp. $99.95. ISBN 3-7643-1511-3.

Papers of one of the twentieth century's leading mathematicians. **(a)**

Writings of Charles S. Peirce: A Chronological Edition. Ed: Christian J. W. Kloesel. Bloomington, IN: Indiana University Press, 1982-1991. *Volume 1, 1857-1866.* xxxv + 698pp. $52.50. ISBN 0-253-37201-1. *Volume 2, 1967-1871.* xlviii + 649pp. $57.50. ISBN 0-253-37202-X. *Volume 3, 1872-1878.* xxxvii + 633pp. $57.50. ISBN 0-253-37203-8. *Volume 4, 1879-1884.* lxx + 698pp. $67.50. ISBN 0-253-37204-6.

First four volumes of a monumental work containing numerous manuscripts never before published and extensive scholarly and editorial notes. **(a)**

George Pólya: Collected Papers. Ed: R. P. Boas, Joseph Hersch, Gian-Carlo Rota. Mathematicians of Our Time;7-8,21-22. Cambridge, MA: MIT Press, 1974-1984. *I: Singularities of Analytic Functions.* xiv + 808pp. $70.00. ISBN 0-262-02104-8. *II:*

Location of Zeros. x + 444pp. $60.00. ISBN 0-262-02103-X. *III: Analysis.* x + 536pp. $65.00. ISBN 0-262-16096-X. *IV: Probability: Combinatorics; Teaching and Learning in Mathematics.* ix + 642pp. $70.00. ISBN 0-262-16097-8.

Each volume concludes with brief commentaries on the papers and a complete Polya bibliography keyed to the four volumes. (**a**)

Collected Papers of Hans Rademacher. Ed: Emil Grosswald. Cambridge, MA: MIT Press, 1974. $135.00 set. ISBN 0-685-03383-X. *Volume I.* xix + 692pp. $70.00. ISBN 0-262-07054-5. *Volume II.* xxi + 638pp. $70.00. ISBN 0-262-07055-3.

These 2 volumes contain all the papers published by Rademacher, either alone or as a joint author, essentially in chronological order. Volume 2 includes list of his writings which are not included in this present collection and a list of dissertations directed by Rademacher. The editor has provided notes for each paper and contributed a biographical sketch. (**a**)

Ramanujan's Notebooks. Srinivasa Ramanujan Aiyangar. Ed: Bruce C. Berndt. New York: Springer-Verlag, 1985-1991. *Part I.* x + 537pp. $76.00. ISBN 0-387-96110-0. *Part II.* xi + 359pp. $79.80. ISBN 0-387-96794-X. *Part III.* xiii + 510pp. $89.80. ISBN 0-387-97503-9.

During the years 1903-1914, Ramanujan recorded many of his mathematical discoveries in notebooks without providing proofs and many of his results were not in the literature. In the 1930s Watson and Wilson began to edit the notebooks, but never completed the task. In 1957 a photostat edition was published in Bombay. These are the first 3 volumes of a projected 5 volume set devoted to the editing of Ramanujan's notebooks. Both a tribute to Ramanujan's genius and Berndt's perseverence. (**a**)

Srinivasa Ramanujan: The Lost Notebook and Other Unpublished Papers. Srinivasa Ramanujan. New York: Springer-Verlag, 1988. xxv + 419pp. $59.50. ISBN 0-387-18726-X.

This book contains reproductions of the "Lost Notebook," several letters between Ramanujan and his colleagues, "Loose Papers," and 117 pages of unpublished work related to various papers. (**a**)

Gesammelte mathematische Werke, wissenschaftlicher Nachlass und Nachtrage. Bernhard Riemann. Ed: Raghavan Narasimhan. Berlin: Springer-Verlag, 1990. vi + 911pp. $120.00. ISBN 03-540-50033-2.

Based on the edition by H. Weber and R. Dedekind, this present edition reproduces these collected works and adds extensive comments by Narasimhan on Riemann's influence in mathematics and other interesting reports on special aspects. Bibliography and secondary literature on Riemann. Expensive. (**a**)

Marcel Riesz: Collected Papers. Ed: Lars Garding, Lars Hormander. New York: Springer-Verlag, 1988. vi + 897pp. $79.00. ISBN 0-387-18115-6.

Almost all of the published papers, largely in French and German. (**a**)

Herbert Robbins: Selected Papers. Ed: T. L. Lai, D. Siegmund. New York: Springer-Verlag, 1985. xli + 518pp. $59.00. ISBN 0-387-96137-2.

A selection of nearly half of Robbins' 133 publications arranged in 3 groups: empirical Bayes methodology, sequential experimentation, and probability and inference. Complete list of his publications. **(a)**

Selected Papers of Abraham Robinson. Ed: H. J. Keisler et al. New Haven, CT: Yale University Press, 1979. *Volume 1: Model Theory and Algebra.* xxxvii + 694pp. $80.00. ISBN 0-300-02071-6. *Volume 2: Nonstandard Analysis and Philosophy.* xlv + 582pp. $70.00. ISBN 0-300-02072-4. *Volume 3: Aeronautics.* xxxii + 270pp. $60.00. ISBN 0-300-02073-2.

This selection contains Robinson's most important papers (98 out of 134). Each volume begins with a superbly written biography by George Seligman, a topical introduction by one of the editors, and frontispiece photographs. Each concludes with a complete bibliography and a list of Robinson's doctoral students. **(a)**

Gesammelte mathematische Abhandlungen. Ludwig Schläfli. Boston: Birkhäuser, 1950-1956. *Band I.* 392pp. $61.95. ISBN 0-8176-0328-X. *Band II.* 381pp. $61.95. ISBN 0-8176-0329-8. *Band III.* 402pp. $61.95. ISBN 0-8176-0330-1.

Schaefli, a great Swiss mathematician, rediscovered the foundations of n-dimensional Euclidean and spherical geometry and developed it in a manner that was not approached by anyone else for half a century. These collected works, which could be improved if arranged more systematically, are profusely annotated by Burkhardt and others. **(a)**

I.J. Schoenberg, Selected Papers. Ed: Carl de Boor. Boston: Birkhäuser, 1988. $150.00 set. ISBN 0-8176-3378-2. *Volume 1.* xviii + 405pp. *Volume 2.* xvi + 441pp.

Sampling of 45 of Schoenberg's 173 papers, spanning the period 1928-87, and covering number theory, geometry, real analysis, complex analysis (v.1) and aspects of spline theory (v.2). Interspersed are brief commentaries by experts who summarize his contributions. Expensive. **(a)**

Collected Papers on Wave Mechanics. E. Schrödinger. New York: Chelsea, 1982. xiii + 207pp. $17.95. ISBN 0-8284-1302-9.

Third English edition of the 1927 German original along with the text of the 1928 "Four Lectures on Wave Mechanics." **(a)**

Gesammelte Mathematische Abhandlungen. Second Edition. Hermann Amandus Schwarz. New York: Chelsea, 1972. xiv + 338 + vii + 370pp. $49.50. ISBN 0-8284-0260-4.

A reprint with corrections of the two 1890 volumes which were themselves a revision, by Schwarz himself, of earlier papers. **(a)**

Selected Papers (1937-1976) of Julian Schwinger. Ed: M. Flato, C. Fronsdal, K. A. Milton. Dordrecht: D. Reidel, 1979. (Available from Kluwer Academic, Norwell, MA). xxvii + 413pp. $29.40; $11.95. ISBN 90-277-0974-2; 90-277-0975-0.

A sixtieth birthday retrospective selection (by Schwinger) of some 50 papers, resulting in the omission of many important papers. Complete bibliography and brief comments on the importance of the papers included. (**a**)

Geometry and Combinatorics: Selected Works of J. J. Seidel. Ed: D. G. Corneil, R. Mathon. San Diego, CA: Academic Press, 1991. xix + 410pp. $69.50. ISBN 0-12-189420-7.

Twenty-eight research and survey papers highlight Seidel's work on the interplay of geometry, combinatorics, and algebra. (**a**)

Collected Papers. Atle Selberg. New York: Springer-Verlag, 1989-1991. *Volume I.* vi + 711pp. $130.00. ISBN 0-387-18389-2. *Volume II.* viii + 251pp. $98.00. ISBN 0-387-50626-8.

Volume 1 ranges from Selberg's early work in analysis to number theory to his later work that enlarged and transfigured the whole concept and structure of arithmetic. Volume 2 contains material (e.g., sieve methods, Eisenstein series, automorphic forms) on which Selberg has lectured recently, but which had not so far appeared in print. Bibliography. Autobiographical note. Expensive. (**a**)

Jean-Pierre Serre: Oeuvres, Collected Papers. Jean-Pierre Serre. New York: Springer-Verlag, 1986. $249 set. ISBN 0-387-15621-6. *Volume I: 1949-1959.* xviii + 596pp. *Volume II: 1960-1971.* iv + 740pp. *Volume III: 1972-1984.* iv + 728pp.

Almost all (132) of Serre's published papers, arranged chronologically, plus a selection of his seminar notes, and summaries of his courses at College de France since 1956. Expensive. (**a**)

Collected Mathematical Papers. Igor R. Shafarevich. New York: Springer-Verlag, 1989. vi + 769pp. $99.00. ISBN 0-387-13618-5.

Contains 43 papers, most in English translation, published between 1943-1984. (**a**)

Collected Mathematical Papers. Henry John Stephen Smith. Ed: J. W. L. Glaisher. New York: Chelsea, 1965. $99.50 set. ISBN 0-8284-0187-X. *Volume I.* xcv + 603pp. *Volume II.* vii + 719pp.

Reprint of 1894 edition. Includes a biographical sketch, recollections, and all of his papers published or in preparation at the time of his death, arranged chronologically. (**a**)

Selecta. Donald C. Spencer. Teaneck, NJ: World Scientific, 1985. $204.00 (set). ISBN 0-317-01208-8. *Volume 1.* xii + 656pp. *Volume 2.* viii + 657pp. *Volume 3.* viii + 441pp.

This set includes most of Spencer's 90 papers, arranged into 10 subject areas, a brief biography by J. J. Kohn, a complete bibliography of his work, and a list of the PhD theses supervised by Spencer. Expensive. (**a**)

Gesammelte Werke. Second Edition. Jacob Steiner. New York: Chelsea, 1971. $75.00 set. ISBN 0-8284-0233-7. *V. I.* viii + 527pp. *V. II.* x + 743pp.

Virtually unaltered reprinting of the collected works of the famous geometer which were originally published in 1881-1882 and edited by Weierstrasse. (**a**)

Selected Mathematical Papers. Su Buchin. New York: Gordon & Breach, 1983. xviii + 411pp. $82.00. ISBN 0-677-31300-4.

Twenty-six papers of Su's work on affine and projective differential geometry are reprinted here, along with a tribute by S. S. Chern, a brief biography by his Chinese colleagues, and a bibliography of his publications. (**a**)

The Collected Mathematical Papers of James Joseph Sylvester. Ed: H. F. Baker. New York: Chelsea, 1973. $195.00 set. ISBN 0-8284-0253-1. *Volume I: 1837-1853*. xii + 659pp. *Volume II: 1854-1873*. xvi + 746pp. *Volume III: 1870-1883*. xv + 697pp. *Volume IV: 1882-1897*. xxxvii + 756pp.

Corrected reprint of the 1904-1912 edition. General index and biographical notice appear in volume IV. Expensive. (**a**)

Gabor Szegö: Collected Papers. Ed: Richard Askey. Boston: Birkhäuser, 1982. $245.00 (set). ISBN 0-8176-3063-5. *Volume 1: 1915-1927*. xx + 857pp. $81.95. ISBN 3-7643-3056-2. *Volume 2: 1927-1943*. x + 869pp. $81.95. ISBN 3-7643-3060-0. *Volume 3: 1945-1972*. x + 880pp. $81.95. ISBN 3-7643-3061-9.

All of Szego's papers and introduced with a short biography, plus a commentary on Szego's life and work, reviews of 3 of his books, and two mathematical commentaries. Expensive. (**a**)

Alfred Tarski, Collected Papers. Ed: Steven R. Givant, Ralph N. McKenzie. Boston: Birkhäuser, 1986. $630.00 set. ISBN 0-8176-3284-0. *Volume 1, 1921-1934*. 659pp. *Volume 2, 1935-1944*. 699pp. *Volume 3, 1945-1957*. 682pp. *Volume 4, 1958-1979*. 757pp.

A complete collection of all Tarski papers, abstracts, and reviews forming the "bedrock" of modern logic. Comprehensive scholarly bibliography. Expensive. (**a**)

The Collected Works of John W. Tukey. Ed: David R. Brillinger, Lyle V. Jones, William S. Cleveland, Colin L. Mallows. Pacific Grove, CA: Wadsworth & Brooks-Cole, 1984-1990. *Volume I: Time Series, 1949-1964*. lxv + 689pp. $57.95. ISBN 0-534-03303-2. *Volume II: Time Series, 1965-1984*. lxvii + 582pp. $52.95. ISBN 0-534-03304-0. *Volume III: Philosophy and Principles of Data Analysis, 1949-1964*. lxviii + 569pp. $57.95. ISBN 0-534-03305-9. *Volume IV: Philosophy and Principles of Data Analysis, 1965-1986*. lxviii + 553pp. $57.95. ISBN 0-534-05101-4. *Volume V: Graphics, 1965-1985*. lxiv + 464pp. $57.95. ISBN 0-534-05102-2. *Volume VI: More Mathematical: 1938-1984*. lxxii + 661pp. $57.95. ISBN 0-534-05103-0.

The first 6 volumes in a series of Tukey's many contributions to statistics. Includes some unpublished papers. (**a**)

Selected Papers of W. T. Tutte. Ed: D. McCarthy, R.G. Stanton. St. Pierre, MB: Charles Babbage Research Centre, 1979. $39.50 set. ISBN 0-9690778-0-7. *V.1*. 438pp. *V.2*. 440pp.

Fifty-three of the most influential papers from the work of one of the world's most prolific and profound graph theorists, honoring him on his 60th birthday. Tutte writes the introductions, giving historical and insightful perspectives. (**a**)

Stanislaw Ulam: Sets, Numbers, and Universes: Selected Works. Ed: W. A. Beyer, J. Mycielski, G. -C. Rota. Mathematicians of Our Time;9. Cambridge, MA: MIT Press, 1974. xxiii + 709pp. $70.00. ISBN 0-262-13094-7.

This book contains 52 articles by the author on mathematical topics in pure mathematics and the application of computers to game theory and physics, together with commentaries by a number of distinguished specialists in these fields. Also included is a 108 item bibliography and a complete reprinting of his book, *A Collection of Mathematical Problems.* **(a)**

Ivan Matveevic Vinogradov: Selected Works. Ed: L. D. Faddeev et al. New York: Springer-Verlag, 1985. xiii + 401pp. $139.00. ISBN 0-387-12788-7.

Sixteen articles plus two monographs on the 'method of trigonometric sums.' Expensive. **(a)**

Collected Works. John Von Neumann. Ed: A. H. Taub. Elmsford, NY: Pergamon, 1961-1963. $1,420.00 set. ISBN 0-08-009566-6. *Volume 1: Logic, Theory of Sets, and Quantum Mechanics.* x + 654pp. *Volume 2: Operators, Ergodic Theory and Almost Periodic Functions in a Group.* ix + 568pp. *Volume 3: Rings of Operators.* ix + 574pp. *Volume 4: Continuous Geometry and Other Topics.* x + 516pp. *Volume 5:Design of Computers, Theory of Automata and Numerical Analysis.* viii + 784pp. *Volume 6: Theory of Games, Astrophysics, Hydrodynamics, and Meteorology.* viii + 538pp.

These 6 volumes contain all the articles published by Von Neumann as well as some of his reports to government agencies and other groups. A useful feature of these well-edited volumes is their indication of unpublished manuscripts found in his files. While papers are generally arranged in chronological order, exceptions have been made in order to preserve a certain unity in some fields. Expensive. **(a)**

Selected Papers in Statistics and Probability. Abraham Wald. Stanford, CA: Stanford University Press, 1969. ix + 702pp. $70.00. ISBN 0-8047-0493-7.

Reprint of 1957 edition. **(a)**

Mathematical Works of J. H. C. Whitehead. Ed: I. M. James. Elkins Park, PA: Franklin Books, 1991. $748.50 set. ISBN 0-08-009873-8. *Vol. I: Differential Geometry.* xxxiii + 361pp. $184.00. ISBN 0-08-009869-X. *Vol. II: Complexes and Manifolds.* xiii + 435pp. $202.31. ISBN 0-08-009870-3. *Vol. III: Homotopy Theory.* xiii + 451pp. $209.75. ISBN 0-08-009871-1. *Vol. IV: Algebraic and Classical Topology.* xiii + 347pp. $167.25. ISBN 0-08-009872-X.

Photocopy reproduction of the 1963 original. "These volumes are believed to contain all the published mathematical work of J. H. C. Whitehead, excluding reviews and lecture notes."—Preface. Expensive. **(a)**

Norbert Wiener: Collected Works with Commentaries. Ed: P. Masani. Mathematicians of Our Time;10,15,20,23. Cambridge, MA: MIT Press, 1976-1985. *Volume I.* xi + 761pp. o.p. ISBN 0-262-23070-4. *Volume II.* xiii + 969pp. $80.00. ISBN 0-262-23092-5. *Volume III.* xiii + 753pp. $70.00. ISBN 0-262-23107-7. *Volume IV.* xx + 1083pp. $85.00. ISBN 0-262-23123-9.

A four volume opus, with papers ranging from cybernetics to harmonic analysis to the Hopf-Wiener integral equation. Papers are grouped by topic and introduced by commentaries linking them to contemporary research. **(a)**

Selected Papers. Jacob Wolfowitz. Ed: Jack C. Kiefer, U. Augustin, Lionel Weiss. New York: Springer-Verlag, 1980. xxiii + 642pp. $79.00. ISBN 0-387-90463-8.

This volume, published to honor Wolfowitz on his 70th birthday, contains 50 of his papers from 1939-1979 and covers a wide range of problems in statistics. A biographical sketch is followed by a thirteen-page discussion of his work. Complete bibliography. **(a)**

Collected Works. Hidehiko Yamabe. New York: Gordon & Breach, 1967. xii + 142pp. $69.00. ISBN 0-677-00610-1.

Eighteen articles written or co-authored by Yamabe from 1950-1960. **(a)**

Selected Papers of Kentaro Yano. Ed: Morio Obata. North-Holland Mathematics Studies;70. New York: Elsevier, 1982. liii + 145pp. $82.00. ISBN 0-444-86495-4.

Yano has selected many of his research papers in differential geometry for this volume and provided a mathematical autobiography. **(a)**

Fuzzy Sets and Applications: Selected Papers by L.A. Zadeh. Ed: R.R. Yager et al. New York: Wiley, 1987. 684pp. $68.95. ISBN 0-471-85710-6.

This volume contains 17 papers by Zadeh on the subject he created, fuzzy sets, beginning with his 1965 paper. **(a)**

Collected Papers. Oscar Zariski. Ed: H. Hironaka, D. Mumford, M. Artin, B. Mazur, J. Lipman, B. Teissier. Mathematicians of Our Time;2,6,12,16. Cambridge, MA: MIT Press, 1972-1979. *V. I: Foundations of Algebraic Geometry and Resolution of Singularities.* xxi + 543pp. $60.00. ISBN 0-262-08049-4. *V. II: Holomorphic Functions and Linear Systems.* xxiii + 505pp. $60.00. ISBN 0-262-01038-0. *V. III: Topology of Curves and Surfaces, and Special Topics in the Theory of Algebraic Varieties.* xxvi + 480pp. $75.00. ISBN 0-262-24021-1. *V. IV: Equisingularity on Algebraic Varieties.* xxvi + 651pp. $75.00. ISBN 0-262-24022-4.

These volumes present Zariski's collected papers, arranged by fields, with editorial introductions that set his work in the entire field of algebraic geometry. Excluded are his 4 books, 2 sets of lecture notes, and 4 early expository articles on other fields of mathematics. Bibliography. **(a)**

Chapter 8

CURRENTLY PUBLISHED JOURNALS OF INTEREST TO MATHEMATICIANS

Nancy D. Anderson

These mathematics journals are arranged alphabetically by keyword. Cross references are provided if the entry is greatly different from the abbreviation used by *Mathematical Reviews.*

These references are composed of the complete title (using cataloged entry), followed by the title abbreviation used in *Mathematical Reviews*, the publisher and place of publication, the ISSN number (when known), and an indication if the title is appropriate for an academic (**a**), public (**p**), or special (**s**) library collection.

Abhandlungen aus dem Mathematischen Seminar der Universität Hamburg. [Abh. Math. Sem. Univ. Hamburg.] Vandenhoeck & Ruprecht, Göttingen. ISSN: 0025-5858 (**a**)

Abstracts of Papers Presented to the American Mathematical Society. [Abstracts Amer. Math. Soc.] Amer. Math. Soc., Providence, RI. ISSN: 0192-5857 (**a,s**)

ACM Transactions on Mathematical Software. [ACM Trans. Math. Software] ACM, New York. ISSN: 0098-3500 (**s**)

Acta Applicandae Mathematicae. [Acta. Appl. Math.] Kluwer Acad. Publ., Boston. ISSN: 0167-8019 (**a,s**)

Acta Arithmetica. [Acta Arith.] Polish Acad. Sci., Inst. Math., Warsaw. ISSN: 0065-1036 (**a**)

Acta Ciencia Indica. Mathematics. [Acta Cienc. Indica Math.] Pragati Prakashan, Meerut. (**a**)

Acta Cybernetica. [Acta Cybernet.] Jozsef Attila Univ., Szeged. ISSN: 0324-721X (**a**)

Acta Informatica. [Acta Inform.] Springer, Berlin-New York. ISSN: 0001-5903 (**s**)

Acta Mathematica. [Acta Math.] Inst. Mittag-Leffler, Djursholm. ISSN: 0001-5962 (**a**)

Acta Mathematica Hungarica. [Acta Math. Hungar.] Akad. Kiadó, Budapest. ISSN: 0236-5294 (**a**)

Acta Mathematica Scientia. English Edition. Shu Hsueh Li Hsueh Pao. See *Shu Hsueh Li Hsueh Pao = Acta Mathematica Scientia.*

Acta Mathematica Sinica. Shu Hsueh Hsueh Pao. See *Shu Hsueh Hsueh Pao = Acta Mathematica Sinica.*

Acta Mathematica Sinica. New Series. [Acta Math. Sinica (N.S.)] Kexue Chubanshe [Science Press], Beijing. ISSN: 0379-7570 (**a**)

Acta Mathematica Universitatis Comenianae. (N.S.) [Acta Math. Univ. Comenian.] Alfa, Bratislava. ISSN: 0862-9544 (**a**)

Acta Mathematica Vietnamica. [Acta Math. Vietnam.] Vien Toan-Hoc, Hanoi. ISSN: 0251-4184 (**a**)

Acta Mathematicae Applicatae Sinica. English Series = Ying Yung Shu Hsueh Hsueh Pao. [Acta. Math. Appl. Sinica (English Ser.)] Kexue Chubanshe [Science Press], Beijing. ISSN: 0168-9673 (**a**)

Acta Mathematicae Applicatae Sinica. Yingyong Shuxue Xuebao. See *Ying Yung Shu Hsueh Hsueh Pao = Acta Mathematicae Applicatae Sinica.*

Acta Polytechnica Scandinavica. Mathematics and Computer Science Series. [Acta Polytech. Scand. Math. Comput. Sci. Ser.] Finn. Acad. Tech. Sci., Helsinki. ISSN: 0355-2713 (**a**)

Acta Scientiarum Mathematicarum. [Acta Sci. Math. (Szeged)] Jozsef Attila Univ., Szeged. ISSN: 0001-6969 (**a**)

Acta Universitatis Carolinae. Mathematica et Physica. [Acta Univ Carolin.-Math. Phys.] Univ. Karlova, Prague. ISSN: 0001-7140 (**a**)

Acta Universitatis Lodziensis Folia Mathematica. [Acta Univ. Lodz. Folia Math.] Univ. Lodz, Lodz. ISSN: 0208-6204 (**a**)

Acta Universitatis Ouluensis. Series A. Scientiae Rerum Naturalium. [Acta Univ. Oulu. Ser. A Sci. Rerum Natur.] Univ. Oulu, Oulu. ISSN: 0355-3191 (**a**)

Acta Universitatis Palackianae Olomucensis. Facultas Rerum Naturalium. Mathematica. [Acta Univ. Palak. Olomuc. Fac. Rerum Natur. Math.] Statni Ped. Naklad., Prague. (**a**)

Acta Universitatis Szegediensis. Acta Scientiarum Mathematicarum. See *Acta Scientiarum Mathematicarum.*

Advances in Applied Mathematics. [Adv. in Appl. Math.] Academic Press, Orlando, FL. ISSN: 0196-8858 (**a**)

Advances in Applied Probability. [Adv. in Appl. Probab.] Appl. Probab. Trust, Sheffield. ISSN: 0001-8678 (**a,s**)

Advances in Mathematics. [Adv. Math.] Academic Press, Duluth, MN. ISSN: 0001-8708 (**a,s**)

Advances in Mathematics. Shuxue Jinzhan. See *Shuxue Jinzhan. Advances in Mathematics.*

Advances in Numerical Methods for Large Sparse Sets of Linear Equations. [Adv. Numer. Methods Large Sparse Sets Linear Equations] Keio Univ., Tokyo. (**a**)

Aequationes Mathematicae. [Aequationes Math.] Birkhäuser, Basel-Boston, MA. ISSN: 0001-9054 (**a,s**)

Afrika Matematika. Series 2. [Afrika Mat. 2] Uniben Press, Benin City. (**a**)

Akademii͡a Nauk Armi͡anskoĭ SSR. Doklady. [Akad. Nauk. Armyan. SSR Dokl.] Akad. Nauk Armyan. SSR, Erevan. ISSN: 0321-1339

Akademiia Nauk Azerbaidzhanskoi SSR, Baku. Doklady. [Akad. Nauk Azerbaidzhan. SSR. Dokl.] "Elm," Baku. ISSN: 0002-3078

Akademiia Nauk Ukrainy. Institut Mathematiki. Preprint. See *Preprint* (*Instytut Matematyky* [*Akademiia Nauk Ukrainskoi RSR*]).

Algebra i Analiz. [Algebra i Analiz] "Nauka" Leningrad. Otdel., Leningrad. ISSN: 0234-0852 (**a**)

Algebra and Logic. [Algebra and Logic] Consultants Bureau, New York-London. ISSN: 0002-5232 (**a**)

Algebra i Logika. [Algebra i Logika] Akad. Nauk SSSR Sibirsk. Otdel., Inst. Mat., Novosibirsk. ISSN: 0373-9252 (**a**)

Algebra Universalis. [Algebra Universalis.] Birkhäuser, Basel-Boston, MA. ISSN: 0002-5240 (**a,s**)

Algebras, Groups and Geometries. [Algebras Groups Geom.] Hadronic Press, Nonantum, MA. ISSN: 0741-9937 (**a**)

Algorithmica. [Algorithmica] Springer, New York-Berlin. ISSN: 0178-4617

The Aligarh Bulletin of Mathematics. [Aligarh Bull. Math.] Aligarh Muslim Univ., Aligarh. (**a**)

The Aligarh Journal of Statistics. [Aligarh J. Statist.] Aligarh Muslim Univ., Aligarh. (**a**)

Alkalmazott Matematikai Lapok. [Alkalmaz. Mat. Lapok] Akad. Kiadó, Budapest. ISSN: 0133-3399 (**a**)

American Journal of Mathematical and Management Sciences. [Amer. J. Math. Management Sci.] Amer. Sci. Press, Columbus, OH. ISSN: 0196-6324 (**a**)

American Journal of Mathematics. [Amer. J. Math.] Johns Hopkins Univ. Press. Baltimore, MD. ISSN: 0002-9327 (**a,p,s**)

The American Mathematical Monthly. [Amer. Math. Monthly] Math. Assoc. America, Washington, DC. ISSN: 0002-9890 (**a,p,s**)

The American Statistician. [Amer. Statist.] Amer. Statist. Assoc., Alexandria, VA. ISSN: 0003-1305 (**a,s**)

Analele Ştiinţifice ale Universiţatii "Al. I. Cuza" din Iaşi. Seria Nouă. Secţiunea I a Matematică. [An. Stiint. Univ. "Al. I. Cuza" Iaşi Sect. I a Mat. (N.S.)] Univ. Al. I. Cuza, Iaşi. ISSN: 0041-9109 (**a**)

Analele Universităţii Bucureşti. Matematică. [An. Univ. Bucureşti. Mat.] Univ. Bucureşti, Bucharest. (**a**)

Analele Universităţii din Craiova. Matematică, Fizică-Chimie. [An. Univ. Craiova Mat. Fiz.-Chim.] Univ. Craiova, Craiova.

Analele Universităţii din Timişoara. Seria Ştiinţe Matematice. [An. Univ. Timişoara Ser. Ştiinţ. Mat.] Univ. Timişoara, Timişoara. (**a**)

Anales del Instituto de Matemáticas. [An. Inst. Mat. Univ. Nac. Autónoma México] Universidad Nacional Autónoma de México. Univ. Nac. Autónoma México, Mexico City. ISSN: 0185-0644 (**a**)

L'Analyse Numérique et la Théorie de L'Approximation. [Anal. Number. Theor. Approx.] Ed. Acad. R.S. Romania, Bucharest. ISSN: 0025-5505 (**a**)

Analysis. [Analysis] Oldenbourg, Munich. ISSN: 0174-4747 (**a**)

Analysis. [Analysis (Oxford)] Blackwell, Oxford. ISSN: 0003-2638

Analysis Mathematica. [Anal. Math.] Akad. Kiado, Budapest. ISSN: 0133-3852 (**a**)

Annales Academiae Scientiarum Fennicae. Series A I. Mathematica. [Ann. Acad. Sci. Fenn. Ser. A I Math.]. Acad. Sci. Fennica, Helsinki. ISSN: 0066-1953 (**a**)

Annales Academiae Scientiarum Fennicae. Series AI. Mathematica Dissertationes. [Ann. Sci. Fenn. Ser. AI. Math. Dissertationes] Acad. Sci. Fennica, Helsinki. ISSN: 0355-0087 (**a**)

Annales de la Faculté des Sciences de Toulouse. Mathématiques. Série 5. [Ann. Fac. Sci. Toulouse Math. (5)] Univ. Toulouse III, Toulouse. ISSN: 0240-2955 (**a**)

Annales de l'Institut Fourier. [Ann. Inst. Fourier (Grenoble)] Univ. Grenoble I, Saint-Martin-d'Hères. ISSN: 0373-0956 (**a**)

Annales de l'Institut Henri Poincaré. Analyse Non Linéaire. [Ann. Inst. H. Poincaré Anal. Non Linéaire.] Gauthier-Villars, Montrouge. ISSN: 0294-1449 (**a**)

Annales de l'Institut Henri Poincaré. Probabilités et Statistique. [Ann. Inst. H. Poincaré Probab. Statist.] Gauthier-Villars, Montrouge. ISSN: 0246-0203 (**a**)

Annales Mathematicae Silesianae. [Ann. Math. Sil.] Uniw. Śląski, Katowice. ISSN: 0208-6336 (**a**)

Annales Polonici Mathematici. [Ann. Polon. Math.] PWN, Warsaw. ISSN: 0006-2216 (**a**)

Les Annales des Sciences Mathématiques du Québec. [Ann. Sci. Math. Quebec] Groupe Cherch. Sci. Math., Montréal, Que. ISSN: 0707-9109 (**a**)

Annales Scientifiques de l'Ècole Normale Supérieure. Quatrième Série. [Ann. Sci. Ecole Norm. Sup. (4)] Gauthier-Villars, Montrouge. ISSN: 0012-9593 (**a**)

Annales Scientifiques de l'Université de Clermont-Ferrand II. Mathématiques. [Ann. Sci. Univ. Clermont-Ferrand II Math.] Univ. Clermont-Ferrand II, Aubière. ISSN: 0249-7042 (**a**)

Annales Scientifiques de l'Université de Clermont-Ferrand II. Probabilités et Applications. [Ann. Sci. Univ. Clermont-Ferrand II Probab. Appl.] Univ. Clermont-Ferrand II, Aubière. ISSN: 0246-1501 (**a**)

Annales Scientifiques de l'Université de Franche-Comté Besançon. Mathématiques. 4e Série. [Ann. Sci. Univ. Franche-Comté Besançon Math. (4)] Univ. Franche-Comté, Besançon. ISSN: 0523-0578 (**a**)

Annales Societatis Mathematicae Polonae. Seria 1. Commentationes Mathematicae. [Comment. Math. Prace Mat.] PWN, Warsaw. ISSN: 0373-8299 (**a**)

Annales Societatis Mathematicae Polonae. Series IV. Fundamenta Informaticae. [Fund. Inform.] IOS, Amsterdam. ISSN: 0169-2968 (**a**)

Annales Universitatis Mariae Curie Skłodowska. Sectio A. Mathematica. [Ann. Univ. Mariae Curie-Skłodowska Sect. A] Uniw. Marii Curie, Lublin. ISSN: 0365-1029 (**a**)

Annales Universitatis Saraviensis. Series Mathematicae. [Ann. Univ. Sarav. Ser. Math.] Univ. Saarlandes, Saarbrücken. ISSN: 0080-5165 (**a**)

Annales Universitatis Scientiarum Budapestinensis de Rolando Eőtvős Nominatae. Sectio Computatorica. [Ann. Univ. Sci. Budapest. Sect. Comput.] Eőotvős Loránd Tudományegyetem, Budapest. ISSN: 0138-9491 (**a**)

Annales Universitatis Scientiarum Budapestinensis de Rolando Eőtvős Nominatae. Sectio Mathematica. [Ann. Univ. Sci. Budapest. Eőtvős Sect. Math.] Eotvos Loránd Tudományegyetem, Budapest. ISSN: 0524-9007 (**a**)

Annales Universitatis Turkuensis. Series A.I. Astronomica-Chemica-Physica-Mathematica. [Ann. Univ. Turku. Ser. A.I.] Univ. Turku, Turku. ISSN: 0082-7002 (**a**)

Annali di Matematica Pura ed Applicata. Serie Quarta. [Ann. Mat. Pura Appl. (4)] Zanichelli, Bologna. ISSN: 0003-4622 (**a**)

Annali della Scuola Normale Superiore di Pisa. Classe di Scienze. Serie IV. [Ann. Scuola Norm. Sup. Pisa Cl. (4)] Scuola Norm. Sup., Pisa. (**a**)

Annali dell'Università di Ferrara. Nuovo Serie. Sezione VII. Scienze Matematiche. [Ann. Univ. Ferrara Sez. VII (N.S.)] Univ. Stud. Ferrara, Ferrara. (**a**)

Annals of Applied Probability. [Ann. Appl. Probab.] Inst. Math. Statist., Hayward, CA. ISSN: 1050-5164 (**a**)

Annals of Differential Equations. Weifen Fangcheng Niankan. [Ann. Differential Equations] Fuzhou Univ., Fuzhou. (**a,s**)

Annals of Global Analysis and Geometry. [Ann. Global Anal. Geom.] Deutscher Verlag Wissensch., Berlin. ISSN: 0232-704X (**a**)

Annals of the Institute of Statistical Mathematics. [Ann. Inst. Statist. Math.] Kluwer, Boston, MA. ISSN: 0020-3157 (**a**)

Annals of the Japan Association for Philosophy of Science. [Ann. Japan Assoc. Philos. Sci.] Japan Assoc. Philos Sci., Tokyo. ISSN: 0453-0691 (**a**)

Annals of Mathematics. Second Series. [Ann of Math. (2)] Princeton Univ. Press, Princeton, NJ. ISSN: 0003-486X (**a,p**)

Annals of Operations Research. [Ann. Oper. Res.] Baltzer, Basel. ISSN: 0254-5330 (**s**)

The Annals of Probability. [Ann. Probab.] Inst. Math. Statist., Hayward, CA. ISSN: 0009-1798 (**a,p,s**)

Annals of Pure and Applied Logic. [Ann. Pure Appl. Logic] North-Holland, Amsterdam-New York. ISSN: 0168-0072 (**a,s**)

The Annals of Statistics. [Ann. Statist.] Inst. Math. Statist., Hayward, CA. ISSN: 0090-5364 (**a,p,s**)

Annuaire de l'Université de Sofia. Faculté de Mathématiques et Mécanique. See *Godishnik na Sofiĭskiĭa Universitet, Fakultet po Matematika i Mekhanika.*

Applicable Analysis. [Appl. Anal.] Gordon & Breach, NY. ISSN: 0003-6811 (**a**)

Applications of Mathematics. [Appl. Math.] Academia, Prague. ISSN: 0862-7940 (**a**)

Applied Mathematical Modelling. [Appl. Math. Modelling] Butterworth- Heinemann, Stoneham, MA. ISSN: 0307-904X (**a,s**)

Applied Mathematics: A Journal of Chinese Universities. See *Kao Teng Ying Yung Shu Hsueh Hsueh = Applied Mathematics: A Journal of Chinese Universities.*

Applied Mathematics and Computation. [Appl. Math. Comput.] Elsevier, NY. ISSN: 0096-3003 (**a,s**)

Applied Mathematics Letters. [Appl. Math. Lett.] Pergamon, Elmsford, NY. ISSN: 0893-9659 (**a,s**)

Applied Mathematics and Mechanics. English Edition. [Appl. Math. Mech. (English Ed.)] Shanghai Univ. Tech., Shanghai. ISSN: 0253-4827

Applied Mathematics and Mechanics. Yinyong Shuxue he Lixue. See *Ying Yun Shu Hsueh Ho Li Hsueh = Applied Mathematics and Mechanics.*

Applied Mathematics and Optimization. [Appl. Math. Optim.] Springer, New York-Berlin. ISSN: 0095-4616 (**a,s**)

Applied Numerical Mathematics. [Appl. Numer. Math.] North-Holland, Amsterdam-New York. ISSN: 0168-9274 (**s**)

Applied Statistics. See *Journal of the Royal Statistical Society. Series C. Applied Statistics.*

Applied Stochastic Models and Data Analysis. [Appl. Stochastic Models Data Anals.] Wiley, Chichester. ISSN: 8755-0024 (**a**)

Approximation Theory and its Applications. [Approx. Theory Appl.] Nanjing Univ., Nanjing. ISSN: 1000-9221 (**a**)

The Arab Journal of Mathematics. [Arab J. Math.] Union Arab Phys. Math., Baghdad. (**a**)

Archimede. [Archimede] Monnier, Florence. ISSN:0390-5543 (**a**)

Archiv der Mathematik. [Arch. Math. (Basel)] Birkhäuser, Basel-Boston, MA. ISSN: 0003-889X (**a,s**)

Archive for Mathematical Logic. [Arch. Math. Logic] Springer, Berlin-New York ISSN: 0933-5846 (**a**)

Archive for History of Exact Sciences. [Arch. Hist. Exact Sci.] Springer, Berlin-New York. ISSN: 0003-9519 (**a**)

Archive for Rational Mechanics and Analysis. [Arch. Rational Mech. Anal.] Springer, Berlin-New York. ISSN: 0003-9527 (**a,s**)

Archivum Mathematicum. [Arch. Math. (Brno)] Univ. J. E. Purkyně, Brno. ISSN: 0044-8753 (**a**)

Arithmetic Teacher. [Arithmetic Teacher] Nat. Council Teachers Math., Reston, VA. ISSN: 0004-136X (**p**)

Arkiv för Matematik. [Ark. Mat.] Inst. Mittag-Leffler, Djursholm. ISSN: 0004-2080 (**a**)

Arkhimedes. [Arkhimedes] Finn. Phys. Soc., Helsinki. ISSN: 0004-1920

Ars Combinatoria. [Ars Combin.] Univ. Manitoba, Winnipeg, MB. ISSN: 0381-7032 (**a**)

Asia-Pacific Journal of Operational Research. [Asia-Pacific J. Oper. Res.] Oper. Res. Soc. Singapore, Singapore. ISSN: 0217-5959

Asymptotic Analysis. [Asymptotic Anal.] Elsevier, Amsterdam. ISSN: 0921-7134 (**a,s**)

Atti della Accademia Nazionale dei Lincei. Classe di Scienze Fisiche, Matematiche e Naturali. Memorie Lincee. Matematica e Applicazioni. Serie IX. See *Memorie Lincee. Matematica e Applicazioni.*

Atti della Accademia Nazionale dei Lincei. Classe di Scienze Fisiche, Matematiche e Naturali. Rendiconti Lincei. Matematica e Applicazioni. Serie IX. See *Rendiconti Lincei. Matematica e Applicazioni.*

Atti del Seminario Matematico e Fisico dell'Universita di Modena. [Atti Sem. Mat. Fis. Univ. Modena] Univ. Modena, Modena. ISSN: 0041-8986 (**a**)

Australasian Journal of Combinatorics. [Australas. J. Combin.] Univ. Queensland, St. Lucia. ISSN: 1034-4942 (**a**)

The Australian Journal of Statistics. [Austral. J. Statist.] Austral. Statist. Publ. Assoc., Canberra. ISSN: 0004-9581 (**a**)

The Australian Mathematical Society. Gazette. See *Gazette (Australian Mathematical Society).*

Azerbaijan SSR Elmlar Akademiiasynyn Khabarlari. Fizika-Tekhnika va Riiaziiiat Elmlari Seriiasy. [Izv. Akad. Nauk Azerbaĭdzhan. SSR Ser. Fiz.-Tekhn. Mat. Nauk] "Èlm," Baku. ISSN: 0002-3108 (**a**)

Bayerische Akademie der Wissenschaften. Mathematisch-Naturwissenschaftliche Klasse. Sitzungsberichte. [Bayer. Akad. Wiss. Math.-Natur. Kl. Sitzungsber.] Bayer. Akad. Wiss., Munich. ISSN: 0340-7586

Bayreuther Mathematische Schriften. [Bayreuth. Math. Schr.] Univ. Bayreuth, Bayreuth. ISSN: 0172-1062 **(a)**

Beiträge zur Algebra und Geometrie. [Beiträge Algebra Geom.] Martin-Luther-Univ. Halle-Wittenberg, Halle. ISSN: 0440-1298 **(a)**

Belgian Journal of Operations Research, Statistics and Computer Science. [Belg. J. Oper. Res. Statist.] Sogesci, Brussels. ISSN: 0770-0512

Bericht. See *Jyvaskylan Yliopisto. Matematiikan Laito. Bericht.*

Biometrical Journal. [Biometrical J.] Akademie-Verlag, Berlin. ISSN: 0323-3847 **(a)**

Biometrics. [Biometrics] Biometric Soc., Washington, DC. ISSN: 0006-341X **(a)**

Biometrie-Praximetrie. [Biometrie-Praximetrie] Soc. Adolphe Quetelet, Gembloux. ISSN: 0006-3436 **(a)**

Biometrika. [Biometrika] Biometrika Trust, London. ISSN: 0006-3444 **(a,s)**

BIT; Computer Science Numerical Mathematics. [BIT] BIT, Copenhagen. ISSN: 0006-3835 **(a,s)**

Boletim da Sociedade Brasileira de Matemática. Nova Série. [Bol. Soc. Brasil. Mat. N.S.] Soc. Brasil. Mat., Rio de Janerio. ISSN: 0100-3569 **(a)**

Boletim da Sociedade Paranaense de Matemática. 2 Série. [Bol. Soc. Paran. Mat. (2)] Soc. Paranaense de Mat., Curitiba. ISSN: 0037-8712 **(a)**

Boletim de Sociedade Portuguesa de Matemática. [Bol. Soc. Port. Mat.] Portuguese Math. Soc., Lisbon.

Boletín de Matemáticas. [Bol. Mat.] Univ. Nac. Colombia, Bogotá. ISSN: 0120-0380 **(a)**

Boletín de la Sociedad Matemática Mexicana. Segunda Serie. [Bol. Soc. Mat. Mexicana (2)] Soc. Mat. Mexicana, Mexico City. **(a)**

Bollettino di Storia delle Scienze Matematiche. [Boll. Storia Sci. Mat.] Compositori, Bologna. ISSN: 0392-4432 **(a)**

Bollettino della Unione Matematica Italiana. Sez. A. Serie VII. [Boll. Un. Mat. Ital. A(7)] Zanichelli, Bologna. **(a)**

Bollettino della Unione Matematica Italiana. Sez. B. Serie VII. [Boll. Un. Mat. Ital. B(7)] Zanichelli, Bologna. **(a)**

Bollettino della Unione Matematica Italiana. Sez. C: Analisi Funzionale e Applicazioni. Ser VI. [Boll. Un. Mat. Ital. C(6)] Zanichelli, Bologna. **(a)**

Bollettino della Unione Matematica Italiana. Sez. D: Algebra e Geometria. Ser VI. [Boll. Un. Mat. Ital. D(6)] Zanichelli, Bologna. **(a)**

Bollettino della Unione Matematica Italiana. Supplemento. [Boll. Un. Mat. Ital. Suppl.] Zanichelli, Bologna. **(a)**

Brazilian Journal of Probability and Statistics. See *Revista Brasileira de Probabilidade e Estatistica.*

Buletinul Academiei de Ştiinţe a R.S.S. Moldovenesti. Matematică = Izvestiia Akademii nauk Moldavskoĭ SSR. Matematika. [Izv. Akad. Nauk Moldav. SSR. Mat.] "Shtiinsta," Kishinev. ISSN: 0236-3089 **(a)**

Buletinul Ştiinţific al Institutului Politehnic Cluj-Napoca. Seria Matematică, Mecanică Aplicată, Construcţii de Maşini. [Bul. Ştiinţ. Inst. Politehn. Cluj-Napoca Ser. Mat. Mec. Apl. Construc. Mas.] Inst. Politehn., Cluj-Napoca.

Buletinul Institutului Politehnic din Iasi. Sectia I, Matematica Mecahica Teoretica Fizica. [Bul. Inst. Politehn. Iasi Sect.] Inst. Politehnic, Iasi. **(a)**

Bulletin of the Allahabad Mathematical Society. [Bull. Allahabad Math. Soc.] Allahabad Math. Soc., Allahabad. ISSN: 0019-5324. **(a)**

Bulletin of the American Mathematical Society. New Series. [Bull. Amer. Math. Soc. (N.S.)] Amer. Math. Soc., Providence, RI. ISSN: 0273-0979 **(a,p,s)**

Bulletin of the Australian Mathematical Society. [Bull. Austral. Math. Soc.] Univ. Queensland Press, St. Lucia. ISSN: 0004-9727 **(a)**

Bulletin of the Calcutta Mathematical Society. [Bull. Calcutta Math. Soc.] Calcutta Math. Soc., Calcutta. ISSN: 0008-0659 **(a)**

Bulletin. Classe des Sciences Mathématiques et Naturelles. Sciences Mathématiques. [Bull. Acad. Serbe Sci. Arts Cl. Sci. Math. Natur] Acad. Serbe Sci. Arts, Belgrade. ISSN: 0001-4184 **(a)**

Bulletin of the Faculty of Science. Assiut University. C. Mathematics. [Bull. Fac. Sci. Assiut Univ. C. Math.] Univ. Assiut, Assiut. **(a)**

Bulletin of the Faculty of Science and Engineering, Chuo University. Series I. Mathematics. [Bull. Fac. Sci. Engrg. Chuo Univ. Ser. I Math.] Chuo Univ., Dept. Math., Fac. Sci. Engrg., Tokyo. **(a)**

Bulletin of the Faculty of Science, Ibaraki University. Series A. Mathematics. [Bull. Fac. Sci. Ibaraki Univ. Ser. A] Ibaraki Univ., Ibaraki-ken. ISSN: 0579-3068 **(a)**

Bulletin of Fukuoka University of Education. Part III. Mathematics, Natural Sciences and Technology. [Bull. Fukuoka Univ. Ed. III] Fukuoka Univ. Education, Fukuoka. ISSN: 0532-811X **(a)**

Bulletin of Information and Cybernetics. [Bull. Inform. Cybernet.] Res. Assoc. Statist. Sci., Fukuoka. ISSN: 0286-522X **(a)**

Bulletin of the Institute of Combinatorics and Its Applications. [Bull. Inst. Combin. Appl.] Inst. Combin., Winnipeg, MB. ISSN: 1183-1278 **(a)**

Bulletin de l'Institut International de Statistique. See *International Statistical Institute. Bulletin.*

Bulletin of the Institute of Mathematics, Academia Sinica. [Bull. Inst. Math. Acad. Sinica] Acad. Sinica, Taipei. ISSN: 0304-9825 **(a)**

Bulletin of the Institute of Mathematics and its Applications. [Bull. Inst. Math. Appl.] Inst. Math. Applications, Southend-on-Sea. **(a,s)**

Bulletin of the Iranian Mathematical Society. [Bull. Iranian Math. Soc.] Iranian Math. Soc., Tehran. **(a)**

Bulletin (Irish Mathematical Society). [Irish Math. Soc. Bull.] Irish Math. Soc., Cork. ISSN: 0790-1690 **(a)**

Bulletin of the Korean Mathematical Society. [Bull. Korean Math. Soc.] Korean Math. Soc., Seoul. **(a)**

Bulletin of the Kyushu Institute of Technology. Mathematics, Natural Science. [Bull. Kyushu Inst. Tech. Math. Natur. Sci.] Kyushu Inst. Tech., Tobata. ISSN: 0454-8221 **(a)**

The Bulletin of the London Mathematical Society. [Bull. London Math. Soc.] London Math. Soc., London. ISSN: 0024-6093 **(a,s)**

Bulletin of the Malaysian Mathematical Society. Second Series. [Bull. Malaysian Math. Soc. (2)] Malaysian Math. Soc., Kuala Lumpur. ISSN: 0126-6705 **(a)**

Bulletin of the Mathematical Association of India. [Bull. Math. Assoc. India] Math. Assoc. India, Kanpur. (**a**)

Bulletin Mathématique de la Société des Sciences Mathématiques de la Republique Socialiste de Roumanie. Nouvelle Série. [Bull. Math. Soc. Sci. Math. R. S. Roumanie (N.S.)] Soc. Sti. Mat. RS Romania, Bucharest. ISSN: 0007-4691 (**a**)

Bulletin of the Polish Academy of Sciences. Mathematics. [Bull. Polish Acad. Sci. Math.] PWN, Warsaw. ISSN: 0239-7269 (**a**)

Bulletin des Sciences Mathématiques. 2e Série. [Bull. Sci. Math (2)] Gauthier-Villars, Montrouge. ISSN: 0007-4497 (**a**)

Bulletin de la Société Franco-Japonaise des Sciences Pures et Appliquées. [Bull. Soc. Franco-Japon. Sci. Pures Appl.] Maison Franco-Japon., Tokyo. ISSN: 0285-3922 (**a**)

Bulletin de la Société Mathématique de Belgique. Série A. [Bull. Soc. Math. Belg. Sér. A] Soc. Math. Belgique, Brussels. ISSN: 0037-9476 (**a**)

Bulletin de la Société Mathématique de Belgique. Série B. [Bull. Soc. Math. Belg. Sér. B] Soc. Math. Belgique, Brussels. ISSN: 0037-9476 (**a**)

Bulletin de la Société Mathématique de France. [Bull. Soc. Math. France] Soc. Math. France, Paris. ISSN: 0037-9484 (**a**)

Bulletin de la Société Mathématique de Grèce. Nouvelle Série. See *Deltion tes Hellenikes Mathematikes Hetaireias.*

Bulletin de la Société Royale des Sciences de Liège. [Bull. Soc. Roy. Sci. Liege] Soc. Roy. Sci. Liege, Liege. ISSN: 0037-9565

Bulletin of Tokyo Gakugei University. Series IV. [Bull. Tokyo Gakugei Univ. (4)] Tokyo Gakugei Univ., Tokyo. ISSN: 0371-6813 (**a**)

Butlleti de la Societat Catalana de Matemàtiques. [Butl. Soc. Catalana Mat.] Soc. Catalana Mat., Barcelona. ISSN: 0214-316X (**a**)

Cahiers du Centre d'Études de Recherche Opérationnelle. [Cahiers Centre Études Rech. Opér.] Inst. Statist. U.L.B., Brussels. ISSN: 0008-9737

Cahiers Mathématiques. [Cahiers Math.] Univ. d'Oran, Oran-Elmenaouar. (**a**)

Cahiers du Séminaire d'Histoire des Mathématiques. See *Séminaire d'Histoire des Mathématiques. Cahiers du Séminaire d'Histoire des Mathématiques.*

Cahiers de Topologie et Géométrie Différentielle Catégoriques. [Cahiers Topologie Géom. Différentielle Catégoriques] Univ. Picardie, Amiens. ISSN: 0008-0004 (**a**)

Calcolo. [Calcolo] Inst. Elab. Inform. CNR, Pisa. ISSN: 0008-0624 (**a**)

Calcutta Statistical Association Bulletin. [Calcutta Statist. Assoc. Bull] Calcutta Statist. Soc., Calcutta. ISSN: 0008-0683 (**a**)

Canadian Journal of Mathematics. Journal Canadien de Mathématiques. [Canad. J. Math.] Canadian Math. Soc., Ottawa, Ont. ISSN: 0008-414X (**a,s**)

The Canadian Journal of Statistics. [Canad. J. Statist.] Statist. Soc. Canada, Ottawa, Ont. ISSN: 0319-5724 (**a**)

Canadian Mathematical Bulletin. [Canad. Math. Bull.] Canadian Math. Soc., Ottawa, Ont. ISSN: 0008-4395 (**a**)

Canadian Mathematical Society. Notes. [Canad. Math. Soc. Notes] Canadian Math. Soc., Ottawa, Ont. ISSN: 0045-5164 (**a,s**)

Caribbean Journal of Mathematics. [Caribbean J. Math.] Univ. West Indies, Barbados. ISSN: 0253-3405 (**a**)

Celestial Mechanics and Dynamical Astronomy. [Celestial Mech. Dynam. Astronom.] Kluwer Acad. Publ., Dordrecht. ISSN: 0923-2958 (**a**)

Chance: New Directions for Statistics and Computing. [Chance: New Directions Statist. Comput.] Springer, NY. ISSN: 0933-2480 (**a**)

Chaos. [Chaos] Amer. Inst. Phys., NY. ISSN: 1054-1500

Chaos, Solitons and Fractals. [Chaos Solitons Fractals] Pergamon, Oxford. ISSN: 0960-0779

Chemometrics and Intelligent Laboratory Systems. [Chemometrics Intelligent Laboratory Systems]. Elsevier, Amsterdam. ISSN: 0169-7439 (**a**)

Chi Suan Shu Hsueh = Mathematica Numerica Sinica. [Math. Numer. Sinica] Kexue Chubanshe [Science Press], Beijing. ISSN: 0254-7791 (**a**)

China Science & Technology Abstracts. Series I. Mathematics, Astronomy, Physics. [China Sci. Tech. Abstracts Ser. I Math. Astronom. Phys.] Kexue Chubanshe [Science Press], Beijing; Pergamon, Elmsford. ISSN: 0254-5179

Chinese Annals of Mathematics. Series A. Shuxue Niankan. Ji A. See *Shu Hsueh Nien K'an. A Chi.*

Chinese Annals of Mathematics. Series B. Shuxue Niankan. Ji B. [Chinese Ann. Math. Ser. B] Shanghai. Sci. Tech., Shanghai. Dist: Baltzer, Basel. ISSN: 0252-9599 (**a**)

Chinese Journal of Applied Probability and Statistics. See *Ying Yung Kai Lu Tung Chi.*

Chinese Journal of Mathematics. [Chinese J. Math.] Nat. Taiwan Univ., Taipei. ISSN: 0379-7570 (**a**)

Chinese Journal of Numerical Mathematics and Applications. [Chinese J. Numer. Math. Appl.] Allerton Press, Allerton, NY. ISSN: 0899-4358 (**a**)

Chinese Quarterly Journal of Mathematics. Shuxue Jikan. [Chinese Quart. J. Math.] Henan Univ., Kaifeng, Henan. (**a**)

Chinese Science Bulletin = Kexue Tongbao. [Kexue Tongbao (English Ed.)] Kexue Chubanshe [Science Press], Beijing; Pergamon Press, Elmsford, NY. ISSN: 0250-7862 (**a**)

Ciencias Matematicas. See *Revista Ciencias Matematicas.*

Collectanea Mathematica. Seminario Matematico de Barcelona. [Collect. Math.] Univ. Barcelona, Barcelona. ISSN: 0010-0757 (**a**)

The College Mathematics Journal. [College Math. J.] Math. Assoc. America, Washington, DC. ISSN: 0746-8342 (**a**)

Colloquium Mathematicum. [Colloq. Math.] PWN, Warsaw. ISSN: 0010-1354 (**a**)

Combinatorica. [Combinatorica] Akad. Kiadó, Budapest. ISSN: 0209-9683 (**a**)

Commentarii Mathematici Helvetici. [Comment. Math. Helv.] Birkhäuser, Basel-Boston, MA. ISSN: 0010-2571 (**a**)

Commentarii Mathematici Universitatis Sancti Pauli. [Comment Math. Univ. St. Paul.] Kinokuniya Book Store, Tokyo. ISSN: 0010-258X (**a**)

Commentationes Mathematicae. Prace Matematyczne. Seria 1. See *Annales Societatis Mathematicae Polonae. Seria 1. Commentationes Mathematicae.*

Commentationes Mathematicae Universitatis Carolinae. [Comment. Math. Univ. Carolin.] Univ. Karlova, Prague. ISSN: 0010-2628 (**a**)

Communications in Algebra. [Comm. Algebra] Dekker, NY. ISSN: 0092-7872 (**a,s**)

Communications of the Association for Computing Machinery. [Comm. ACM] ACM, New York. ISSN: 0001-0782 (**s**)

Communications of the Dublin Institute for Advanced Studies. Series A. [Comm. Dublin Inst. Adv. Stud. Ser. A] Dublin Inst. Adv. Stud., Dublin. ISSN: 0070-7414 (**a**)

Communications de la Faculté des Sciences de l'Université d'Ankara. Série 1A: Mathematics and Statistics. [Comm. Fac. Sci. Univ. Ankara Ser. A1 Math. Statist.] Univ. Anakara, Ankara. ISSN: 0251-0871 (**a**)

Communications in Mathematical Physics. [Comm. Math. Phys.] Springer, Berlin-New York. ISSN: 0010-3616 (**a,s**)

Communications in Partial Differential Equations. [Comm. Partial Differential Equations] Dekker, NY. ISSN: 0360-5302 (**a,s**)

Communications on Pure and Applied Mathematics. [Comm. Pure Appl. Math.] Wiley, NY. ISSN: 0010-3640 (**a,s**)

Communications in Statistics. Simulation and Computation. [Comm. Statist. Simulation Comput.] Dekker, NY. ISSN: 0361-0918 (**a,s**)

Communications in Statistics. Stochastic Models. [Comm. Statist. Stochastic Models] Dekker, NY. ISSN: 0882-0287 (**a,s**)

Communications in Statistics. Theory and Methods. [Comm. Statist. Theory Methods] Dekker, NY. ISSN: 0361-0926 (**a,s**)

Complex Systems. [Complex Systems] Complex Systems Publ., Champaign, IL. ISSN: 0891-2531 (**a,s**)

Complex Variables. Theory and Application. [Complex Variables Theory Appl.] Gordon & Breach, New York. ISSN: 0278-1077 (**a**)

Compositio Mathematica. [Compositio Math.] Kluwer, Dordrecht. ISSN: 0010-437X (**a**)

Comptes Rendus de l'Académie Bulgare des Sciences. See *Doklady Bolgarskoĭ Akademiĭ Nauk = Comptes Rendus de l'Académie Bulgare des Sciences.*

Comptes Rendus Mathématiques de l'Académie des Sciences. Mathematical Reports of the Academy of Sciences. [C. R. Math Rep. Acad. Sci. Canada] Univ. Waterloo, Ont. ISSN: 0706-1994 (**a**)

Comptes Rendus des Seances de l'Académie des Sciences. Série I. Mathématique. [C. R. Acad. Sci. Paris Sér. I Math.] Gauthier-Villars, Montrouge. ISSN: 0249-6291 (**a**)

Computational Complexity. [Comput. Complexity] Birkhäuser, Basel. ISSN: 1016-3328 (**a**)

Computational Geometry: Theory and Applications. [Comput. Geom.] Elsevier, Amsterdam. ISSN: 0925-7721 (**a**)

Computational Mathematics and Mathematical Physics. [Comput. Math. and Math. Phys.] Pergamon, Oxford-Elmsford, New York. ISSN: 0965-5425

Computational Statistics & Data Analysis. [Comput. Statist. Data Anal.] North-Holland, Amsterdam-New York. ISSN: 0167-9473 (**a**)

Computational Statistics Quarterly. [CSQ-Comput. Statist. Quart.] Physica, Heidelberg. ISSN: 0723-712X (**a**)

Computer Aided Geometric Design. [Comput. Aided Geom. Design] Elsevier, Amsterdam. ISSN: 0167-8396

Computer and Control Abstracts. [CCA] IEE, Hitchin. ISSN: 0036-8113

Computers and Mathematics with Applications. [Comput. Math. Appl.] Pergamon, Oxford. ISSN: 0898-1221 **(a,s)**

Computing. [Computing] Springer, Vienna. ISSN: 0010-485X **(a,p,s)**

Computing Reviews. [CR] ACM, NY. ISSN: 0010-4884 **(s)**

Conferenze del Seminario di Mathematica dell'Universita di Bari. [Confer. Sem. Mat. Univ. Bari] Laterza, Bari. **(a,s)**

Congressus Numerantium. [Congr. Numer.] Utilitas Math., Winnipeg, MB. ISSN: 0384-9864 **(a)**

Constructive Approximation. [Constr. Approx.] Springer, New York-Berlin. ISSN: 0176-4276 **(a,s)**

Control and Cybernetics. [Control Cybernet.] PWN, Warsaw. ISSN: 0424-8569

Control Theory & Applications. Kongzhi Lilun yu Yingyong. [Control Theory Appl.] Huanan Ligong Daxue Chubanshe, Guangzhou. ISSN: 1000-8152

Crux Mathematicorum. [Crux Mathematicorum] Canadian Math. Soc., Ottawa, Ont. ISSN: 0705-0348 **(a)**

Cryptologia. [Cryptologia] Rose Hulman Inst. Tech., Terre Haute, IN. ISSN: 0161-1194 **(a)**

CSQ-Computational Statistics Quarterly. See *Computational Statistics Quarterly.*

Cuadernos de Estadistica Matemática. Série A. Probabilidad. [Cuadern. Estadist. Mat. Sér. A. Probab.] Fac. Cienc. Mat., Univ. de Granada, Granada. ISSN: 0211-853X **(a)**

Current Index to Statistics. [Current Index Statist.] Amer. Statist. Assoc., Alexandria, VA. ISSN: 0364-1228 **(a,s)**

Current Mathematical Publications. [CMP] Amer. Math. Soc., Providence, RI. ISSN: 0361-4794 **(a,s)**

CWI Quarterly. [CWI Quarterly] Centrum Wisk. Inform., Amsterdam. ISSN: 0922-5366 **(a)**

Czechoslovak Mathematical Journal. [Czechoslovak Math. J.] Academia, Prague. ISSN: 0011-4642 **(a)**

Deltion tēs Hellēnikēs Mathematikēs Hetaireias. [Bull. Soc. Math. Greece (N.S.)] Greek Math. Soc., Athens. **(a)**

Demonstratio Mathematica. [Demonstratio Math.] Politech. Warszaw., Warsaw. ISSN: 0420-1213 **(a)**

Designs, Codes and Cryptography. [Des. Codes Cryptogr.] Kluwer Acad. Publ., Hingham, MA. ISSN: 0925-1022 **(a)**

Diagrammes. [Diagrammes] Univ. Paris VII, Paris. ISSN: 0224-3911 **(a)**

Differential Equations. [Differential Equations] Consultants Bureau, New York-London. ISSN: 0012-2661 **(a)**

Differential Geometry and Its Applications. [Differential Geom. Appl.] North-Holland, Amsterdam. ISSN: 0926-2245 **(a)**

Differential and Integral Equations. [Differential Integral Equations] Ohio Univ., Athens, OH. ISSN: 0893-4983 **(a)**

Differen͡tsial'na͡ia Geometri͡ia Monogoobraziĭ Figur. [Differentsial'naya Geom. Mnogoobraz. Figur.] Kaliningrad. Gos. Univ., Kaliningrad. ISSN: 0321-4796 **(a)**

Differents͡ial'nye Uravneni͡ia. [Differentsial'nye Uravneniya] "Nauka i Tekhnika," Minsk. ISSN: 0374-0641 (**a**)

Differents͡ial'nye Uraveneni͡ia i ikh Primenenie. [Differentsial'nye Uravneniya i Primenen.] Akad. Nauk Litovsk. SSR, Inst. Mat. Kibernet., Vilnius. ISSN: 0132-7801 (**a**)

Dinamika Sistem. [Dinamika Sistem] Gorkov. Gos. Univ., Gorki. ISSN: 0130-318X (**a**)

Discrete Applied Mathematics. [Discrete Appl. Math.] North-Holland, Amsterdam-New York. ISSN: 0166-218X (**a,s**)

Discrete & Computational Geometry. [Discrete Comput. Geom.] Springer, New York-Berlin. ISSN: 0179-5376 (**a**)

Discrete Mathematics. [Discrete Math.] North-Holland, Amsterdam-New York. ISSN: 0012-365X (**a,s**)

Discrete Mathematics and Its Applications. [Discrete Math. Appl.] VSP, Utrecht. ISSN: 0924-9265 (**a**)

Discussiones Mathematicae. [Discuss. Math.] Higher College Engrg., Zielona Góra. (**a**)

Diskretna͡ia Matematika. [Diskret. Mat.] Nauka, Moscow. ISSN: 0234-0860 (**a**)

Dissertationes Mathematicae. Rozprawy Matematyczne. [Dissertationes Math. (Rozprawy Mat.)] PWN, Warsaw. ISSN: 0012-3862 (**a**)

Doğa. Türk Matematik Dergisi. Doğa. Turkish Journal of Mathematics. [Doğa Mat.] Turk. Bilimsel Tek. Araştirma Kurumu, Ankara. ISSN: 1010-7622 (**a**)

Doklady Akademi͡ia Nauk Armi͡anskoĭ SSR. See *Akademi͡ia Nauk Armi͡anskoĭ SSR. Doklady*.

Doklady Akademi͡ia Nauk Azerbaĭdzhanskoĭ SSR. See *Akademi͡ia Nauk Azerbaĭdzhanskoĭ SSR. Doklady*.

Doklady Akademiĭ Nauk SSSR. [Dokl. Akad. Nauk SSR] "Nauka," Moscow. ISSN: 0002-3264 (**a**)

Doklady Akademi͡ia Nauk Ukrainskoĭ SSR. Matematika, Estestvoznanie, Tekhnicheskie Nauki. [Dokl. Akad. Nauk Ukrain. SSR] "Naukova Dumka," Kiev. ISSN: 0868-8044 (**a**)

Doklady Bolgarskoĭ Akademiĭ Nauk = Comptes Rendus de l'Académie des Sciences. [C.R. Acad. Bulgare Sci.] Bulgar. Acad. Sci., Sofia. ISSN: 0366-8688 (**a**)

Doklady Rasshirennykh Zasedaniĭ Seminara Instituta Prikladnoĭ Matematiki im. I.N. Vekua. [Dokl. Rasshiren. Zasedan. Sem. Inst. Prikl. Mat. I.N. Vekua] Tbilis. Gos. Univ., Tbilisi. (**a**)

Dongbei Shuxue. Northeastern Mathematical Journal. See *Tung Pei Shu Hsueh = Northeastern Mathematical Journal*.

Dopovidī Akademiï Nauk Ukraïns'koï RSR. Seriia A. [Dopovidi Akad. Nauk Ukrain. RSR SER. A] "Naukova Dumka," Kiev. ISSN: 0002-3531 (**a**)

Dresdner Reihe Forschung. [Dresdner Reihe Forsch.] Pädagogische Hochschule Dresden, Dresden. ISSN: 0233-0164 (**a**)

Duke Mathematical Journal. [Duke Math. J.] Duke Univ. Press, Durham, NC. ISSN: 0012-7094 (**a,s**)

Dynamic Systems and Applications. [Dynamic Systems Appl.] Dynamic, Atlanta, GA. ISSN: 1056-2176 (**a**)

Dynamics and Control. [Dynamics Control] Kluwer Acad. Publ., Hingham, MA. ISSN: 0925-4668 **(a)**

Dynamics And Stability of Systems. [Dynamics Stability Systems] Oxford Univ. Press., Oxford. ISSN: 0268-1110

Econometric Reviews. [Econometric Rev.] Dekker, NY. ISSN: 0747-4938 **(a)**

Educational Studies in Mathematics. [Ed. Stud. Math.] Kluwer Acad. Publ., Dordrecht. ISSN: 0013-1954 **(a)**

The Egyptian Statistical Journal. [Egyptian Statist. J.] Univ. Cairo, Giza/Cairo. ISSN: 0542-1748

Elemente der Mathematik. [Elem. Math.] Birkhäuser, Basel-Boston, MA. ISSN: 0013-6018 **(a)**

L'Enseignement Mathématique. IIe Série. [Enseign. Math. (2)] Enseignement Math., Geneva. ISSN: 0013-8584 **(a)**

Ergodic Theory and Dynamical Systems. [Ergodic Theory Dynamical Systems] Cambridge Univ. Press, New York. ISSN: 0143-3857 **(a)**

ESN Information Bulletin: Reports on Current European / Middle Eastern Science. [ESN Inform. Bull. Rep. Curr. European / Middle Eastern Sci.] Off. Naval. Res., New York. **(a)**

Estadística Española. [Estadíst. Española] Inst. Nac. Estadíst., Madrid. ISSN: 0014-1151 **(a)**

Eureka. [Eureka] Cambridge Univ. Math. Soc., Cambridge. ISSN: 0071-2248 **(a)**

European Journal of Applied Mathematics. [European J. Appl. Math.] Cambridge Univ. Press, New York. ISSN: 0956-7925 **(a,s)**

European Journal of Combinatorics. [European J. Combin.] Academic Press, London-New York. ISSN: 0195-6698 **(a)**

Expositiones Mathematicae. [Exposition. Math.] Bibliographisches Inst., Mannheim. ISSN: 0723-0869 **(a)**

Extracta Mathematica. [Extracta Math.] Univ. Extremadura, Bibl. Dep. Mat., Badajoz. ISSN: 0213-8743 **(a)**

Facta Universitatis. Series: Mathematics and Informatics. [Facta Univ. Ser. Math. Inform.] Univ. Nis, Nis. ISSN: 0352-9665 **(a)**

Fasciculi Mathematici. [Fasc. Math.] Politech. Poznańska, Poznań. ISSN: 0044-4413 **(a)**

The Fibonacci Quarterly. [Fibonacci Quart.] Fibonacci Assoc., Santa Clara, CA. ISSN: 0015-0517 **(a)**

Finite Elements in Analysis and Design. [Finite Elem. Anal. Des.] Elsevier, Amsterdam. ISSN: 0168-874X

Fiziko-Matematichesko Spisanie. [Fiz.-Mat. Spis. B″lgar. Akad. Nauk.] Bulgar. Acad. Sci. Sofia. ISSN: 0015-3265 **(a)**

Forum Mathematicum. [Forum Math.] de Gruyter, Berlin. ISSN: 0933-7741 **(a)**

Fu-Chien Shih Fan Ta Hsueh Hsueh Pao. Tzu Jan Ko Hsueh Pan. [Fujian Shifan Daxue Xuebao Ziran Kexue Ban] Fujian Normal Univ., Fuzhou. ISSN: 1000-5277 **(a)**

Fujian Shifan Daxue Xuebao Ziran Kexue Ban. See *Fu-Chien Shih Fan Ta Hsueh Hsueh Pao. Tzu Jan Ko Hsueh Pan.*

Functional Analysis and its Applications. [Functional Anal. Appl.] Consultants Bureau, New York-London. ISSN: 0016-2663 **(a,s)**

Functiones et Approximatio Commentarii Mathematici. [Funct. Approx. Comment. Math.] Mickiewicz Univ. Press, Poznań. ISSN: 0208-6573 (**a**)

Fundamenta Informatica. See *Annales Societatis Mathematicae Polonae. Series IV. Fundamenta Informatica.*

Fundamenta Mathematicae. [Fund. Math.] PWN, Warsaw. ISSN: 0016-2736 (**a**)

Funkcialaj Ekvacioj. [Funkcial. Ekvac.] Math. Soc. Japan, Kobe. ISSN: 0532-8721 (**a**)

Funkt͡sional'nyĭ Analiz i ego Prilozhenii͡a. [Funktsional. Anal. i Prilozhen.] "Nauka," Moscow. ISSN: 0374-1990 (**a**)

Fuzzy Mathematics = Mohu Shuxue. [Fuzzy Math.] Huazhong Gong Xueyuan Chubanshe, Wuhan. (**a**)

Fuzzy Sets and Systems. [Fuzzy Sets and Systems] North-Holland, Amsterdam-New York. ISSN: 0165-0114 (**a,s**)

Ganit. [Ganit] Bangladesh Math. Soc., Dacca. (**a**)

Gaṇita. [Ganita] Bharata Ganita Parisad, Lucknow. (**a**)

Gaṇita Bhāratī. Bulletin of the Indian Society for History of Mathematics. [Ganita Bharati] Birla Inst. Tech., Ranchi. ISSN: 0970-0307 (**a**)

Gaṇita Sandesa = Gaṇita Sandesh. [Ganita Sandesh] Government College, Dept. Math., Ajmer. ISSN: 0970-9169 (**a**)

Gaṇita Sandesh. See *Gaṇita Sandesa.*

Gaoxiao Yingyong Shuxue Xuebao. Applied Mathematics. A Journal of Chinese Universities. See *Kao Teng Ying Yung Shu Hsueh Hsueh = Applied Mathematics: A Journal of Chinese Universities.*

Gazeta Matematica. [Gat. Mat.] Soc. de Stiinte Mat., Bucharest. ISSN: 0016-5441 (**a**)

Gazette (*Australian Mathematical Society*). [Austral. Math. Soc. Gaz.] Austral. Math. Soc., St. Lucia. ISSN: 0311-0729 (**a**)

Gazette des Mathématiciens. [Gazette Mathématiciens] Soc. Math. France, Montrouge. ISSN: 0224-8999 (**a**)

Geometriae Dedicata. [Geom. Dedicata] Kluwer Acad. Publ., Dordrecht. ISSN: 0046-5755 (**a**)

Geometric and Functional Analysis. [Geom. Functional Anal.] Birkhäuser, Basel. ISSN: 1016-443X (**a**)

Glasgow Mathematical Journal. [Glasgow Math. J.] Scottish Academic Press, Edinburgh. ISSN: 0017-0895 (**a**)

Glasnik Matematicki. Serija III. [Glas. Mat. Ser. III] Soc. Math. Phys. S. R. Croatia, Zagreb. (**a**)

Godishnik na Sofiĭskii͡a Universitet, Fakultet po Matematika i Mekhanika. [Annuaire Univ. Sofia Fac. Math. Mec.] Univ. Sofia, Sofia. ISSN: 0205-0808 (**a**)

Godishnik na Visshite Uchebni Zavedenii͡a Prilozhna Matematika. [Godishnik Vissh. Uchebn. Zaved. Prilozhna Mat.] "Tekhnika," Sofia. (**a**)

Graphs and Combinatorics. [Graphs Combin.] Springer, Tokyo. ISSN: 0911-0119 (**a**)

Hardy-Ramanujan Journal. [Hardy-Ramanujan J.] Tata Inst. Fund. Res., Bombay. (**a**)

Hiroshima Mathematical Journal. [Hiroshima Math. Journal] Hiroshima Univ., Hiroshima. ISSN: 0018-2079 (**a**)

Historia Mathematica. [Historia Math.] Academic Press, Orlando, FL. ISSN: 0315-0860 (**a**)

Hokkaido Mathematical Journal. [Hokkaido Math. J.] Hokkaido Univ., Sapporo. ISSN: 0385-4035 (**a**)

Honam Mathematical Journal. [Honam Math. J.] Cholla and Cheju Math. Soc., Kwangju. (**a**)

Houston Journal of Mathematics. [Houston J. Math.] Univ. Houston, Houston, TX. ISSN: 0362-1588 (**a,s**)

Hsi T'ung K'o Hsueh Yu Shu Hsueh = Journal of Systems Science and Mathematical Sciences. Xitong Kexue Yu Shuxue. [J. Systems Sci. Math. Sci.] Kexue Chubanshe [Science Press], Beijing. ISSN: 1000-0577 (**a**)

Hunan Annals of Mathematics. Hunan Shuxue Niankan. [Hunan Ann. Math.] Hunan Univ., Changsha. (**a**)

IAPQR Transactions. [IAPQR Trans.] Indian Assoc. Productivity, Quality and Reliability, Calcutta.

Illinois Journal of Mathematics. [Illinois J. Math.] Univ. Illinois Press, Champaign, IL. ISSN: 0019-2089 (**a**)

IMA Journal of Applied Mathematics. [IMA J. Appl. Math.] Oxford Univ. Press, Oxford. ISSN: 0272-4960 (**a,s**)

IMA Journal of Mathematical Control and Information. [IMA J. Math. Control Inform.] Oxford Univ. Press, Oxford. ISSN: 0272-4960 (**a**)

IMA Journal of Mathematics Applied in Business and Industry. [IMA J. Math. Appl. Bus. Indust.] Inst. Math. Appl., Southend-on-Sea. ISSN: 0953-0061 (**a**)

IMA Journal of Mathematics Applied in Medicine and Biology. [IMA J. Math. Appl. Med. Biol.] Oxford Univ. Press, London-New York. ISSN: 0265-0746 (**a**)

IMA Journal of Numerical Analysis. [IMA J. Numer. Anal.] Inst. Math. Appl., Southend-on-Sea. ISSN: 0272-4979 (**a,s**)

Indagationes Mathematicae. [Indag. Math. (N.S.)] North-Holland, Amsterdam. ISSN: 0023-3358 (**a**)

Indian Journal of Mathematics. [Indian J. Math.] Allahabad Math. Soc., Allahabad. ISSN: 0019-5324 (**a**)

Indian Journal of Pure and Applied Mathematics. [Indian J. Pure Appl. Math.] Indian Nat. Sci. Acad., New Delhi. ISSN: 0019-5588 (**a**)

Indiana University Mathematics Journal. [Indiana Univ. Math. J.] Indiana Univ., Bloomington, IN. ISSN: 0022-2518 (**a,s**)

Industrial Mathematics. [Indust. Math.] Indust. Math. Soc., Roseville, MI. ISSN: 0019-8528

Informatica. [Informatica] Lith. Acad. Sci., Inst. Math. Cybernet., Vilnius. ISSN: 0868-4952

Information and Computation. [Inform. and Comput.] Academic Press, Duluth, MN. ISSN: 0890-5401 (**a,s**)

Information and Control. [Inform. and Control (Shenyang)] Chinese Automat. Soc., Shenyang. ISSN: 1002-0411

Information Processing Letters. [Inform. Process. Lett.] North-Holland, Amsterdam-New York. ISSN: 0020-0190 (**s**)

Institut des Hautes Ètudes Scientifiques. Publications Mathématiques. See *Publications Mathématiques* (*Institut des Hautes Ètudes Scientifiques*).

Institute of Mathematical Statistics Bulletin. [Inst. Math. Statist. Bull.] Inst. Math. Statist., Hayward, CA. ISSN: 0146-3942 (**a**)

Insurance: Mathematics & Economics. [Insurance Math. Econom.] North-Holland, Amsterdam-New York. ISSN: 0167-6687

Integral Equations and Operator Theory. [Integral Equations Operator Theory] Birkhäuser, Basel-Boston, MA. ISSN: 0378-620X (**a,s**)

International Journal of Algebra and Computation. [Internat. J. Algebra Comput.] World Sci. Publishing, Singapore. ISSN: 0218-1967 (**a**)

International Journal of Bifurcation and Chaos in Applied Sciences and Engineering. [Int. J. Bifur. Chaos Appl. Sci. Engrg.] World Sci. Publishing, Singapore.ISSN: 0218-1274 (**a,s**)

International Journal of Computational Geometry and Applications. [Internat. J. Comput. Geom. Appl.] World Sci. Publishing, Singapore. ISSN: 0218-1959 (**a**)

International Journal of Computer Mathematics. [Internat. J. Comput. Math.] Gordon & Breach, NY. ISSN: 0020-7160

International Journal of Game Theory. [Internat. J. Game Theory] Physica, Heidelberg. ISSN: 0020-7276 (**a**)

International Journal of Mathematical Education in Science and Technology. [Internat. J. Math. Ed. Sci. Tech.] Taylor & Francis, London. ISSN: 0020-739X

International Journal of Mathematics. [Int. J. Math.] World Sci. Publishing, Singapore. ISSN: 0129-167X (**a**)

International Journal of Mathematics and Mathematical Sciences. [Internat. J. Math. Math. Sci.] Univ. Central Florida, Orlando, FL. ISSN: 0161-1712 (**a,p**)

International Mathematical News. See *Internationale Mathematische Nachrichten.*

International Mathematics Research Notices. [Internat. Math. Res. Notices] Duke Univ. Press, Durham, NC (Included in Duke Math. J.) (**a**)

International Statistical Institute. Bulletin. [Bull. Inst. Internat. Statist.] Internat. Statist. Inst., Voorburg. (**a**)

International Statistical Review. [Internat. Statist. Rev.] Longman, London-New York. ISSN: 0306-7734 (**a**)

Internationale Mathematische Nachrichten. [Internat. Math News] Österreich. Math. Gesellsch., Vienna. ISSN: 0020-7926 (**a**)

Inventiones Mathematicae. [Invent. Math.] Springer, Berlin-New York. ISSN: 0020-9910 (**a**)

Inverse Problems. [Inverse Problems] Inst. Phys., London. ISSN: 0266-5611 (**s**)

Irish Mathematical Society Bulletin. See *Bulletin* (*Irish Mathematical Society*)

Islamabad Journal of Sciences. Journal of Mathematics and Sciences. [Islamabad J. Sci.] Quaid-i-Azam Univ., Islamabad. ISSN: 0304-5218 (**a**)

Israel Journal of Mathematics. [Israel J. Math.] Weizmann, Jerusalem. ISSN: 0021-2172 (**a**)

Issledovaniia po Prikladnoĭ Matematike. [Issled. Prikl. Mat.] Kazan. Gos. Univ., Kazań. ISSN: 0205-8758 (**a**)

İstanbul Üniversitesi. Fen Fakültesei. Matematik Dergisi. University of Istanbul. Faculty of Science. The Journal of Mathematics. [İstanbul Üniv. Fen Fak. Mat. Derg.] Univ. Istanbul, Istanbul. (**a**)

Istoriko-Matematicheskie Issledovaniia. [Istor.-Mat. Issled.] "Nauka," Moscow. ISSN: 0136-0949 (**a**)

Itogi Nauki i Tekhniki. Algebra, Topologiia, Geometriia. [Itogi Nauki i Tekhniki. Ser. Algebra, Topologiia, Geom.] Akad. Nauk SSSR, Vsesoiuz. Inst. Nauchn. i Tekhn. Inform., Moscow. ISSN: 0202-7445 (**a**)

Itogi Nauki i Tekhniki. Matematicheskiĭ Analiz. [Itogi Nauki i Tekhniki. Ser. Mat. Anal.] Akad. Nauk SSSR, Vsesoiuz. Inst. Nauchn. i Tekhn. Inform., Moscow. ISSN: 0202-7453 (**a**)

Itogi Nauki i Tekhniki. Problemy Geometrii. [Itogi Nauki i Tekhniki. Ser. Probl. Geom.] Akad. Nauk SSSR, Vsesoiuz. Inst. Nauchn. i Tekhn. Inform., Moscow. ISSN: 0202-7461 (**a**)

Itogi Nauki i Tekhniki. Sovremennye Problemy Matematiki. [Itogi Nauki i Tekhniki. Ser. Sovrem. Probl. Mat.] Akad. Nauk SSSR, Vsesoiuz. Inst. Nauchn. i Tekhn. Inform., Moscow. ISSN: 0202-747X (**a**)

Izvestiia Akademii Nauk Armianskoĭ SSR. Seriia Matematika. [Izv. Akad. Nauk Armyan. SSR Ser. Mat.] Akad. Nauk Armyan. SSR, Erevan. ISSN: 0002-3043 (**a**)

Izvestiia Akademii Nauk Azerbaĭdzhanskoĭ SSR. Seriia Fiziko-Tekhnicheskikh i Matematicheskikh Nauk. See *Azarbaijan SSR Elmlar Akademiĭasynyn Khabarlari. Fizika-Tekhnika va Riĭaziĭat Elmlari Seriĭasy.*

Izvestiia Akademii Nauk Kazakhskoĭ SSR. Seriia Fiziko-Matematicheskaia. [Izv. Akad. Nauk Kazakh. SSR Ser. Fiz.-Mat.] "Nauka" Kazakh. SSR, Alma-Ata. ISSN: 0002-3191

Izvestiia Akademii Nauk Moldavskoĭ SSR. Matematika. See *Buletinul Academiei de Ştiinţe a R.S.S. Moldovenesti. Matematika* = Izvestiia Akademii nauk Moldavskoĭ *SSR. Matematika.*

Izvestiia Akademii Nauk SSSR. Seriia Matematicheskaia. [Izv. Akad. Nauk SSSR Ser. Mat.] "Nauka," Moscow. ISSN: 0373-2436 (**a**)

Izvestiia Severo-Kavkazkogo Nauchnogo TSentra Vyssheĭ Shkoly. Estestvenny Nauki. [Izv. Severo-Kavkaz. Nauchn. Tsentra Vyssh. Shkoly Estest. Nauk.] Rostov. Gos. Univ., Rostov-on-Don. ISSN: 0321-3005 (**a**)

Izvestiia Vysshikh Uchebnykh Zavedeniĭ. Matematika. [Izv. Vyssh. Uchebn. Zaved. Mat.] Kazan. Gos. Univ., Kazan. ISSN: 0021-5446 (**a**)

Jahrbuch Kurt-Gödel-Gesellschaft. [Jahrb. Kurt-Gödel-Ges.] Österr. Akad. Wiss., Vienna. (**a**)

Jahrbuch Überblicke Mathematik. [Jahrb. Überbl. Math.] Friedr. Vieweg & Sohn, Braunschweig. ISSN: 0172-8512 (**a**)

Jahresbericht der Deutschen Mathematiker-Vereinigung. [Jahresber. Deutsch. Math.-Verein.] Teubner, Stuttgart. ISSN: 0012-0456 (**a**)

James Cook Mathematical Notes. [James Cook Math. Notes] B.C. Rennie, Burnside, S. Australia.

Japan Journal of Industrial and Applied Mathematics. [Japan J. Indust. Appl. Math.] Kinokuniya, Tokyo. ISSN: 0916-7005 (**a**)

Japanese Journal of Mathematics. New Series. [Japan. J. Math. (N.S.)] Math. Soc. Japan, Tokyo. ISSN: 0289-2316 (**a**)

Jñānābha. [Jnanabha] Vijnana Parishad of India, Orai. ISSN: 0304-9892 (**a**)

Journal of Algebra. [J. Algebra] Academic Press, Duluth, MN. ISSN: 0021-8693 **(a,s)**

Journal of Algebraic Combinatorics. [J. Algebraic Combin.] Kluwer, Hingham, MA. ISSN: 0925-9899 **(a)**

Journal of Algebraic Geometry. [J. Algebraic Geom.] Univ. Press, Hong Kong. ISSN: 1056-3911 **(a)**

Journal of Algorithms. [J. Algorithms] Academic Press, Duluth, MN. ISSN: 0196-6774 **(a,s)**

Journal of the American Mathematical Society. [J. Amer. Math. Soc.] Amer. Math. Soc., Providence, RI. ISSN: 0894-0347 **(a,p,s)**

Journal of the American Statistical Association. [J. Amer. Statist. Assoc.] Amer. Statist. Assoc., Alexandria, VA. ISSN: 0162-1459 **(a,p,s)**

Journal d'Analyse Mathématique. [J. Analyse Math.] Weizmann, Jerusalem. ISSN: 0021-7670 **(a)**

Journal of Applied Mathematics and Mechanics. [J. Appl. Math. Mech.] Pergamon, Oxford-Elmsford, NY. ISSN: 0021-8928 **(s)**

Journal of Applied Mathematics and Stochastic Analysis. [J. Appl. Math. Stochastic Anal.] Univ. Pittsburgh Bradford, Bradford, PA. ISSN: 1048-9533 **(a)**

Journal of Applied Probability. [J. Appl. Probab.] Appl. Probab. Trust, Sheffield. ISSN: 0021-9002 **(a,s)**

Journal of Applied Statistics. [J. Appl. Statist.] Carfax, Abingdon, UK. ISSN: 0266-4763 **(a,s)**

Journal of Approximation Theory. [J. Approx. Theory] Academic Press, Orlando, FL. ISSN: 0021-9045 **(a,s)**

Journal of the Association for Computing Machinery. [J. Assoc. Comput. Mach.] ACM, New York. ISSN: 0004-5411 **(a,s)**

Journal of the Australian Mathematical Society. Series A, Pure Mathematics and Statistics. [J. Austral. Math. Soc. Ser. A] Austral. Math. Soc., St. Lucia. ISSN: 0263-6115 **(a)**

Journal of the Australian Mathematical Society. Series B, Applied Mathematics. [J. Austral. Math. Soc. Ser. B] Austral. Math. Soc., St. Lucia. ISSN: 0334-2700 **(a)**

Journal of Automated Reasoning. [J. Automat. Reason.] Kluwer Acad. Publ., Dordrecht. ISSN: 0168-7433 **(s)**

Journal of the Bihar Mathematical Society. [J. Bihar Math. Soc.] Lok Sahityagar, Bhagalpur.

Journal of Biomathematics. Shengwu Shuxue Xuebao. See *Sheng We Shu Hsueh Hsueh Pao = Journal of Biomathematics.*

Journal of Biopharmaceutical Statistics. [J. Biopharm. Statist.] Dekker, New York. ISSN: 1054-3406 **(a)**

Journal of Business & Economic Statistics. [J. Busin. Econom. Statist.] Amer. Statist. Assoc., Alexandria, VA. ISSN: 0735-0015

Journal of Classification. [J. Classification] Springer, New York-Berlin. ISSN: 0176-4268 **(a)**

Journal of Combinatorial Mathematics and Combinatorial Computing. [J. Combin. Math. Combin. Comput.] Charles Babbage Res. Centre, Winnipeg, MB. ISSN: 0835-3026 **(a)**

Journal of Combinatorial Theory. Series A. [J. Combin. Theory Ser. A] Academic Press, Orlando, FL. ISSN: 0097-3165 **(a)**

Journal of Combinatorial Theory. Series B. [J. Combin. Theory. Ser. B] Academic Press, Orlando, FL. ISSN: 0095-8956 **(a)**

Journal of Combinatorics, Information & System Sciences. [J. Combin. Inform. System Sci.] Forum Interdisciplinary Math., Delhi. ISSN: 0250-9628 **(a)**

Journal of Complexity. [J. Complexity] Academic Press, Duluth, MN. ISSN: 0885-064X **(a)**

Journal of Computational and Applied Mathematics. [J. Comput. Appl. Math.] North-Holland, Amsterdam-New York. ISSN: 0377-0427 **(a,s)**

Journal of Computational Mathematics. [J. Comput. Math.] VSP, Zeist. ISSN: 0254-9409 **(a)**

Journal of Computer and System Sciences. [J. Comput. System Sci.] Academic Press, Duluth, MN. ISSN: 0022-0000 **(s)**

Journal of Cryptology. [J. Cryptology] Springer, New York-Berlin. ISSN: 0933-2790

Journal of Differential Equations. [J. Differential Equations] Academic Press, Orlando, FL. ISSN: 0022-0396 **(a,s)**

Journal of Differential Geometry. [J. Differential Geom.] Lehigh Univ., Bethlehem, PA. ISSN: 0022-040X **(a,s)**

Journal of Dynamics and Differential Equations. [J. Dynamics Differential Equations] Plenum, NY. ISSN: 1040-7294 **(a,s)**

Journal of the Faculty of Science. University of Tokyo. Section IA. Mathematics. [J. Fac. Sci. Univ. Tokyo Sect. IA Math.] Univ. Tokyo Press, Tokyo. ISSN: 0040-8980 **(a)**

Journal of Fujian Teachers University. Natural Science Edition. See *Fu-Chien Shih Fan Ta Hsueh Hsueh Pao. Tzu Jan Ko Hsueh Pan.*

Journal of Functional Analysis. [J. Funct. Anal.] Academic Press, Duluth, MN. ISSN: 0022-1236 **(a,s)**

Journal of Geometric Analysis. [J. Geom. Anal.] CRC, Boca Raton, FL. ISSN: 1050-6926 **(a)**

Journal of Geometry. [J. Geom.] Birkhäuser, Basel-Boston, MA. ISSN: 0047-2468 **(a)**

Journal of Geometry and Physics. [J. Geom. Phys.] Elsevier, Amsterdam. ISSN: 0393-0440 **(a,s)**

Journal of Global Optimization. [J. Global Optim.] Kluwer Acad. Publ., Hingham, MA. ISSN: 0925-5001 **(a,s)**

Journal of Graph Theory. [J. Graph Theory.] Wiley, New York. ISSN: 0364-9024 **(a,s)**

Journal of the Indian Academy of Mathematics. [J. Indian Acad. Math.] Indian Acad. Math., Indore. ISSN: 0970-5120 **(a)**

The Journal of the Indian Mathematical Society. New Series. [J. Indian Math. Soc. (N.S.)] Indian Math. Soc., Delhi. ISSN: 0019-5839 **(a)**

Journal of the Indian Society of Agricultural Statistics. [J. Indian Soc. Agricultural Statist.] Indian Soc. Agricultural Statist., New Delhi. **(a)**

Journal of the Indian Society of Statistics and Operations Research. [J. Indian Soc. Statist. Oper. Res.] Aligarh Muslim Univ., Aligarh. ISSN: 0250-9636

Journal of the Indian Statistical Association. [J. Indian Statist. Assoc.] Indian Statist. Assoc., Poona. ISSN: 0537-2585 **(a)**

Journal of Information & Optimization Sciences. [J. Inform. Optim. Sci.] Anal. Publ, Delhi. ISSN: 0252-2667 **(a)**

Journal of Institute of Mathematics & Computer Sciences (Mathematics Series). [J. Inst. Math. Comput. Sci. Math. Ser.] Inst. Math. Comput. Sci., Calcutta. **(a)**

Journal of Integral Equations and Applications. [J. Integral Equations Appl.] Rocky Mountain Math. Consortium, Tempe, AZ. ISSN: 0163-5549 **(a,s)**

Journal of the Japan Statistical Society = Nihon Tokei Gakkai shi. [J. Japan Statist. Soc.] Japan Statist. Soc., Tokyo. ISSN: 0389-5602 **(a)**

Journal of the Japanese Society of Computational Statistics. [J. Japanese Soc. Comput. Statist.] Japan. Soc. Comput. Stat., Okayama. ISSN: 0915-2350 **(a)**

Journal of Karadeniz Technical University, Faculty of Arts and Sciences. Series of Mathematics-Physics. [J. Karadeniz Tech. Univ. Fac. Arts Sci. Ser. Math.-Phys.] Karadeniz Tek. Univ., Trabzon. **(a)**

Journal of Knot Theory and its Ramifications. [J. Knot Theory Ramifications] World Sci. Publishing, Singapore. ISSN: 0218-2165 **(a)**

Journal of the Korean Mathematical Society. [J. Korean Math. Soc.] Korean Math. Soc., Seoul. ISSN: 0304-9914 **(a)**

Journal of the Korean Statistical Society. See *Tonggyehak Yongu.*

Journal of Logic and Computation. [J. Logic Comput.] Oxford Univ. Press, Eynsham, England. ISSN: 0955-792X **(a)**

Journal of Logic Programming. [J. Logic Programming] Elsevier, NY. ISSN: 0743-1066 **(a,s)**

The Journal of the London Mathematical Society. Second Series. [J. London Math. Soc. (2)] London Math. Soc., London. ISSN: 0024-6107 **(a,s)**

Journal of Mathematical Analysis and Applications. [J. Math. Anal. Appl.] Academic Press, Duluth, MN. ISSN: 0022-247X **(a,s)**

Journal of Mathematical Biology. [J. Math. Biol.] Springer, New York-Berlin. ISSN: 0303-6812

Journal of Mathematical Economics. [J. Math. Econom.] North-Holland, Amsterdam-New York. ISSN: 0304-4068 **(a)**

Journal of Mathematical and Physical Sciences. [J. Math. Phys. Sci.] Indian Instit. Thec., Madras. ISSN: 0047-2557 **(a)**

Journal of Mathematical Physics. [J. Math. Phys.] Amer. Inst. Physics, New York. ISSN: 0022-2488 **(a,s)**

Journal of Mathematical Research and Exposition. [J. Math. Res. Exposition] Inst. Appl. Math., Dalian Inst. Tech., Dalian. ISSN: 1000-341X **(a)**

Journal of Mathematical Sciences. [J. Math. Sci.] Soc. of Math. Sciences, Delhi. ISSN: 0449-2757 **(a)**

Journal of the Mathematical Society of Japan. [J. Math. Soc. Japan] Math. Soc. Japan, Tokyo. ISSN: 0025-5645 **(a)**

Journal of Mathematics of Kyoto University. [J. Math. Kyoto Univ.] Kyoto Univ., Kyoto. ISSN: 0023-608X **(a,s)**

Journal of Mathematics. Shu Hsueh Tsa Chih. See *Shu Hsueh Tsa Chih. Journal of Mathematics.*

Journal of Mathematics, Tokushima University. [J. Math. Tokushima Univ.] Tokushima Univ., Tokushima. ISSN: 0075-4293 **(a)**

Journal de Mathématiques Pures et Appliquées. Neuvième Série. [J. Math. Pures Appl. (9)] Gauthier-Villars, Montrouge. ISSN: 0021-7824 (**a**)

Journal of Multivariate Analysis. [J. Multivariate Anal.] Academic Press, Duluth, MN. ISSN: 0047-259X (**a,s**)

Journal of National Academy of Mathematics, India. [J. Nat. Acad. Math. India] Nat. Acad. Math. India, Gorakhpur. (**a**)

Journal of Natural Geometry. [J. Natur. Geom.] Math. Res. Unit, Birbeck Coll., Univ. London, London. ISSN: 0963-2654

The Journal of Natural Sciences and Mathematics. [J. Natur. Sci. Math.] Government College, Lahore. ISSN: 0022-2941 (**a**)

Journal of the Nigerian Mathematical Society. [J. Nigerian Math. Soc.] Nigerian Math. Soc., Ibadan. ISSN: 0189-8965 (**a**)

The Journal of Non-Classical Logic. [J. Non-Classical Logic] Univ. Estad. Campinas, São Paulo. ISSN: 0102-3411 (**a**)

Journal of Nonlinear Science. [J. Nonlinear Sci.] Springer, Secaucus, NJ. ISSN: 0938-8974 (**a,s**)

Journal of Number Theory. [J. Number Theory] Academic Press, Orlando, Fl. ISSN: 0022-314X (**a,s**)

Journal on Numerical Methods and Computer Applications. Shuxue Jisuan Yu Jisuanji Yingyong. See *Shu Chih Chi Suan Yu Chi Suan Chi Ying Yung.*

Journal of Official Statistics. [J. Official Statist.] Statist. Sweden, Stockholm. ISSN: 0282-423X (**a**)

Journal of Operator Theory. [J. Operator Theory] Nat. Inst. Sci. Tech. Creation, Bucharest. ISSN: 0379-4024 (**a**)

Journal of Optimization Theory and Applications. [J. Optim. Theory Appl.] Plenum, NY. ISSN: 0022-3239 (**a,s**)

Journal of Orissa Mathematical Society. [J. Orissa Math. Soc.] Orissa Math. Soc., Bhubaneswar. (**a**)

Journal of Partial Differential Equations. [J. Partial Differential Equations] Internat. Acad. Publ., Beijing; Pergamon, Oxford. ISSN: 1000-940X (**a**)

Journal of Philosophical Logic. [J. Philos. Logic.] Kluwer Acad. Publ., Dordrecht. ISSN: 0022-3611

Journal of Pure and Applied Algebra. [J. Pure Appl. Algebra] North-Holland, Amsterdam-New York. ISSN: 0022-4049 (**a,s**)

Journal of Pure Mathematics. [J. Pure. Math.] Calcutta Univ., Calcutta. (**a**)

Journal of the Ramanujan Mathematical Society. [J. Ramanujan Math. Soc.] Ramanujan Math. Soc., Annamalainagar. ISSN: 0970-1249 (**a**)

Journal of Recreational Mathematics. [J. Recreational Math.] Baywood, Amityville, NY. ISSN: 0022-412X (**a**)

Journal für die Reine und Angewandte Mathematik. [J. Reine Angew. Math.] de Gruyter, Berlin-New York. ISSN: 0075-4102 (**a**)

Journal of the Royal Statistical Society. Series A. General. [J. Roy. Statist. Soc. Ser. A] Roy. Statist. Soc., London. ISSN: 0035-9238 (**a,s**)

Journal of the Royal Statistical Society. Series B. Methodological. [J. Roy. Statist. Soc. Ser. B] Roy. Statis. Soc., London. ISSN: 0035-9246 (**a,s**)

Journal of the Royal Statistical Society. Series C. Applied Statistics. [J. Roy. Statist. Soc. Ser. C] Roy. Statist. Soc. London. ISSN: 0035-9254 **(a,s)**

Journal of Saitama University, Faculty of Education. Mathematics and Natural Science. [J. Saitama Univ. Fac. Ed. Math. Natur. Sci.] Saitama Univ., Urawa. ISSN: 0387-9313

Journal of Soviet Mathematics. [J. Soviet Math.] Consultants Bureau, New York-London. ISSN: 0094-4104 **(a)**

Journal of Statistical Computation and Simulation. [J. Statist. Comput. Simulation] Gordon & Breach, New York. ISSN: 0094-9655 **(a,s)**

Journal of Statistical Planning and Inference. [J. Statist. Plann. Inference] North-Holland, Amsterdam-New York. ISSN: 0378-3758 **(a,s)**

Journal of Statistical Research. [J. Statist. Res.] Univ. Dacca, Dacca. ISSN: 0020-3165 **(a)**

Journal of Symbolic Computation. [J. Symbolic Comput.] Academic Press, London-New York. ISSN: 0747-7171 **(a,s)**

The Journal of Symbolic Logic. [J. Symbolic Logic] Assoc. Symbol. Logic, Providence, RI. ISSN: 0022-4812 **(a,s)**

Journal of Systems Science and Mathematical Sciences. Xitong Kexue Yu Shuxue. See *Hsi T'ung K'o Hsueh Yu Shu Hsueh.*

Journal of the Tensor Society of India. [J. Tensor Soc. India] Tensor Soc. of India, Lucknow. **(a)**

Journal of Theoretical Probability. [J. Theoret. Probab.] Plenum, NY. ISSN: 0894-9840 **(a)**

Journal of Time Series Analysis. [J. Time Ser. Anal.] Blackwell, Oxford. ISSN: 0143-9782 **(a)**

Journal of Tsuda College. [J. Tsuda College] Tsuda College, Tokyo. ISSN: 0287-7805 **(a)**

Journal of Undergraduate Mathematics. [J. Undergrad. Math.] Guilford College, Greensboro, NC. ISSN: 0022-5339 **(a)**

Jyväskylän Yliopisto. Matematiikan Laitos. Bericht. [Bericht] Univ. Jyväskylä, Jyväskylä. ISSN: 0075-4641 **(a)**

K-Theory. [K-Theory] Kluwer Acad. Publ., Boston, MA. ISSN: 0920-3036 **(a)**

Kao Teng Ying Yung Shu Hsueh Hsueh Pao = Applied Mathematics: A Journal of Chinese Universities. [Appl. Math. J. Chinese Univ.] Hangzhou Univ., Zhejiang. ISSN:1000-4424 **(a)**

Kobe Journal of Mathematics. [Kobe J. Math.] Kobe Univ., Kobe. ISSN: 0289-9051 **(a)**

Kodai Mathematical Journal. [Kodai Math. J.] Tokyo Inst. Tech., Tokyo. ISSN: 0386-5991 **(a)**

Korean Journal of Applied Statistics. Ungyong T'ongkye Yonku. [Korean J. Appl. Stat.] Korean Statist. Soc., Seoul. **(a)**

Kumamoto Journal of Mathematics. [Kumamoto J. Math.] Kumamoto Univ., Kumamoto. ISSN: 0914-675X **(a)**

Kvant. [Kvant] "Nauka," Moscow. ISSN: 0130-2221 **(a)**

Kyungpook Mathematical Journal. [Kyungpook Math. J.] Kyungpook Nat. Univ., Taegu. **(a)**

Latviĭskiĭ Matematicheskiĭ Ezhegodnik. [Latv. Mat. Ezhegodnik] "Zinatne," Riga. ISSN: 0321-2270 (**a**)

Leningradskiĭ Gosudarstvennyĭ Ordena Lenina Universitet imeni A. A. Zhdanova. Uchenye Zapiski. Seriia Matematicheskikh Nauk. See *Seriia Matematicheskikh Nauk.*

Letters in Mathematical Physics. [Lett. Math. Phys.] Kluwer Acad. Publ., Dordrecht. ISSN: 0377-9017 (**a,s**)

Libertas Mathematica. [Libertas Math.] Amer. Roman. Acad. Arts Sci. Publ., Los Angeles, CA. ISSN: 0278-5307 (**a**)

Linear Algebra and its Applications. [Linear Algebra Appl.] Elsevier, NY. ISSN: 0024-3795 (**a,s**)

Linear and Multilinear Algebra. [Linear and Multilinear Algebra] Gordon & Breach, New York. ISSN: 0308-1087 (**a**)

Lithuanian Mathematical Journal. [Lithuanian Math. J.] Consultants Bureau, New York-London. ISSN: 0363-1672 (**a**)

Litovskiĭ Matematicheskiĭ Sbornik. [Litovsk. Mat. Sb.] "Mokslas," Vilnius. ISSN: 0132-2818 (**a**)

Logique et Analyse. Nouvelle Série. [Logique et Anal. (N.S.)] CNR de Logique, Brussels. ISSN: 0024-5836

Manuscripta Mathematica. [Manuscripta Math.] Springer, Berlin-New York. ISSN: 0025-2611 (**a**)

Match. [Match] Max-Planck-Inst. Kohlenforschung, Mulheim. ISSN: 0340-6253

Matemática Aplicada e Computacional. [Mat. Apl. Comput.] Soc. Brasil. Mat. Apl. Comput., Rio de Janeiro. ISSN: 0101-8205 (**a**)

Le Matematiche. [Matematiche (Catania)] Univ. Catania, Catania. (**a**)

Matematicheskaia Fizika i Nelineĭnaia Mekhanika. [Mat. Fiz. Nelinein. Mekh.] "Naukova Dumka," Kiev. ISSN: 0233-7568 (**a**)

Matematicheskaia Logika i ee Primeneniia. [Mat. Logika Primenen.] Akad. Nauk Litovsk. SSR, Inst. Mat. i Kibernet., Vilnius. ISSN: 0132-7801

Matematicheskie Issledovaniia. [Mat. Issled.] "Shtiintsa," Kishinev. (**a**)

Matematicheskie Metody i Fiziko-Mekhanicheskie Poliа. [Mat. Metody i Fiz.-Mekh. Polia] "Naukova Dumka," Kiev. ISSN: 0130-9420 (**a**)

Matematicheskie Metody v Sotsial'nykh Naukakh. [Mat. Metody v Sotsial. Nauk.] Akad. Nauk Litovsk. SSR, Inst. Mat. I Kibernet., Vilnius. ISSN: 0136-3700

Matematicheskie Voprosy Kibernetiki i Vychislitel'noĭ Tekhniki. [Mat. Voprosy Kibernet. Vychisl. Tekhn.] Akad. Nauk Armyan. SSR, Erevan. ISSN: 0131-4645

Matematicheskie Zametki. [Mat. Zametki] "Nauka," Moscow. ISSN: 0025-567X (**a**)

Matematicheskie Zapiski. [Ural. Gos. Univ. Mat. Zap.] Ural. Gos. Univ., Sverdlovsk. (**a**)

Matematicheskiĭ Sbornik. [Mat. Sb.] "Nauka," Moscow. ISSN: 0368-8666 (**a**)

Matematichki Bilten. [Mat. Bilten] Sojuz Drush. Mat. Inform. Maked., Skopje. ISSN: 0025-5165 (**a**)

Matematicki Vesnik. [Mat. Vesnik] Drustvo Mat. Fiz. Astronoma Srbije, Belgrade. ISSN: 0025-5165 (**a**)

Matematicko Fizicki List za Učenike Srednjih Skola. [Mat. Fiz. List Za Ucen. Sred. Skola] Izd. Drustvo Mat. i Fiz. SRH, Zagreb. ISSN: 0543-0038

Matematika. [Matematika (Zagreb)] Skolska Knjiga, Zagreb.

Matematikai Lapok. [Mat. Lapok] Akad. Kiadó, Budapest. ISSN: 0025-519X (**a**)

Matematiskfysiske Meddelelser udgivet af Det Kongelige Danske Videnskabernes Selskab. [Mat.-Fys. Medd. Danske Vid. Selsk.] Roy. Danish Acad. Sci. Letters, Copenhagen. ISSN: 0023-3323

Matematyka. [Zeszyty Nauk. Politech. Gdansk. Mat.] Politechniki Gdanskiej, Gdansk. ISSN: 0072-0380 (**a**)

Matematyka. [Zeszyty Nauk. Politech. Łódź. Mat.] Łódź Tech. Univ., Łódź. ISSN: 0137-2572 (**a**)

Matematyka. [Zeszyty Nauk. Wyż. Szkoły Ped. w Opolu Mat.] Wydawn. Wyższej Skoły Ped. Powst. Śląsk. Opolu, Opole. (**a**)

Matematyka Stosowana. See *Roczniki. Seria III: Matematyka Stosowana.*

Mathematica. [Mathematica (Cluj)] Acad. Rep. RSR, Cluj-Napoca. ISSN: 0025-5505 (**a**)

Mathematica Applicata. Yingyong Shuxue. See *Ying Yung Shu Hsueh.*

Mathematica Balkanica. New Series. [Math. Balkanica (N.S.)] Bulgar. Acad. Sci., Sofia. ISSN: 0205-3217 (**a**)

Mathematica Bohemica. [Math. Bohem.] Academia, Prague. ISSN: 0862-7959 (**a**)

Mathematica Japonica. [Math. Japon.] Assoc. Math. Japon., Osaka. ISSN: 0025-5513 (**a**)

Mathematica Numerica Sinica. Jisuan Shuxue. See *Chi Suan Shu Hsueh = Mathematica Numerica Sinica.*

Mathematica Pannonica. [Math. Pannon.] Univ. Miskolc Math. Inst., Egyetemvaros. ISSN: 0865-2090 (**a**)

Mathematica Scandinavica. [Math. Scand.] Aarhus Univ., Aarhus. ISSN: 0025-5521 (**a**)

Mathematica Slovaca. [Math. Slovaca] Slovak Acad. Sci., Math. Inst. ISSN: 0139-9918 (**a**)

Mathematicae Notae. [Math. Notae] Univ. Nac. Rosario, Rosario. ISSN: 0025-553X (**a**)

Mathematical Biosciences. [Math. Biosci.] Elsevier, NY. ISSN: 0025-5564

Mathematical Chronicle. [Math. Chronicle] Univ. Auckland, Auckland. ISSN: 0481-1155 (**a**)

Mathematical and Computer Modelling. [Math. Comput. Modelling] Pergamon, Oxford-Elmsford, New York. ISSN: 0895-7177 (**a**)

The Mathematical Gazette. [Math. Gaz.] Math. Assoc., Leicester. ISSN: 0025-5572 (**a**)

The Mathematical Intelligencer. [Math. Intelligencer] Springer, Berlin-New York. ISSN: 0343-6993 (**a,s**)

Mathematical Journal of Okayama University. [Math. J. Okayama Univ.] Okayama Univ., Okayama. ISSN: 0030-1566 [**a**)

Mathematical Medley. [Math. Medley] Singapore Math. Soc., Singapore. ISSN: 0217-2976 (**a**)

Mathematical Methods in the Applied Sciences. [Math. Methods Appl. Sci.] Teubner, Stuttgart. ISSN: 0170-4214 (**a,s**)

Mathematical Modelling. [Math. Modelling] Pergamon, Elmsford, NY. ISSN: 0270-0255 (**a,s**)

Mathematical Models & Methods in Applied Sciences. [Math. Models Methods Appl. Sci.] World Sci. Publishing, Singapore. ISSN: 0218-2025 (**a**)

Mathematical Notes of the Academy of Sciences of the USSR. [Math. Notes] Consultants Bureau, New York-London. ISSN: 0001-4346 (**a**)

Mathematical Population Studies. [Math. Population Stud.] Gordon & Breach, London. ISSN: 0889-8480 **(a)**

Mathematical Proceedings of the Cambridge Philosophical Society. [Math. Proc. Cambridge Philos. Soc.] Cambridge Univ. Press, Cambridge-New York. ISSN: 0305-0041 **(a,s)**

Mathematical Programming. [Math. Programming] North-Holland, Amsterdam-New York. ISSN: 0025-5610 **(a,s)**

Mathematical Reports of College of General Education, Kyushu University. [Math. Rep. Kyushu Univ.] Kyushu Univ., Fukuoka. ISSN: 0287-9980 **(a)**

Mathematical Reviews. [MR] Amer. Math. Soc., Providence, RI. ISSN: 0025-5629 **(a,s)**

The Mathematical Scientist. [Math. Sci.] Austral. Math. Soc., St. Lucia. ISSN: 0312-3685 **(a)**

Mathematical Social Sciences. [Math. Social Sci.] North-Holland, Amsterdam. ISSN: 0165-4896 **(a)**

Mathematical Structures in Computer Science. [Math. Structures Comput. Sci.] Cambridge Univ. Press, New York. ISSN: 0960-1295 **(a)**

Mathematical Systems Theory. [Math. Systems Theory] Springer, New York-Berlin. ISSN: 0025-5661 **(a,s)**

Mathematics Colloquium of the University of Cape Town. [Math. Colloq. Univ. Cape Town] Univ. Cape Town, Cape Town. ISSN: 0304-9922 **(a)**

Mathematics of Computation. [Math. Comp.] Amer. Math. Soc., Providence, RI. ISSN: 0025-5718 **(a,s)**

Mathematics and Computer Education. [Math. Comput. Ed.] MATYC J., Old Bethpage, NY. ISSN: 0730-8639

Mathematics and Computers in Simulation. [Math. Comput. Simulation] North-Holland, Amsterdam-New York. ISSN: 0378-4754 **(s)**

Mathematics of Control, Signals, and Systems. [Math. Control Signals Systems] Springer, New York-Berlin. ISSN: 0932-4194 **(a)**

The Mathematics Education. [Math. Ed. (Siwan)] Math. Ed., Siwan. ISSN: 6047-6269

Mathematics Journal of Toyama University. [Math. J. Toyama Univ.] Toyama Univ., Toyama. ISSN: 0916-6009 **(a)**

Mathematics Magazine. [Math. Mag.] Math. Assoc. America, Washington, DC. ISSN: 0025-570X **(a,p)**

Mathematics of Operations Research. [Math. Oper. Res.] Inst. Management Sci., Providence, RI. ISSN: 0364-765X **(s)**

Mathematics in Practice and Theory. Shuxue de Shijian yu Renshi. [Math. Practice Theory] Kexue Chubanshe [Science Press], Beijing. **(a)**

The Mathematics Student. [Math. Student] Indian Math. Soc., Delhi. ISSN: 0025-5742

Mathematics Teacher. [Math. Teacher] Nat. Council Teachers Math., Reston, VA. ISSN: 0025-5769 **(p)**

Mathematics Today. [Math. Today] Vaidya Sanman Nidhi, Ahmedabad. **(a)**

Mathematics of the USSR-Izvestiya. [Math. USSR-Izv.] Amer. Math. Soc., Providence, RI. ISSN: 0025-5726 **(a,s)**

Mathematics of the USSR-Sbornik. [Math. USSR-Sb.] Amer. Math. Soc., Providence, RI. ISSN: 0025-5734 **(a,s)**

Mathematika. [Mathematika] Univ. London, London. ISSN: 0025-5793 **(a)**

Mathématiques Informatique et Sciences Humaines. [Math. Inform. Sci. Humaines] École Hautes Études Sci. Soc., Paris. ISSN: 0995-2314

Mathematische Annalen. [Math. Ann.] Springer, Berlin-New York. ISSN: 0025-5831 **(a,s)**

Mathematische Nachrichten. [Math. Nachr.] Akademie-Verlag, Berlin. ISSN: 0025-584X **(a)**

Mathematische und Naturwissenschaftliche Unterricht. [Math. Naturwiss. Unterricht] Ferd. Dummler, Bonn. ISSN: 0025-5866 **(a)**

Mathematische Semesterberichte. [Math. Semesterber.] Vandenhoeck & Ruprecht, Göttingen. ISSN: 0720-728X **(a)**

Mathematische Zeitschrift. [Math. Z.] Springer, Berlin-New York. ISSN: 0025-5874 **(a,s)**

Mathesis. [Mathesis] Univ. Nac. Autónoma México, Mexico City. ISSN: 0185-6200

Matimyás Matematika. [Matimyás Mat.] Univ. Philippines, Quezon City. ISSN: 0115-6926 **(a)**

The Matrix and Tensor Quarterly. [Matrix Tensor Quart.] Tensor Soc. Great Britain, London. ISSN: 0025-5998 **(a)**

Memoirs of the American Mathematical Society. [Mem. Amer. Math. Soc.] Amer. Math. Soc., Providence, RI. ISSN: 0065-9266 **(a,p,s)**

Memoirs of the Faculty of General Education, Kumamoto University. Natural Sciences. Mathematics. [Mem. Fac. Gen. Ed. Kumamoto Univ. Natur. Sci.] Kumamoto Univ., Kumamoto ISSN: 0286-5769 **(a)**

Memoirs of the Faculty of Science, Kochi University. Series A. Mathematics. [Mem. Fac. Sci. Kochi Univ. Ser. A Math.] Kochi Univ., Kochi. ISSN: 0389-0252 **(a)**

Memoirs of the Faculty of Science, Kyushu University. Series A. Mathematics. [Mem. Fac. Sci. Kyushu Univ. Ser. A] Kyushu Univ., Fukuoka. ISSN: 0373-6385 **(a)**

Memorie Lincei. Matematica e Applicazioni. [Atti Accad. Naz. Lincei Cl. Sci Fiz. Mat. Natur. Mem. (9) Mat. Appl.] Accad. Naz. Lincei, Rome. ISSN: 1120-6357 **(a)**

Menemui Matematik. [Menemui Mat.] Malaysian Math. Soc., Kuala Lumpur. ISSN: 0126-9003

Metody Diskretnogo Analiza. [Metody Diskret. Analiz.] Akad. Nauk Sibirsk. Otdel., Inst. Mat., Novosibirsk. ISSN: 0136-1228 **(a)**

Metody Vychisleniĭ. [Metody Vychisl.] Leningrad. Univ., Leningrad. ISSN: 0131-2146 **(a)**

Metrika. [Metrika] Physica, Vienna. ISSN: 0026-1335 **(a)**

Metron. [Metron] Univ. degli Studi di Roma, Rome. ISSN: 0026-1424 **(a)**

The Michigan Mathematical Journal. [Michigan Math. J.] Univ. Michigan, Ann Arbor, MI. ISSN: 0026-2285 **(a)**

Miscelánea Matemática. [Miscelánea Mat.] Soc. Mat. Mexicana, Mexico City. **(a)**

Mitteilungen der Gesellschaft für Angewandte Mathematik und Mechanik. [Mitt. Ges. Angew. Math. Mech.] Ges. Angew. Math. Mech., Essen. ISSN: 0936-7195 **(a)**

Mitteilungen der Mathematische Gesellschaft der DDR. [Mitt. Math. Ges. DDR] Math. Gesellsch. DDR, Berlin. **(a)**

Mitteilungen der Mathematischen Gesellschaft in Hamburg. [Mitt. Math. Ges. Hamburg] Math. Gesellsch., Hamburg. ISSN: 0340-4358 **(a)**

Mitteilungen aus dem Mathem. Seminar Giessen. [Mitt. Math. Sem. Giessen] Univ. Giessen, Giessen. ISSN: 0373-8221 **(a)**

Mitteilungen (*Schweizerische Vereinigung der Versicherungsmathematiker*). [Schweiz. Verein. Versicherungsmath. Mitt.] Stampfli & Cie, Bern.

Modeling, Identification and Control. [MIC-Modeling Identification Control] Roy. Norwegian Council Sci. Indust. Res., Oslo. ISSN: 0332-7353

Monatshefte für Mathematik. [Monatsh. Math.] Springer, Vienna. ISSN: 0026-9255 (**a**)

Monografias da Sociedade Paranaense de Matemática. [Monograf. Soc. Paran. Mat.] Soc. Paranaense de Mat., Curitiba. ISSN: 0102-3292 (**a**)

Moscow University Computational Mathematics and Cybernetics. [Moscow Univ. Comput. Math. Cybernet.] Allerton Press, Allerton, N.Y. ISSN: 0278-6419

Moscow University Mathematics Bulletin. [Moscow Univ. Math. Bull.] Allerton Press, Allerton, NY. ISSN: 0027-1322 (**a,s**)

Nachrichten der Akademie der Wissenschaften in Göttingen. II, Mathematisch-Physikalische Klasse. [Nachr. Akad. Wiss. Göttingen Math.-Phys Kl. II] Vandenhoeck & Ruprecht, Göttingen. ISSN: 0065-5295 (**a**)

Nagoya Mathematical Journal. [Nagoya Math. J.] Nagoya Univ., Nagoya. ISSN: 0027-7630 (**a**)

Nanjing Daxue Xuebao. Shuxue Bannian Kan. See *Shu Hsueh Pan Nien Kan. Mathematical biquarterly.*

Natural Science Report of the Ochanomizu University. [Natur. Sci. Rep. Ochanomizu Univ.] Ochanomizu Univ., Tokyo. ISSN: 0029-8190 (**a**)

Naval Research Logistics. [Naval Res. Logist.] Wiley, NY. ISSN: 0028-1441 (**a,s**)

Nelineinye Granichnye Zadachi. [Nelinein. Granich. Zadachi] "Naukova Dumka," Kiev. ISSN: 0236-0497 (**a**)

The Nepali Mathematical Sciences Report. [Nepali Math. Sci. Rep.] Tribhuvan Univ., Kathmandu.

New Zealand Mathematical Society Newsletter. [New Zealand Math Soc. Newslett.] New Zealand Math. Soc., Wellington. ISSN: 0110-0025 (**a**)

The New Zealand Mathematics Magazine. [New Zealand Math. Mag.] Auckland Math. Assoc., Auckland. ISSN: 0549-0510 (**a**)

New Zealand Statistician. [New Zealand Statist.] New Zealand Statist. Assoc., Wellington. ISSN: 0111-9176 (**a**)

Nieuw Archief voor Wiskunde. Vierde Serie. [Nieuw Arch. Wisk. (4)] Math. Centrum, Amsterdam. ISSN: 0028-9825 (**a**)

Nieuw Tijdschrift voor Wiskunde. [Nieuw Tijdschr. Wisk.] Kluwer, Dordrecht. ISSN: 0028-985X

Nihonkai Mathematical Journal. [Nihonkai Math. J.] Niigata Univ., Dept. Math., Niigata. (**a**)

Nonlinear Analysis. [Nonlinear Anal.] Pergamon, Oxford-Elmsford, New York. ISSN: 0362-546X (**a,s**)

Nonlinear Science Today. [Nonlinear Sci. Today] Springer, Secaucus, NJ. ISSN: 0938-9008 (**a**)

Nonlinearity. [Nonlinearity] Inst. Phys., Bristol. ISSN: 0951-7715 (**a,s**)

Normat. [Normat] Universitetsforlaget, Oslo. ISSN: 0029-1412 (**a**)

Northeastern Mathematical Journal. See *Tung Pei Shu Hsueh = Northeastern Mathematical Journal.*

Notas de Matemática. [Notas Mat.] Univ. de los Andes, Mérida. **(a)**

Notas de la Sociedad de Matemática de Chile. [Notas Soc. Mat. Chile] Soc. Mat. Chile, Santiago. ISSN: 0716-1298 **(a)**

Note di Matematica. [Note Mat.] Univ. Stud. Lecce, Lecce. **(a)**

Notices of the American Mathematical Society. [Notices Amer. Math. Soc.] Amer. Math. Soc., Providence, RI. ISSN: 0002-9920 **(a,s)**

Notre Dame Journal of Formal Logic. [Notre Dame J. Formal Logic] Univ. Notre Dame Press, South Bend, IN. ISSN: 0029-4527 **(a,s)**

Novoe v Zhizni, Nauke, Tekhnike. Seriia: Matematika, Kibernetika. [Novoe Zhizni Nauk. Tekhn. Ser. "Mat. Kibernet."] "Znanie," Moscow.

Numerical Algorithms. [Numer. Algorithms] Baltzer, Basel. ISSN: 1017-1398

Numerical Functional Analysis and Optimization. [Numer. Funct. Anal. Optim.] Dekker, NY. ISSN: 0163-0563 **(a,s)**

Numerical Mathematics. Gaodeng Xuexiao Jisuan Shuxue Xuebao. [Numer. Math. J. Chinese Univ.] Nanjing Univ., Nanjing. ISSN: 1000-081X **(a)**

Numerical Methods for Partial Differential Equations. [Numer. Methods Partial Differential Equations] Wiley, NY. ISSN: 0749-159X **(a,s)**

Numerische Mathematik. [Numer. Math.] Springer, Berlin-New York. ISSN: 0029-599X **(a,s)**

Obzornik za Matematiko in Fiziko. [Obzornik Mat. Fiz.] Drustvo Mat. Fiz. Astronom. SRS, Ljubljana. ISSN: 0473-7466

Operations Research. [Oper. Res.] Oper. Res. Soc. Amer., Baltimore, MD. ISSN: 0030-364X **(s)**

Optimal Control: Applications & Methods. [Optimal Control Appl. Methods] Wiley, Chichester. ISSN: 0143-2087

Optimization. [Optimization] Akademie-Verlag, Berlin. ISSN: 0323-3898 **(a,s)**

Optimizatsiia. [Optimizatsiya] Akad. Nauk SSSR Sibirsk. Otdel., Inst. Mat., Novosibirsk. ISSN: 0134-3998 **(a)**

Opuscula Mathematica. [Opuscula Math.] Akad. Górniczo-Hutnicza S. Staszice Krakow., Krakow. ISSN: 0137-6535 **(a)**

OR Spektrum. [OR Spektrum] Springer, New York-Berlin. ISSN: 0171-6468

Order. [Order] Kluwer Acad. Publ., Hingham, MA. ISSN: 0167-8094 **(a)**

Osaka Journal of Mathematics. [Osaka J. Math.] Osaka Univ., Osaka. ISSN: 0030-6126 **(a)**

Otbor i Obrabotka Informatsiĭ. [Otbor i Obrabotka Informatsiĭ] "Naukova Dumka," Kiev. ISSN: 0474-8662

Pacific Journal of Mathematics. [Pacific J. Math.] Pacific J. Math., Carmel Valley, CA. ISSN: 0030-8730 **(a,p,s)**

Pakistan Journal of Statistics. [Pakistan J. Statist.] Pakistan J. Statist., Lahore. ISSN: 1012-9367 **(a)**

Panamerican Mathematical Journal. [Panamer. Math. J.] Univ. Central Florida, Dept. Math., Orlando, FL. **(a)**

Pentagon. [Pentagon] Kappa Mu Epsilon, Macomb, IL. ISSN: 0031-4870 **(a,p)**

Periodica Mathematica Hungarica. [Period. Math. Hungar.] Akad. Kiadó, Budapest. ISSN: 0031-5303 **(a)**

Philosophia Mathematica. Second Series. [Philos. Math. (2)] Old Dominion Univ., Norfolk, VA. ISSN: 0031-8019 **(a)**

Philosophical Transactions of the Royal Society of London. Series A. Mathematical and Physical Sciences. [Philos. Trans. Roy. Soc. London Ser. A] Royal Soc. London, London. ISSN: 0080-4614 **(s)**

Pi Mu Epsilon Journal. [Pi Mu Epsilon J.] Pi Mu Epsilon, St. Paul, MN. ISSN: 0031-952X **(a)**

PLISKA Studia Mathematica Bulgarica. [PLISKA Stud. Math. Bulgar.] Bulgar. Acad. Sci., Sofia. ISSN: 0204-9805 **(a)**

Pokroky Matematiky, Fyziky a Astronomie. [Pokroky Mat. Fyz. Astronom.] Academia, Prague. ISSN: 0032-2423 **(a)**

Polish Academy of Sciences. Institute of Philosophy and Sociology. Bulletin of the Section of Logic. [Polish Acad. Sci. Inst. Philos. Sociol. Bull. Sect. Logic] Polish Acad. Sci., Warsaw. ISSN: 0138-0680 **(a)**

Portugaliae Mathematica. [Portugal. Math.] Portuguese Math. Soc., Lisbon. ISSN: 0032-5155 **(a)**

Praxis der Mathematik. [Praxis Math.] Aulis/Deubner, Cologne. ISSN: 0032-7042 **(a)**

Preprint (*Instytut Matematyky* (*Akademii͡a Nauk RSR*)). [Akad. Nauk Ukrainy Inst. Mat. Preprint] Inst. Math. Akad. Nauk Ukrainy, Kiev. **(a)**

Prikladnai͡a Matematika i Mekhanika. [Prikl. Mat. Mekh.] "Nauka," Moscow. ISSN: 0032-8235

Primenenie Teorii Veroi͡atnosteĭ i Matematicheskoĭ Statistiki. [Primenen. Teor. Veroi͡atnost. i Mat. Statist.] Akad. Nauk Litovsk. SSR, Inst. Mat. i Kibernet., Vilnius. ISSN: 0207-1169 **(a)**

Probability in the Engineering and Informational Sciences. [Probab. Engrg. Inform. Sci.] Cambridge Univ. Press, New Rochelle, NY. ISSN: 0269-9648 **(a)**

Probability and Mathematical Statistics. [Probab. Math. Statist.] PWN, Wydawn. Uniw. Wroclaw., Wroclaw. ISSN: 0208-4147 **(a)**

Probability Theory and Related Fields. [Probab. Theory Related Fields] Springer, Berlin-New York. ISSN: 0178-8051 **(a)**

Problems of Control and Information Theory. Problemy Upravlenii͡a i Teorii Informatsiĭ. [Problems Control Inform. Theory/Problemy Upravlen. Teor. Inform.] Akad. Kiadó, Budapest. ISSN: 0370-2529

Problemy Matematyczne. [Problemy Mat.] Wydawn Uczel. WSP, Bydogoszcz. **(a)**

Problemy Peredachi Informat͡siĭ. [Problemy Peredachi Informatsii] "Nauka," Moscow. ISSN: 0555-2923 **(a)**

Proceedings of the American Mathematical Society. [Proc. Amer. Math. Soc.] Amer. Math. Soc., Providence, RI. ISSN: 0002-9939 **(a,p,s)**

Proceedings of the Edinburgh Mathematical Society. Series II. [Proc. Edinburgh Math. Soc. (2)] Scottish Academic Press, Edinburgh. ISSN: 0013-0915 **(a)**

Proceedings of the Institute of Statistical Mathematics. See *Tokei Suri Kenkyujo Iho.*

Proceedings of the Japan Academy. Series A, Mathematical Sciences. [Proc. Japan Acad. Ser. A Math. Sci.] Japan Acad., Tokyo. ISSN: 0386-2194 **(a)**

Proceedings of the London Mathematical Society. Third Series. [Proc. London Math. Soc. (3)] Oxford Univ. Press, London-New York. ISSN: 0024-6115 **(a,s)**

Proceedings of the Mathematical and Physical Society of Egypt. [Proc. Math. Phys. Soc. Egypt] Nat. Inform. Documentation Centre, Cairo. ISSN: 1110-0613 (**a**)

Proceedings. Mathematical Sciences. [Proc. Indian Acad. Sci. Math. Sci.] Indian Acad. Sci., Bangalore. ISSN: 0370-0089 (**a**)

Proceedings of the Mathematical Society. [Proc. Math. Soc.] Banaras Hindu Univ., Math. Soc., Varanasi. (**a**)

Proceedings of the Royal Irish Academy. Section A. [Proc. Roy. Irish Acad. Sect. A] Roy. Irish Acad., Dublin. ISSN: 0035-8975 (**a**)

Proceedings of the Royal Society of Edinburgh. Section A. Mathematics. [Proc. Roy. Soc. Edinburgh Sect. A] Roy. Soc. Edinburgh, Edinburgh. ISSN: 0308-2105 (**a,s**)

Proceedings of the Steklov Institute of Mathematics. [Proc. Steklov Inst. Math.] Amer. Math. Soc., Providence, RI. ISSN: 0081-5438 (**a,s**)

Programmirovanie ĖVM. [Programmirovanie ĖVM] Lietuvos TSR ma Mat. ir Kibernet. Inst., Vilnius. (**a**)

Progress of Mathematics. [Progr. Math. (Varanasi)] Acad. Progr. Math., Varanasi. (**a**)

Proyecciones. Revista de Matemática. [Proyecciones] Univ. Catól. Norte, Fac. Cienc., Dep. Mat., Antofagasta, Chile. ISSN: 0716-0917 (**a**)

Pubblicazioni del Dipartimento di Metodi e Modelli Matematici per le Scienze Applicate, Università degli Studi di Roma"La Sapienza." See *Quaderni (Università degli Studi di Roma"La Sapienza." Dipartimento di Metodi e Modelli Matematici per le Scienze Applicate).*

Publicaciones Matemáticas del Uruguay. [Publ. Mat. Uruguay] Univ. Republica, Montevideo. (**a**)

Publicacions Matemàtiques. [Publ. Mat.] Univ. Autónoma Barcelona, Barcelona. ISSN: 0214-1493 (**a**)

Publications du Centre de Recherches en Mathématiques Pures. Série I. [Publ. Centre Rech. Math. Pures (I) (Neuchâtel)] Univ. Neuchâtel, Neuchâtel. ISSN: 0259-4897 (**a**)

Publications de l'Institut Mathématique. Nouvelle Série. [Publ. Inst. Math. (Beograd) (N.S.)] Inst. Math., Belgrade. ISSN: 0034-5318 (**a**)

Publications de l'Institut de Statistique de l'Université de Paris. [Publ. Inst. Statist. Univ. Paris] Univ. Paris VI, Paris. (**a**)

Publicationes Mathematicae. [Publ. Math. Debrecen] Kossuth Lajos Tudoman., Debrecen. ISSN: 0033-3883 (**a**)

Publications Mathématiques (Institut des Hautes Études Scientifiques). [Inst. Hautes Études Sci. Publ. Math.] Presses Univ. France, Paris. ISSN: 0073-8301 (**a**)

Publications of the Research Institute for Mathematical Sciences. [Publ. Res. Inst. Math. Sci.] Kyoto Univ., Kyoto. ISSN: 0034-5318 (**a,s**)

Publikacije Elektrotehnickog Fakulteta. Serija: Matematika. [Univ. Beograd. Publ. Elektrotehn. Fak. Ser. Mat.] Univ. Belgrade, Belgrade. ISSN: 0353-8893 (**a**)

Punime Matematike. [Punime Mat.] Univ. Kosoves Prishtine, Prishtina. ISSN: 0352-5759 (**a**)

Punjab University Journal of Mathematics. [Punjab Univ. J. Math. Lahore] Punjab, Lahore. (**a**)

Pure and Applied Mathematics. Chuncui Shuxue yu Yingyong Shuxue. [Pure Appl. Math.] Northwest Univ., Xian. (**a**)

Pure and Applied Mathematika Sciences. [Pure Appl. Math. Sci.] Math. Sci. Soc., Saharanpur. ISSN: 0379-3168 (**a**)

Pure Mathematics and Applications. Series A. [Pure Math. Appl. Ser. A] Budapest Univ. Econom., Inst. Math. Comput. Sci., Budapest. (**a**)

Pure Mathematics and Applications. Series B. [Pure Math. Appl. Ser. B] Budapest Univ. Econom., Inst. Math. Comput. Sci., Budapest. (**a**)

Pure Mathematics and Applications. Series C. [Pure Math. Appl. Ser. C] Budapest Univ. Econom., Inst. Math. Comput. Sci., Budapest. (**a**)

Pure Mathematics Manuscript. [Pure Math. Manuscript] Calcutta Univ., Calcutta. (**a**)

Quaderni dell Mathesis di Cosenza. [Quad. Mathesis Cosenza] Luciano, Rome. (**a**)

Quaderni (Università degli Studi di Roma"La Sapienza." Dipartimento di Metodi Matematici per le Scienze Applicate) [Pubbl. Dip. Metod. Model. Mat. Sci. Appl. Univ. Stud. Roma Quad.] Univ. Roma, Rome. (**a**)

Quaestiones Mathematicae. [Quaestiones Math.] South African Math. Soc., Potchefstroom. ISSN: 0379-9468 (**a**)

Quantum. [Quantum] Nat. Sci. Teachers Assoc., Washington, DC. ISSN: 1048-8820 (**p**)

Quarterly of Applied Mathematics. [Quart. Appl. Math.] Amer. Math. Soc., Providence, RI. ISSN: 0033-569X (**a,p,s**)

The Quarterly Journal of Mathematics. Oxford Series. Second Series. [Quart. J. Math. Oxford Ser. (2)] Oxford Univ. Press, London-New York. ISSN: 0033-5606 (**a,s**)

The Quarterly Journal of Mechanics and Applied Mathematics. [Quart. J. Mech. Appl. Math.] Oxford Univ. Press, Oxford. ISSN: 0033-5614 (**a,s**)

Qüestiió. [Questiio] Univ. Politec. Catalunya, Barcelona. ISSN: 0210-8054 (**a**)

Questions and Answers in General Topology. [Questions Answers Gen. Topology] Osaka Kyoiku Univ., Osaka. (**a**)

Queueing Systems. Theory and Applications. [Queueing Systems Theory Appl.] Baltzer, Basel. ISSN: 0257-0130 (**a**)

Radovi Matematički. [Rad. Mat.] Akad. Nauk. i Umjetnost. Bosne i Hercegovine, Sarajevo. ISSN: 0352-6100 (**a**)

RAIRO Informatique Théorique et Applications. [RAIRO Inform. Théor. Appl.] Dunod, Paris. ISSN: 0988-3754

RAIRO Modélisation Mathématique et Analyse Numérique. [RAIRO Modél. Math. Anal. Numér.] Dunod, Paris. ISSN: 0399-0516 (**a**)

RAIRO Recherche Opérationnelle. [RAIRO Rech. Opér.] Dunod, Paris. ISSN: 0399-0559

Ranchi University Mathematical Journal. [Ranchi Univ. Math. J.] Ranchi Univ., Ranchi. ISSN: 0079-9602 (**a**)

Random Structures & Algorithms. [Random Structures Algorithms] Wiley, NY. ISSN: 1042-9832 (**a**)

Real Analysis Exchange. [Real Anal. Exchange] Western Illinois Univ., Macomb, IL. ISSN: 0147-1937 (**a**)

Rebrape. See *Revista Brasileira de Probabilidade e Estatistica.*

Referativnyĭ Zhurnal. Matematika. [RZhMat] Akad. Nauk SSSR, Vsesoyuz. Inst. Nauchn. i Tekhn. Inform., Moscow. ISSN: 0034-2467 (**a**)

Rendiconti della Accademia Nazionale delle Scienze detta dei XL. Serie V. Memorie di Matematica. Parte I. [Rend. Accad. Naz. Sci. XL Mem. Mat. (5)] Accad. Naz. Sci. XL, Rome. ISSN: 0392-4106 (**a**)

Rendiconti del Circolo Matematico di Palermo. Serie II. [Rend. Circ. Mat. Palermo (2)] Circ. Mat. Palermo, Palermo. ISSN: 0009-725X (**a**)

Rendiconti del Circolo Matematico di Palermo. Serie II. Supplemento. [Rend. Circ. Mat. Palermo (2) Suppl.] Circ. Mat. Palermo, Palermo. ISSN: 0009-725X (**a**)

Rendiconti dell'Istituto di Matematica dell'Università di Trieste. [Rend. Istit. Mat. Univ. Trieste] Univ. degli Studi Trieste, Trieste. ISSN: 0049-4704 (**a**)

Rendiconti Lincei. Matematica e Applicazioni. [Atti Accad. Naz. Lincei Cl. Sci. Fis. Mat. Natur. Rend. (9) Mat. Appl.] Accad. Naz. Lincei, Rome. ISSN: 1120-6330 (**a**)

Rendiconti di Matematica e della sue Applicazioni. Serie VII. [Rend. Mat. Appl. (7)] Univ. Studi Roma"La Sapienza," Rome. ISSN: 0034-4427 (**a**)

Rendiconti del Seminario della Facoltà di Scienze della Università di Cagliari. [Rend. Sem. Fac. Sci. Univ. Cagliari] Univ. Cagliari, Cagliari.

Rendiconti del Seminario Matematico. [Rend. Sem. Mat. Univ. Politec. Torino] Univ. Politec. Torino, Dir. Sem. Mat., Turin. ISSN: 0373-1243 (**a**)

Rendiconti del Seminario Matematico e Fisico di Milano. [Rend. Sem. Mat. Fis. Milano.] Univ. degli Studi Milano, Milan. ISSN: 0370-7377 (**a**)

Rendiconti del Seminario Matematico dell'Università di Padova. [Rend. Sem. Mat. Univ. Padova] CEDAM, Padova. (**a**)

Rendiconto dell'Accademia delle Scienze Fisiche e Matematiche. Serie IV. [Rend. Accad. Sci. Fis. Mat. Napoli (4)] Giannini & Figli, Naples. ISSN: 0370-3568

Report (Institut Mittag-Leffler) [Rep. Inst. Mittag-Leffler] Inst. Mittag-Leffler, Djursholm. (**a**)

Reporte de Investigación del Instituto de Matemática, Cibernética y Computación. [Rep. Investigación Inst. Mat. Cibernet. Comput.] Acad. Cienc. Cuba, Havana. ISSN: 0138-8916

Reports of the Faculty of Science and Engineering, Saga University. Mathematics. [Rep. Fac. Sci. Engrg. Saga Univ. Math.] Saga Univ., Saga. ISSN: 0387-8538 (**a**)

Reports of Mortality and Morbidity Experience. [Rep. Mortality Morbidity Experience] Soc. Actuaries, Itasca, IL.

Reports on Mathematical Logic. [Rep. Math. Logic.] PWN, Warsaw. ISSN: 0137-2904 (**a**)

Reports on Mathematical Physics. [Rep. Math. Phys.] Pergamon, Oxford-Elmsford, New York. ISSN: 0034-4877 (**a,s**)

Reports of Statistical Application Research, Union of Japanese Scientists and Engineers. [Rep. Statist. Appl. Res. Un. Japan. Sci. Engrs.] Union Japan. Sci. Engin., Tokyo. ISSN: 0034-4842 (**a**)

Research Memoir of the Statistical Bureau. See *Tokeikyoku Kenkyu Iho.*

Research Report (Keio Gijuku Daigaku. Dept. of Mathematics). [Res. Rep. Keio Univ. Dept. Math.] Keio Univ., Yokohama. (**a**)

Resultate der Mathematik. [Resultate Math.] Birkhäuser, Boston, MA. ISSN: 0378-6218 (**a**)

Reviews in Mathematical Physics. [Rev. Math. Phys.] World Sci. Publishing, Singapore. ISSN: 0129-055X (**a**)

Revista Brasileira de Probabilidade e Estatística. [Rebrape] Assoc. Brasileira Estatist., São Paulo. (**a**)

Revista Ciencias Matemáticas. [Cienc. Mat. (Havana)] Univ. Habana, Havana. ISSN: 0256-5374 (**a**)

Revista Colombiana de Matemáticas. [Rev. Colombiana Mat.] Soc. Colombiana Mat., Bogota. ISSN: 0034-7426 (**a**)

Revista Integración. Temas de Matemáticas. [Rev. Integr. Temas Mat.] Univ. Indust. Santander, Bucaramanga. ISSN: 0120-419X (**a**)

Revista Internacional de Métodos Numéricos para Cálculo y Diseño en Ingeniería. [Rev. Internac. Métod. Numér. Cálc. Diseñ. Ingr.] Univ. Politéc. Catalunya, Barcelona. ISSN: 0213-1315

Revista de Matemática e Estatística. [Rev. Mat. Estatíst.] Univ. Estad. Paulista, São Paulo. ISSN: 0102-0811 (**a**)

Revista Matemática Iberoamericana. [Rev. Mat. Iberoamericana] Consejo Sup. Inv. Cient., Madrid. ISSN: 7294-1985 (**a**)

Revista Matemática de la Universidad Complutense de Madrid. [Rev. Mat. Univ. Complut. Madrid] Editorial Univ. Complutense Madrid, Madrid. ISSN: 0214-3577 (**a**)

Revista de Matemáticas Aplicadas. [Rev. Mat. Apl.] Univ. Chile, Santiago. ISSN: 0716-5803 (**a**)

Revista. Serie A, Matemáticas y Física Teórica. [Univ. Nac. Tucumán Rev. Ser. A] Univ. Nac. Tucumán, Tucumán. ISSN: 0080-2360 (**a**)

Revista de la Unión Matemática Argentina. [Rev. Un. Mat. Argentina] Union Mat. Argentina, Buenos Aires. ISSN: 0041-6932 (**a**)

Revue de Mathématiques Speciales. [Rev. Math. Speciales] Vuibert, Paris. ISSN: 0035-1504

Revue Roumaine de Mathématiques Pures et Appliquées. [Rev. Roumaine Math. Pures Appl.] Ed. Acad. R.S. Romania, Bucharest. ISSN: 0035-3965 (**a**)

Revue de Statistique Appliquée. [Rev. Statist. Appl.] Univ. Paris VI, Paris. ISSN: 0035-175X (**a**)

Riazi: The Journal of Karachi Mathematical Association. [Riazi J. Karachi Math. Assoc.] Univ. Karachi, Karachi. ISSN: 0255-7568 (**a**)

La Ricerca: Matematiche Pure ed Applicate. [Ricerca: Mat. Pure Appl.] ESA Editrice, Rome. (**a**)

Ricerche di Matematica. [Ricerche Mat.] Univ. Degli Studi di Napoli, Naples. (**a**)

Rivista di Matematica Pura ed Applicata. [Riv. Mat. Pura Appl.] Aviani, Tricesimo. (**a**)

Rivista di Matematica per le Scienze Economiche e Sociali. [Riv. Mat. Sci. Econom.] Assoc. Mat. Appl. Sci. Econom. Soc. (A.M.A.S.E.S.), Milan. (**a**)

Rivista di Matematica della Università di Parma. Serie IV. [Riv. Mat. Univ. Parma (4)] Univ. Parma, Parma. ISSN: 0035-6298 (**a**)

Rivista di Statistica Applicata. [Riv. Statist. Appl.] Ed. Sci. Tec. Europee, Milano. ISSN: 0036-6549 (**a**)

The Rocky Mountain Journal of Mathematics. [Rocky Mountain J. Math.] Rocky Mountain Math. Consortium, Tempe, AZ. ISSN: 0035-7596 (**a**)

Roczniki Polskiego Towarzystwa Matematycznego. Seria II. Wiadomości Matematyczne. [Wiadom. Mat.] PWN, Warsaw. ISSN: 0373-8302

Roczniki. Seriia III: Matematyka Stosowana. [Mat. Stos.] PWN, Warsaw. ISSN: 0137-2890 (**a**)

Rostocker Mathematisches Kolloquium. [Rostock. Math. Kolloq.] Wilhelm-Pieck-Univ., Rostock. ISSN: 0138-3248 (**a**)

Russian Journal of Numerical Analysis and Mathematical Modelling. [Russian J. Numer. Anal. Math. Modelling] VSP, Zeist. ISSN: 0927-6467

Russian Mathematical Surveys. [Russian Math. Surveys] London Math. Soc., London. ISSN: 0036-0279 (**a,s**)

Ryuku Mathematical Journal. [Ryuku Math. J.] Univ. Ryukyus, Dept. Math., Okinawa. (**a**)

Sächsische Akademie der Wissenschaften, Leipzig. Mathematisch-Naturwissenschaftliche Klasse. Sitzungsberichte. [Sitzungsber. Sächs. Akad. Wiss. Leipzig Math. Natur. Kl.] Akademie-Verlag, Berlin. ISSN: 0371-327X

St. Petersburg Mathematical Journal. [St. Petersburg Math. J.] Amer. Math. Soc., Providence, RI. ISSN: 1061-0022 (**a**)

Saitama Mathematical Journal. [Saitama Math. J.] Saitama Univ., Urawa. ISSN: 0289-0739 (**a**)

Sankhyā. Series A. [Sankhya Ser. A] Statist. Publ. Soc., Calcutta. ISSN: 0581-572X (**a**)

Sankhyā. Series B. [Sankhya Ser. B] Statist. Publ. Soc., Calcutta. ISSN: 0581-5738 (**a**)

Scandinavian Actuarial Journal. [Scand. Actuar. J.] Almqvist & Wiksell, Stockholm. ISSN: 0346-1238 (**a**)

Scandinavian Journal of Statistics. Theory and Applications. [Scand. J. Statist.] Scand. Univ. Press, Oslo. ISSN: 0303-6898 (**a**)

Schweizerische Vereinigung der Versicherungsmathematiker. Mitteilungen. See *Mitteilungen (Schweizerische Vereinigung der Versicherungsmathematiker)*

Science in China (Scientia Sinica). Series A. Mathematics, Physics, Astronomy & Technological Sciences [Sci. China Ser. A] Kexue Chubanshe [Science Press], Beijing; Pergamon, Elmsford. ISSN: 0253-5831 (**a**)

Science Reports of the Yokohama National University. Section I. Mathematics, Physics, Chemistry. [Sci. Rep. Yokohama Nat. Univ. Sect. I Math. Phys. Chem.] Yokohama Nat. Univ., Yokohama. ISSN: 0513-5613 (**a**)

Scientific Papers of the College of Arts and Sciences, the University of Tokyo. [Sci. Papers College Arts Sci. Univ. Tokyo] Univ. Tokyo, Tokyo. ISSN: 0289-7520 (**a**)

Selecta Mathematica Sovietica. [Selecta Math. Soviet.] Birkhäuser Boston, Boston, MA. ISSN: 0272-9903 (**a**)

Semigroup Forum. [Semigroup Forum] Springer, New York-Berlin. ISSN: 0037-1912 (**a**)

Séminaire d'Analyse Convexe. Travaux du Séminaire d'Analyse Convexe. [Sém. Anal. Convexe] Univ. Sci. Tech. Languedoc, Montpellier. (**a**)

Séminaire Gaston Darboux de Géométrie et Topologie. [Sém. Gaston Darboux Géométrie Topologie] Univ. Sci. Tech. Languedoc, Montpellier. (**a**)

Séminaire d'Histoire des Mathématiques. Cahiers du Séminaire d'Histoire des Mathématiques. [Cahiers Sém. Hist. Math.] Univ. Pierre et Marie Curie, Paris. (**a**)

Séminaire de Théorie des Nombres de Bordeaux. Série 2. [Sém. Théor. Nombres Bordeaux (2)] Univ. Bordeaux I, Talence. ISSN: 0989-5558 (**a**)

Seminar. Institut Prikladnoĭ Matematiki. Doklady. See *Doklady. Seminar Rasshirennykh Zasedaniĭ Seminara Instituta Prikladnoĭ Matematiki im. I.N. Vekua.*

Seminar on Mathematical Sciences. [Sem. Math. Sci.] Keio Univ., Yokohama. (**a**)

Seminar Sophus Lie: Darmstadt, Erlangen, Greifswald, Leipzig. [Sem. Sophus Lie] Heldermann, Berlin. ISSN: 0940-2268 (**a**)

Seminar po Vektornomu i Tenzornomu Analizu. Trudy Seminara po Vektornomu i Tenzornomu Analizu s ikh Prilozhenii͡ami k Geometrii͡, Mekhanike i Fizike. [Trudy Sem. Vektor. Tenzor. Anal.] Moskov. Gos. Univ., Moscow, ISSN: 0204-3165 (**a**)

Sequential Analysis. [Sequential Anal.] Dekker, NY. ISSN: 0747-4946 (**a**)

Serdica. [Serdica] Bulgar. Acad. Sci., Sofia. ISSN: 0204-4110 (**a**)

Serii͡a Matematicheskikh Nauk. [Leningrad. Gos. Univ. Uchen. Zap. Ser. Mat. Nauk] Leningrad. Univ., Leningrad. ISSN: 0136-8141 (**a**)

Sheng We Shu Hsueh Hsueh Pao = Journal of Biomathematics. [J. Biomath.] J. Biomath. Office, Hefei. (**a**)

Shu Chih Chi Suan Yu Chi Suan Chi Ying Yung = Journal of Numerical Methods and Computer Applications. [J. Numer. Methods Comput. Appl.] Kexue Chubanshe [Science Press], Beijing. ISSN: 1000-3266 (**a**)

Shu Hsueh Hsueh Pao = Acta Mathematica Sinica. [Acta Math. Sinica] Kexue Chubanshe [Science Press], Beijing. ISSN: 0583-1431 (**a**)

Shu Hsueh Li Hsueh Pao = Acta Mathematica Scientia. [Acta Math. Sci. (English Ed.)] Acta Math. Sci. Press, Wuhan. ISSN: 0252-9602 (**a**)

Shu Hsueh Nien K'an. A Chi. = Chinese Annals of Mathematics. Ser. A. [Chinese Ann. Math. Ser. A] Shanghai Sci. Tech., Shanghai. ISSN: 1000-8314 (**a**)

Shu Hsueh Pan Nien Kan = Mathematical Biquarterly. [Nanjing Daxue Xuebao. Shuxue Bannian Kan] Nanjing Univ., Nanjing. (**a**)

Shu Hsueh Tsa Chih = Journal of Mathematics. [J. Math. (Wuhan)] Wuhan Univ., Wuhan. (**a**)

Shuxue Jinzhan = Advances in Mathematics. [Adv. in Math. (China)] Beijing Daxue Chubanshe, Beijing. ISSN: 1000-0917 (**a**)

SIAM Journal on Applied Mathematics. [SIAM J. Appl. Math.] SIAM, Philadelphia, PA. ISSN: 0036-1399 (**a,s**)

SIAM Journal on Computing. [SIAM J. Comput.] SIAM, Philadelphia, PA. ISSN: 0097-5397 (**a,s**)

SIAM Journal on Control and Optimization. [SIAM J. Control Optim.] SIAM, Philadelphia, PA. ISSN: 0363-0129 (**a,s**)

SIAM Journal on Discrete Mathematics. [SIAM J. Discrete Math.] SIAM, Philadelphia, PA. ISSN: 0895-4801 (**a,s**)

SIAM Journal on Mathematical Analysis. [SIAM J. Math. Anal.] SIAM, Philadelphia, PA. ISSN: 0036-1410 (**a,s**)

SIAM Journal on Matrix Analysis and Applications. [SIAM J. Matrix Anal. Appl.] SIAM, Philadelphia, PA. ISSN: 0895-4798 **(a,s)**

SIAM Journal on Numerical Analysis. [SIAM Journal on Numerical Analysis] SIAM, Philadelphia, PA. ISSN: 0036-1429 **(a,s)**

SIAM Journal on Optimization. [SIAM J. Optimization] SIAM, Philadelphia, PA. ISSN: 1052-6234 **(a,s)**

SIAM Journal on Scientific and Statistical Computing. [SIAM J. Sci. Statist. Comput.] SIAM, Philadelphia, PA. ISSN: 0196-5204 **(a,s)**

SIAM News. [SIAM News]. SIAM, Philadelphia, PA. **(a,s)**

SIAM Review. [SIAM Rev.] SIAM, Philadelphia, PA. ISSN: 0036-1445 **(a,s)**

Siberian Advances in Mathematics. [Siberian Adv. Math.] Allerton, New York. ISSN: 1055-1344 **(a)**

Siberian Mathematical Journal. [Siberian Math. J.] Consultants Bureau, New York-London. ISSN: 0037-4466 **(a)**

Sibirskiĭ Matematicheskiĭ Zhurnal. [Sibirsk. Mat. Zh.] "Nauka" Sibirsk. Otdel., Novosibirsk. ISSN: 0037-4474 **(a)**

Simon Stevin. [Simon Stevin] Natuur-en Geneeskundige Vennootschap, Ghent. ISSN: 0037-5454 **(a)**

Sitzungsberichte der Sächsischen Akademie der Wissenschaften zu Leipzig. Mathematisch-Naturwissenschaftliche Klasse. See *Sächsische Akademie der Wissenschaften, Leipzig. Mathematisch-Naturwissenschaftliche Klasse.*

Soochow Journal of Mathematics. [Soochow J. Math.] Soochow Univ., Taipei. ISSN: 0250-3255 **(a)**

South African Statistical Journal. [South African Statist. J.] South African Statist. Assoc., Pretoria. ISSN: 0038-271X **(a)**

Southeast Asian Bulletin of Mathematics. [Southeast Asian Bull. Math.] World Sci. Publishing, Singapore. ISSN: 0129-2021 **(a)**

Soviet Journal of Contemporary Mathematical Analysis. [Soviet J. Contemporary Math. Anal.] Allerton, NY. ISSN: 0735-2719

Soviet Mathematics. Doklady. [Soviet Math. Dokl.] Amer. Math. Soc., Providence, RI. ISSN: 0197-6788 **(a)**

Soviet Mathematics. (*Iz. VUZ*). [Soviet Math. (Iz. VUZ)] Allerton Press, Allerton, NY. ISSN: 0917-7156 **(a)**

Statistica. [Statistica (Bologna)] Coop. Lib. Univ., Bologna. ISSN: 0390-590X **(a)**

Statistica Neerlandica. [Statist. Neerlandica] Vereniging voor Statist., Rotterdam. ISSN: 0039-0402 **(a)**

Statistica Sinica. [Statist. Sinica] Statist. Sinica, Univ. Manitoba, Dept. Statist., Winnipeg, MB. ISSN: 1017-0405 **(a)**

Statistical Science. [Statist. Sci.] Inst. Math. Statist., Hayward, CA. ISSN: 0883-4237 **(a,p)**

Statistical Software Newsletter. [Statist. Software Newsl.] Gesellschaft fur Strahlen-und Umweltforschung, Neuherberg. ISSN: 0173-5896 **(a)**

Statistical Theory and Method Abstracts. [STMA] Longman, London-New York. ISSN: 0039-0518 **(a)**

Statisticheskie Problemy Upravleniia͡. [Statist. Problemy Upravleniya] Akad. Nauk Litovsk. SSR, Inst. Mat. i Kibernet., Vilnius. ISSN: 0132-5302

Statistician. [Statistician] Carfax, Abingdon, UK. ISSN: 0039-0526 (**a**)

Statistics. [Statistics] Akademie-Verlag, Berlin. ISSN: 0323-3944 (**a**)

Statistics and Computing. [Statist. Comput.] Chapman & Hall, London. ISSN: 0960-3174 (**a**)

Statistics & Decisions. [Statist. Decisions] Oldenbourg, Munich. ISSN: 0721-2631 (**a**)

Statistics & Probability Letters. [Statist. Probab. Lett.] North-Holland, Amsterdam-New York. ISSN: 0167-7152 (**a**)

Statistique et Analyse des Données. [Statist. Anal. Données] Assoc. Statist. Util., Paris. ISSN: 0750-7364

Statistische Hefte. [Statist. Hefte] Springer, New York-Berlin. ISSN: 0039-0631 (**a**)

Stats. [Stats] Amer. Statist. Assoc., Alexandria, VA. (**a**)

Stochastic Analysis and Applications. [Stochastic Anal. Appl] Dekker, NY. ISSN: 0736-2994 (**a,s**)

Stochastic Processes and their Applications. [Stochastic Process. Appl.] North-Holland, Amsterdam-New York. ISSN: 0304-4149 (**a,s**)

Stochastica. [Stochastica] Univ. Politec. Barcelona, Barcelona. ISSN: 0210-7821 (**a**)

Stochastics and Stochastics Reports. [Stochastics Stochastics Rep.] Gordon and Breach, London. ISSN: 0090-9491 (**a**)

Structural Topology. Topologie Structurale. See *Topologie Structurale = Structural Topology*.

Studia Logica. [Studia Logica] Ossolineum, Wroclaw. ISSN: 0039-3215 (**a**)

Studia Mathematica. [Studia Math.] PWN, Warsaw. ISSN: 0039-3223 (**a**)

Studia Scientiarum Mathematicarum Hungarica. [Studia Sci. Math. Hungar.] Akad. Kiadó, Budapest. ISSN: 0081-6906 (**a**)

Studia Universitatis Babeş-Bolyai. Mathematică. [Studia Univ. Babeş-Bolyai Math.] Univ. "Babeş-Bolyai," Cluj-Napoca. (**a**)

Studies in Applied Mathematics. [Stud. Appl. Math.] Elsevier, NY. ISSN: 0022-2526 (**a,s**)

Studii şi Cercetări Matematice. [Stud. Cerc. Mat.] Ed. Acad. R.S. România, Bucharest. ISSN: 0039-4068 (**a**)

Sûgaku. [Sûgaku] Math. Soc. Japan, Tokyo. ISSN: 0039-470X (**a**)

Sugaku Expositions. [Sugaku Expositions] Amer. Math. Soc., Providence, RI. ISSN: 0898-9583 (**a,s**)

Su-hak. [Su-hak] Acad. Sci., P'yōngyang.

Survey Methodology. [Survey Meth.] Stat. Canada, Ottawa, ON. ISSN: 0714-0045 (**a**)

Surveys on Mathematics for Industry. [Surveys Math. Indust.] Springer, Vienna. ISSN: 0938-1953

SUT Journal of Mathematics. [SUT J. Math.] Tokyo Science Univ., Tokyo. ISSN: 0496-6597 (**a**)

System Science and Mathematical Sciences. [Systems Sci. Math. Sci.] Science Press, Beijing; Allerton, NY. ISSN: 1000-9590.

Systems & Control Letters. [Systems Control Letters] North-Holland, Amsterdam; New York. ISSN: 0167-6911

Szigma. (*Matematikai-Kőzgazdasági Folyóirat*.) [Szigma—Mat.-Kozgazdasagi Folyoirat] Akad. Kiadó, Budapest. ISSN: 0039-8128

Tamkang Journal of Mathematics. [Tamkang J. Math.] Tamkang Univ., Tamsui. ISSN: 0049-2930 (**a**)

Tap chí Toán học. [Tap chí Toán học] Math. Soc., Hanoi. ISSN: 0866-7179 (**a**)

Teaching Statistics. [Teaching Statist.] Longman, Harlow, UK. ISSN: 0141-982X

Technometrics. [Technometrics] Amer. Statist. Assoc., Alexandria, VA. ISSN: 0040-1706 (**a,s**)

Tensor. [Tensor (N.S.] Tensor Soc., Chigasaki. ISSN: 0040-3504 (**a**)

Teoriia Funktsii, Funktsional'nyi Analiz i ikh Prilozheniia. [Teor. Funktsii Funktsional. Anal. i Prilozhen.] "Osnova," Kharkov. ISSN: 0321-4427 (**a**)

Teoriia Veroiatnostei i Matematicheskaia Statistika. [Teor. Veroyatnost. i Mat. Statist.] "Vishcha Shkola," Kiev. ISSN: 0131-6982

Teoriia Veroiatnostei i ee Primeneniia. [Teor. Veroyatnost. i Primenen.] "Nauka," Moscow. ISSN: 0040-361X (**a**)

Teoreticheskaia i Matematicheskaia Fizika. [Teoret. Mat. Fiz.] "Nauka," Moscow. ISSN: 0564-6162

Theoretic Papers. [Theoret. Papers] Blindern Theoret. Res. Team, Oslo. ISSN: 0801-3128 (**a**)

Theoretical Computer Science. [Theoret. Comput. Sci.] North-Holland, Amsterdam-New York. ISSN: 0304-3975 (**a**)

Theoretical and Mathematical Physics. [Theoret. and Math. Phys.] Consultants Bureau, New York-London. ISSN: 0040-5779 (**s**)

Theory of Probability and its Applications. [Theory Probab. Appl.] SIAM, Philadelphia, PA. ISSN: 0040-585X (**a,s**)

Theory of Probability and Mathematical Statistics. [Theory Probab. Math. Statist.] Amer. Math. Soc., Providence, RI. ISSN: 0094-9000 (**a**)

The Tôhoku Mathematical Journal. Second Series. [Tôhoku Math. J. (2)] Tôhoku Univ., Sendai. ISSN: 0040-8735 (**a**)

Tokei Suri Kenkyujo Iho. [Proc. Inst. Statist. Math.] Inst. Statist. Math., Tokyo. ISSN: 0563-685X (**a**)

Tokeikyoku Kenkyu Iho. [Res. Mem. Stat. Bureau] Stat. Bureau and Stat. Center, Tokyo. ISSN: 0446-5849 (**a**)

Tokyo Journal of Mathematics. [Tokyo J. Math.] Kinokuniya Book Store, Tokyo. ISSN: 0387-3870 (**a**)

Tonggyehak Yongu. [J. Korean Statist. Soc.] Korean Statist. Soc., Seoul. (**a**)

Topologie Structurale = Structural Topology. [Structural Topology] Univ. Québec, Montreal, QUE. ISSN: 0026-9171 (**a**)

Topology. [Topology] Pergamon, Oxford-Elmsford, New York. ISSN: 0040-9383 (**a**)

Topology and its Applications. [Topology Appl.] North-Holland, Amsterdam-New York. ISSN: 0166-8641 (**a**)

Topology Proceedings. [Topology Proc.] Auburn Univ., Auburn, AL. ISSN: 0146-4124 (**a**)

Trabajos de Estadistica. [Trabajos Estadist.] Estadist., Málaga. ISSN: 0213-8190

Transactions of the American Mathematical Society. [Trans. Amer. Math. Soc.] Amer. Math. Soc., Providence, RI. ISSN: 0002-9947 (**a,s**)

Transactions of the Moscow Mathematical Society. [Trans Moscow Math. Soc.] Amer. Math. Soc., Providence, RI. ISSN: 0077-1554 (**a,s**)

Trudy Geometricheskogo Seminara. (*Kazanskiĭ Gosudarstvennyĭ Universitet*). [Trudy Geom. Sem. Kazan. Univ.] Kazań. Univ., Kazan. ISSN: 0206-0329 (**a**)

Trudy (*Institut Vychislitel'noĭ Matematiki im. N.I. Muskhelishvili*). [Trudy Inst. Vychisl. Mat. Akad. Nauk Gruzin. SSR] "Metsniereba," Tbilisi. ISSN: 0135-0730 (**a**)

Trudy Instituta Matematiki. [Trudy Inst. Mat. (Novosibirsk)] "Nauka" Sibirsk. Otdel., Novosibirsk. ISSN: 0208-0060 (**a**)

Trudy Instituta Prikladnoĭ Matematiki imeni I.N. Vekua. [Tbiliss. Gos. Univ. Inst. Prikl. Mat. Trudy] Tbilis, Gos. Univ., Tbilisi. ISSN: 0367-2637 (**a**)

Trudy Leningradskogo Matematicheskogo Obshchestva. [Trudy Leningrad. Mat. Obshch.] Leningrad. Univ., Leningrad. ISSN: 0868-6688 (**a**)

Trudy Moskovskogo Matematicheskogo Obshchestva. [Trudy Moskov. Mat. Obshch.] Moskov. Gos. Univ., Moscow. ISSN: 0134-8663 (**a**)

Trudy Ordena Lenina i Ordena Okti͡abryskoĭ Revoli͡utsiĭ Matematicheskogo Instituta imeni V.A. Steklova. [Trudy Mat. Inst. Steklov.] "Nauka," Moscow. ISSN: 0371-9085 (**a**)

Trudy Seminara imeni I. G. Petrovskogo. [Trudy Sem. Petrovsk.] Moskov. Gos. Univ., Moscow. ISSN: 0321-2971 (**a**)

Trudy Seminara po Kraevym Zadacham. [Trudy Sem. Kraev. Zadacham] Kazan. Gos. Univ., Kazan. ISSN: 0132-3652 (**a**)

Trudy Seminara po Vektornomu i Tenzornomu Analizu s ikh Prilozheni͡ami k Geometrii, Mekhanike i Fizike. See *Seminar po Vektornomu i Tenzornomu Analizu. Trudy Seminara po Vektornomu i Tenzornomu Analizu s ikh Prilozheni͡ami k Geometrii, Mekhanike i Fizike.*

Trudy Tbilisskogo Ordena Trudovogo Krasnogo Znameni Matematicheskogo Instituta imeni A.M. Razmadze. [Trudy Tbiliss. Mat. Inst. Razmadze Akad. Nauk Gruzin. SSR] "Metsniereba," Tbilisi. ISSN: 0135-0781 (**a**)

Tsukuba Journal of Mathematics. [Tsukuba J. Math.] Univ. Tsukuba, Ibaraki. ISSN: 0387-4982 (**a**)

Tung Pei Shu Hsueh = Northeastern Mathematical Journal. [Dongbei Shuxue] Jilin Univ., Changchun. ISSN: 1000-1778 (**a**)

Ukrainian Mathematical Journal. [Ukrainian Math. J.] Consultants Bureau, New York-London. ISSN: 0041-5995 (**a**)

Ukrainskiĭ Geometricheskiĭ Sbornik. [Ukrain. Geom. Sb.] "Osnova," Kharkov. ISSN: 0135-6992 (**a**)

Ukrainskiĭ Matematicheskiĭ Zhurnal. [Ukrain. Mat. Zh.] "Naukova Dumka," Kiev. ISSN: 0041-6053 (**a**)

Universidad Nacional de Tucumán. Facultad de Ciencias Exactas y Tecnologia. Revista. Serie A. Matemáticas y Fisíca Teórica. See *Revista. Serie A, Matemáticas y Fisíca Teórica.*

Universitatis Iagellonicae Acta Mathematica. [Univ. Iagel. Acta. Math.] PWN, Warsaw. ISSN: 0083-4386 (**a**)

Univerzitet u Beogradu. Publikacije Elektrotehnickog Fakulteta. Serija. Matematika. See *Publikacije Elektrotehnickog Fakulteta. Serija: Matematika.*

Univerzitet u Novom Sadu. Zbornik Radova Prirodno-Matematičkog Fakulteta. Serija za Matematika. See *Zbornik Radova Prirodno-Matematičkog Fakulteta. Serija za Matematika.*

Upravlíaemye Sistemy. [Upravlyaemye Sistemy] Akad. Nauk SSSR Sibirsk. Otdel., Inst. Mat., Novosibirsk. ISSN: 0566-7275

Ural'skiĭ Gosudarstvennyĭ Universitet im A.M. Gor'kogo. Ural'skoe Matematicheskoe Obshchestvo. Matematicheskie Zapiski. See *Matematicheskie Zapiski.*

Uspekhi Matematicheskikh Nauk. [Uspekhi Mat. Nauk] "Nauka," Moscow. ISSN: 0042-1316 (**a**)

Utilitas Mathematica. [Utilitas Math.] Utilitas Math., Winnipeg, Man. ISSN: 0315-3681 (**a**)

Vestnik Khar'kovskogo Gosudarstvennogo Universiteta. Seriía Mekhaniko-Matematicheskaía. Zapiski Mekhaniko-Matematicheskogo Fakul'teta Khar'kovskogo Matematicheskogo Obshchestva. [Vestnik Khar'kov. Gos. Univ.] "Vishcha Shkola," Kharkov. ISSN: 0453-8048

Vestnik Leningrad University. Mathematics. [Vestnik Leningrad Univ. Math.] Allerton Press, Allerton, NY. ISSN: 0146-924X

Vestnik Leningradskogo Universiteta. Matematika, Mekhanika, Astronomiía. [Vestnik Leningrad. Univ. Mat. Mekh. Astronom.] Leningrad. Univ., Leningrad. ISSN: 0024-0850 (**a**)

Vestnik Moskovskogo Universiteta. Seriía I. Matematika, Mekhanika. [Vestnik Moskov. Univ. Ser. I Mat. Mekh.] Moskov. Gos. Univ., Moscow. ISSN: 0201-7385 (**a**)

Vestnik Moskovskogo Universiteta. Seriía XV. Vychislitel'naía Matematika i Kibernetika. [Vestnik Moskov. Univ. Ser. XV Vychisl. Mat. Kibernet.] Moskov. Gos. Univ., Moscow. ISSN: 0137-0782 (**a**)

Vestsī Akademīī Navuk BSSR. Seryyi Fīzīka-Matèmatychnykh Navuk. [Vestsi Akad. Navuk BSSR Ser. Fiz.-Mat. Navuk] Navuka i Tèkhnika, Minsk. ISSN: 0002-3574 (**a**)

The Vikram Mathematical Journal. [Vikram Math. J.] Vikram Univ., Ujjain. (**a**)

Vīsnyk Kyivs'kono Unīversytetu. Serīía Matematyky ta Mekhanīky. [Visnik Kiïv. Univ. Ser. Mat. Mekh.] "Vishcha Shkola," Kiev. ISSN: 0201-7601 (**a**)

Vychislitel'naía i Prikladnaía Matematika. [Vychisl. Prikl. Mat. (Kiev)] "Lybid," Kiev. ISSN: 0321-4117 (**a**)

Wiadomości Matematyczne. See *Roczniki Polskiego Towarzystwa Matematycznego Seria II. Wiadomosci Matematiyczne.*

Wissenschaftliche Zeitschrift der Friedrich-Schiller-Universität. Mathematisch-Naturwissenschaftliche Reihe. [Wiss. Z. Friedrich-Schiller-Univ. Jena Math.-Natur. Reihe] Friedrich-Schiller Univ., Jena. ISSN: 0043-6836 (**a**)

Wissenschaftliche Zeitschrift der Karl-Marx-Universität Leipzig. Mathematisch-Naturwissenschaftliche Reihe. [Wiss. Z. Karl-Marx-Univ. Leipzig Math.-Natur. Reihe] Karl-Marx-Univ., Leipzig. ISSN: 0300-0540

Wissenschaftliche Zeitschrift der Pädagogischen Hochschule Erfurt/Mühlhausen. Mathematische-Naturwissenschaftliche Reihe. [Wiss. Z. Pädagog. Hochsch. Erfurt/Mühlhausen Math.-Natur. Reihe] Pädagog. Hochsch. Erfurt/Mühlhausen, Erfurt.

Ying Yun Shu Hsueh Ho Li Hsueh = Applied Mathematics and Mechanics. [Appl. Math. Mech.] Appl. Math. Mech., Chongqing.

Ying Yung Kai Lu Tung Chi. = Chinese Journal of Applied Probability and Statistics [Chinese J. Appl. Probab. Statist.] East China Normal Univ., Shanghai. (**a**)

Ying Yung Shu Shueh = Mathematica Applicata. [Math Appl.] Huazhong Univ. Sci. Tech., Wuhan. (**a**)

Ying Yung Shu Hsueh Hsueh Pao = Acta Mathematicae Applicatae Sinica. [Acta Math. Appl. Sinica] Kexue Chubanshe (Science Press), Beijing. (**a**)

Yokohama Mathematical Journal. [Yokohama Math. J.] Yokohama City Univ., Yokohama. ISSN: 0044-0523 (**a**)

ZAMP. [Z. Angew. Math. Phys.] Birkhäuser, Basel-Boston, MA. ISSN: 0044-2275 (**a,s**)

Zapiski Nauchnykh Seminarov LOMI. [Zap. Nauchn. Sem. Leningrad. Otdel. Mat. Inst. Steklov. (LOMI)] "Nauka" Leningrad. Otdel., Leningrad. ISSN: 0206-8540 (**a**)

Zastosowania Matematyki. [Zastos. Mat.] PWN, Warsaw. ISSN: 0044-1899 (**a**)

Zbornik Radova Filozofskog Fakulteta u Nišu. Serija Matematika. [Zb. Rad.] Univ. Niš, Niš. (**a**)

Zbornik Radova Prirodno-Matematičkog Fakulteta Serija za Matemati. [Zb. Rad. Prirod.-Mat. Fak. Ser. Mat.] Univ. Novi Sad, Inst. Math., Novi Sad. ISSN: 0352-0900 (**a**)

Zbornik Radova (Univerzitet "Svetozar Markovic" u Kragujevac. Prirodno-Matematički Fakultet.) [Zb. Rad. (Kragujevac)] Svetozar Marković Univ., Kragujevac. (**a**)

Zeitschrift für Analysis und ihre Anwendungen. [Z. Anal. Anwendungen] Deutscher Verlag Wissensch., Berlin. ISSN: 0232-2064 (**a**)

Zeitschrift für Angewandte Mathematik und Mechanik. [Z. Angew. Math. Mech.] Akademie-Verlag, Berlin. ISSN: 0044-2267 (**a,s**)

Zeitschrift für Angewandte Mathematik und Physik. See *ZAMP.*

Zeitschrift für Mathematische Logik und Grundlagen der Mathematik. [Z. Math. Logik Grundlag. Math.] Deutscher Verlag Wissensch., Berlin. ISSN: 0044-3050 (**a**)

Zentralblatt für Mathematik und ihre Grenzgebiete. [Zbl] Springer, Berlin-New York. ISSN: 0044-4235 (**a**)

Zeitschrift für Operations Research. [Z. Oper. Res.] Physica, Heidelberg. ISSN: 0340-9422 (**a**)

Zeszyty Naukowe Geometria. [Zeszyty Nauk. Geom.] Politech. Poznanska, Poznan. ISSN: 0239-488X (**a**)

Zeszyty Naukowe Politechniki Łódzkiej. Matematyka. See *Matematyka.*

Zeszty Naukowe Politechniki Rzeszówskiej. Matematyka i Fizyka. [Zeszyty Nauk. Politech. Rzeszówskiej. Mat. Fiz.] Politech. Rzeszowska, Rzeszów. ISSN: 0209-2670

Zeszyty Naukowe Wyższej Szkoły Pedagogicznej w Opolu. Matematyka. See *Matematyka.*

Zhurnal Vychislitel'noĭ Matematiki i Matematicheskoĭ Fiziki. [Zh. Vychisl. Mat. i Mat. Fiz.] "Nauka," Moscow. ISSN: 0044-4669 (**a**)

ABOUT THE CONTRIBUTORS

Nancy D. Anderson is an Associate Professor of Library Administration and has been Mathematics Librarian at the University of Illinois at Urbana-Champaign for the past twenty-one years. She is author of *French Mathematical Seminars: A Union List* and co-author of "Mathematics Research Libraries: A 1990 Snapshot." She is currently a member of two American Mathematical Society's committees: the Committee on the Library (co-chair) and the Committee on Electronic Products and Services. She is a former chair of the Special Library Association's Physics-Astronomy-Mathematics Division. She has contributed numerous articles on mathematics librarianship to the *P-A-M Bulletin*.

Lois M. Pausch is an Associate Professor of Library Administration and Geology Librarian at the University of Illinois at Urbana-Champaign. She was formerly mathematical statistics subject specialist in the Mathematics Library and was also, for many years, the cordinator of the science cataloging. She is the author of several articles on bibliographical instruction.

Willis F. Cunningham is Assistant Social Sciences Librarian at Oklahoma State University. Formerly he was a graduate assistant in the Mathematics Librarian at the University of Illinois at Urbana-Champaign.

Gary Davidoff is head of the Mathematics/ComputerScience/Nuclear and General Physics Library at Argonne National Laboratory in Argonne, Illinois.

Mary B. Harper is the current graduate assistant in the Mathematics Library at the University of Illinois at Urbana-Champaign. Her background is in educational administration and evaluation.

Allen D. Merry is currently working as a reference librarian at the Queensborough Public Library, Flushing, New York. Formerly he was a graduate assistant in the Mathematics Library at the University of Illinois at Urbana-Champaign.

Robert Thornhill is Extension Librarian at the Siskiyou County Public Library in Yreka, California. He was formerly the science and business librarian at the Chicago Public Library main branch.

AUTHOR INDEX

(All entries printed in **boldface type** refer to text chapters 2-5)

SUBJECT INDEX

(All entries printed in **boldface type** refer to text chapters 2-5)